Armed Jews in the Americas

Jewish Latin America

ISSUES AND METHODS

VOLUME 13

The titles published in this series are listed at *brill.com/jlam*

Armed Jews in the Americas

Edited by

Raanan Rein
David M.K. Sheinin

BRILL

LEIDEN | BOSTON

Cover image: https://www.torontopubliclibrary.ca/detail.jsp?Entt=RDMDC-TSPA_0050191F&R=DC-TSPA _0050191F
Credit: Phil Givens at the Police Gun Range, Mario Geo, no date, Toronto Star Photo Archive, tspa_0050191f Toronto Public Library.

Library of Congress Cataloging-in-Publication Data

Names: Rein, Raanan, 1960- editor. | Sheinin, David, editor.
Title: Armed Jews in the Americas / edited by Raanan Rein, David M.K. Sheinin.
Description: Leiden ; Boston : Brill, 2021. | Series: Jewish Latin America, 2211-0968 ; volume 13 | Includes index. | Summary: "A Jewish weapons manufacturer during the American Civil War, a Jewish-Canadian chair of the Metropolitan Toronto Police Board, and Jewish-Argentine guerrilla fighters-these are some of the individuals discussed in this first-of-its-kind volume. It brings together some of the best new works on armed Jews in the Americas. Links between Jews and their ties to weapons are addressed through multiple cultural, political, social, and ideological contexts, thus breaking down longstanding, stilted myths in many societies about Jews and weaponry. Anti-Semitism and Jewish self-defense, Jewish volunteers in the Spanish Civil War and in the 1948 Arab-Israeli war, and Jewish-American gangsters as ethnic heroes form part of the little-researched topic of Jews and arms in the Americas"—Provided by publisher.
Identifiers: LCCN 2021016916 (print) | LCCN 2021016917 (ebook) | ISBN 9789004462533 (hardback) | ISBN 9789004462540 (ebook)
Subjects: LCSH: Jews—History, Military. | Jewish soldiers—Western Hemisphere—History. | Jews—United States—History. | Firearms—Western Hemisphere—History. | Firearms—United States—History. | Firearms industry and trade—Western Hemisphere—History. | Firearms industry and trade—United States—History. | Jews—Legal status, laws, etc.—United States—History.
Classification: LCC DS111.6 .A76 2021 (print) | LCC DS111.6 (ebook) | DDC 970.004/924—dc23
LC record available at https://lccn.loc.gov/2021016916
LC ebook record available at https://lccn.loc.gov/2021016917

Typeface for the Latin, Greek, and Cyrillic scripts: "Brill". See and download: brill.com/brill-typeface.

ISSN 2211-0968
ISBN 978-90-04-46253-3 (hardback)
ISBN 978-90-04-46254-0 (e-book)

In memory of

Joseph Sheinin (1928–2007), structural engineer and member of the Communist Party of Canada in the 1940s and 1950s. In the 1960s, Lukens Canada Ltd. hired him to solve a design flaw in the enormous mechanical press at their Rexdale, Ontario, plant, where nuclear submarine outer hulls were molded for the US Navy.

•••

Nehama Rein (1930–2020), who served in the Haganah, the main paramilitary organization in Jewish Palestine during British Mandatory rule. In 1948, she joined the recently-established Israeli navy, where she met Raanan's father.

Contents

Acknowledgements

Publication of this book was made possible thanks to the generous support of the Elías Sourasky Chair of Iberian and Latin American Studies and the S. Daniel Abraham Center for International and Regional Studies, both at Tel Aviv University.

We also express our thanks to Omri Elmaleh and Maayan Nahari for their able assistance, to Beryl Belsky for her excellent copyediting of the text, and to our colleagues at Brill, especially Katelyn Chin and Erika Mandarino, for their efficiency and longstanding support of this book series.

Notes on Contributors

Lukas Böckmann
studied Ibero-American History, Romance Languages and Literature and Philosophy at the universities of Cologne, Universidad Veracruzana, and Universidad de Buenos Aires. He completed his Diploma in 2013 and worked as an academic assistant at the Academy Project "European Traditions—Encyclopedia of Jewish Cultures." He is currently a doctoral candidate in Leipzig at the Leibniz Institute for Jewish History and Culture–Simon Dubnow. His project "A Paradise on Earth" analyzes the hidden religious roots of the Latin American guerrilla movements of the 1960s and the New Left milieu surrounding it.

Larry D. Gragg
is Curators' Distinguished Teaching Professor Emeritus at Missouri University of Science and Technology where he served as chair of the history and political science department for seventeen years. He has won more than twenty teaching awards, including the University of Missouri President's Award for Outstanding Teaching. He is the author of ten books including *"Bright Light City": Las Vegas in Popular Culture* (2013); *Benjamin "Bugsy" Siegel: The Gangster, the Flamingo, and The Making of Modern Las Vegas* (2015); and *Becoming America's Playground: Las Vegas in the 1950s* (2019).

Gabriela Jonas Aharoni
teaches in the School of Audio and Visual Arts, Communication and Liberal Arts and Sciences Department of Sapir College (Israel). Her book, *Argentinian Telenovelas. Southern Sagas Rewrite Social and Political Reality*, was published by Sussex Academic Press (2015). Jonas Aharoni has also published several articles in academic journals, such as "Jewish Identities in Argentinian Television Fiction: The case of Graduados," in *Jewish Film and New Media*, and "Globalización e identidades plurales en las telenovelas argentinas de fines de siglo," in *E.I.A.L.* Her chapter, "The Other as Mirror: Representation of Jews and Palestinians on Argentinian and Chilean Television Screens," was included in the volume *The New Ethnic Studies: Jewish Latin Americans in a Comparative Perspective*, Brill, 2017.

Adrián Krupnik
obtained his PhD in history at Tel Aviv University. After studying sociology at the University of Buenos Aires, he received an MA in Israel Studies at Haifa

University. Adrián Krupnik's research stands at the intersection of Latin American and Israel studies by uniting two strands of modern Jewish history: development and change in a Diaspora community whose members integrated successfully into the host society, and the evolving relationship of that community with Israel. During his sojourn at Universität Potsdam he completed his first book manuscript, "Between Two Homelands: Jewish-Argentine Return Migration from Israel, 1948–2006." Krupnik's new project is "German-Argentine and Jewish-Argentine Youth in Radical Times: Ethno-Political Identities and Transnational Bonds in Comparative Perspective, 1966–1976."

Michael Petrou
is a historian and journalist. He is an adjunct professor of history at Carleton University in Canada and is editor-in-chief of *Open Canada*, an online magazine about Canada's place in the world. Petrou was the 2018 Martin Wise Nieman Fellow at Harvard University and was previously a Chevening Scholar at Saint Antony's College, University of Oxford, where he completed a DPhil in modern history. Petrou is the author of *Renegades: Canadians in the Spanish Civil War* and *Is This Your First War? Travels through the Post-9/11 Islamic World*. He has won the Ottawa Book Award and three Canadian National Magazine Awards.

Stephanie M. Pridgeon
is assistant professor of Hispanic Studies at Bates College and the author of *Revolutionary Visions: Jewish Life and Politics in Latin American Film* (University of Toronto Press), as well as several articles on Latin American film, literature, and cultural studies.

Raanan Rein
is the Elias Sourasky Professor of Latin American and Spanish History and former vice president of Tel Aviv University. He is also head of the S. Daniel Abraham Center for International and Regional Studies. Rein is the author and editor of more than forty books and numerous articles and book chapters. He is a member of Argentina's National Academy of History, and former president of the Latin American Jewish Studies Association (LAJSA). The Argentine government awarded him the title of Commander in the Order of the Liberator San Martin for his contribution to Argentine culture.

David M.K. Sheinin
is professor and director of the History Graduate Program at Trent University. He is the winner of the university's Distinguished Research Award (2017) and

his most recent book is *Race and Transnationalism in the Americas* (co-edited with Benjamin Bryce, University of Pittsburgh Press, 2021).

Barry L. Stiefel
PhD is an Associate Professor in the Historic Preservation & Community Planning program at the College of Charleston, which is part of the Department of Art and Architectural History. He is interested in researching how the sum of local preservation efforts affects regional, national, and multinational policies within the field of cultural resource management and natural heritage conservation, with a particular focus on Jewish heritage. Dr. Stiefel has published numerous books and articles.

Amy Weiss
is an Assistant Professor of Judaic Studies and History and holds the Maurice Greenberg Chair of Judaica Studies at the University of Hartford. She was previously a Thomas and Elissa Ellant Katz Fellow at the University of Pennsylvania's Herbert D. Katz Center for Advanced Judaic Studies. Her research and publications focus on the intersections of American religion, Israeli culture, and Jewish-Protestant relations. Most recently, her articles have appeared in the journals *American Jewish History*, *Holocaust and Genocide Studies*, and *Israel Studies*. She is currently writing a book manuscript on the evolving interfaith alliances between American Jews and evangelicals. Weiss received her PhD from the Hebrew and Judaic Studies and History departments at New York University.

CHAPTER 1

Introduction: A Curious Mix of Jews and Weapons

David M.K. Sheinin and Raanan Rein

In addition to bringing together some of the best new works on armed Jews in the Americas, a subject not addressed previously in an edited volume, this book demystifies the links between Jews and their ties to weapons. Demystification is not the objective of individual chapters or of the collection itself, but as studies that tackle the book's theme through multiple cultural, political, social, and ideological contexts, they break down longstanding myths rooted in many societies about Jews and weaponry.

Armed Jews, or more often, the hazy intersection of armed Jews and the *fantasy* of armed Jews, have long made others nervous. In early 2021, there was outrage in and out of the US Congress over a series of conspiracy theories advocated by a first-term member of the House of Representatives, Marjorie Taylor Greene (R-Georgia). The most bizarre of the conspiracies Greene touted was that Jews had triggered the disastrous 2018 California wildfires by means of a space laser. It got stranger still. Greene's inclusion of the Rothschilds among the secretive co-conspirators evoked centuries of anti-Semitic bile and was a reminder that the fear of armed Jews is only one aspect of a broad series of discriminatory tropes dating back to the original fantasy of violent Jewish sin, the supposed Jewish killing of Christ, a doctrine renounced by the Catholic Church in the early 1960s.[1] The US House of Representatives Democratic Party majority voted to oust Taylor from her committee assignments in response to her egregious declarations. At that point, quipped satirist Andy Borowitz, she "created a committee of her own, the House Committee on Semitic Aerospace Weaponry."[2]

Greene's death ray rant was also reminiscent of the power of the armed-Jew-as-dangerous fiction that reached a pinnacle of European anti-Semitic

1 Zack Beauchamp, "Marjorie Taylor Greene's Space Laser and the Age-Old Problem of Blaming the Jews," *Vox*, January 30, 2021, https://www.vox.com/2256258/marjorie-taylor-greene-jewish-space-laser-anti-semitism-conspiracy-theories.

2 Andy Borowitz, "Defiant Marjorie Taylor Greene Creates Own House Committee on Semitic Aerospace Weaponry," *The New Yorker*, February 5, 2021, https://www.newyorker.com/humor/borowitz-report/defiant-marjorie-taylor-greene-creates-own-house-committee-on-semitic-aerospace-weaponry.

 | DOI:10.1163/9789004462540_002

paranoia in the late-nineteenth-century prosecution (and attendant persecution) of Captain Alfred Dreyfus in what came to be known as *l'Affaire Dreyfus*. Never far from the minds of many Europeans in the weeks before the 1938 *Reichskristallnacht*, German authorities arrested unknown numbers of Jewish citizens for *lawful* gun ownership—as though *they* were the impending menace to society.[3]

Like other aspects of anti-Semitism, including the association of Jews with both the vanguard of Communist revolution and the pinnacle of capitalist exploitation, the discriminatory trope of Jews and guns has often been contradictory. In 2018, the journalist Adam C. Smith wondered aloud whether, in their support for Florida Republican gubernatorial candidate Ron DeSantis, the National Rifle Association (NRA) was alerting Floridians to the threat of gun control advocates, or to the more specific threat of Jewish gun control advocates. "Don't let [former New York Mayor] Michael Bloomberg and [Billionaire social justice advocate] George Soros control Florida!" screamed an NRA flyer in support of DeSantis.[4] In American popular culture, Jewish gun owners might be a threat, but some regard Jewish opponents of gun ownership as equally dangerous.

In his chapter in this volume, "David Lopez Jr. and Weapons Manufacturing during the American Civil War," Barry L. Stiefel underscores that sometimes—and without anti-Semitic intent—Jews linked to weapons and violence were at the very least, on the wrong side of history. By the outbreak of the American Civil War in 1860, David Lopez Jr. had become an accomplished and highly skilled weapons manufacturer, capable also of their repair. It remains unclear whether his support for the Confederacy was a matter of political conviction or simply a function of his having been a pillar of the Charleston, South Carolina, Jewish community at the time, and a man engaged in technical and business activities not normally associated with Jewish-Americans. Whatever the case, Lopez played a key role in arming the Confederate war machine in defense of secession and slavery.

According to a parallel narrative frequently—and incongruously—posited, Jews dislike guns and are reluctant gun owners. A 2012 story in the *Washington Jewish Week* focused on "pro-gun Jews" as a rarity:

---

3 David C. Jones, "'A Beastly Affair': Visual Representations of Animality and the Politics of the Dreyfus Affair," *Canadian Journal of History 46*, no. 1 (2011): 35–62; Stephen P. Halbrook, *Gun Control in the Third Reich: Disarming the Jews and "Enemies of the State"* (Oakland: The Independent Institute, 2013), 145–48.

4 Adam C. Smith, "Is the NRA Warning Floridians about Gun Control Advocates or Jewish Gun Control Advocates?" *Tampa Bay Times*, November 3, 2018, 3.

> Day by day, the stereotype that Jews don't hunt or even own guns is fading. They aren't the Jewish gangsters of old like Meyer Lansky and Bugsy Siegel, nor do they enjoy killing … They are just regular people who want to protect their families.[5]

This was a theme in the US media after the October 2018 massacre at the Tree of Life synagogue in Pittsburgh, with various reports of the putative novelty of Jewish-Americans buying firearms for the first time. No statistical evidence was presented, however, demonstrating a marked increase in Jewish gun ownership, or proving that Jews had been reluctant gun owners in the past.[6] In Europe, with the recent surge in anti-Semitism, the supposed Jewish rejection of gun ownership has served as a backdrop for calls in the Jewish community for Jewish Europeans to take their defense into their own hands. In 2015, Brussels-based rabbi Menachem Margolin, director of the European Jewish Association, called on European countries to pass legislation to allow "designated people" in Jewish communities to own weapons. In part, Margolin's suggestion came in answer to what he called Israeli Prime Minister Binyamin Netanyahu's "Pavlovian" calls on Jews to immigrate to Israel: "A call to make aliyah is not the solution to anti-Semitic terror," Margolin reasoned. Guns were the answer.[7]

Over the course of the twentieth century, and despite holdouts like Marjorie Taylor Green, in many Western cultures non-Jewish popular notions of a threat to society when Jews were armed faded to an extent. Americans sometimes replaced or interspersed the danger narrative with powerful counter-narratives of Jewish triumph under arms, where the threat of Jewish violence was not so much eliminated as romanticized, normalized, and rendered appealing. The inverse of the Jews, guns, and danger narrative is the idea that if only Jews were more like (presumably non-Jewish) NRA members, they might avoid violence directed their way. In 2015, though likely unaware of the pre-*Reichskristallnacht* detention of Jewish gun owners in Germany, Republican presidential candidate and future US Secretary of Housing and Urban Development Ben Carson told a national television audience in the United States that had Jewish Europeans owned guns, the Holocaust might have been "greatly diminished."[8] And in the

5 Suzanne Pollak, "Pro-gun Jews," *Washington Jewish Week*, December 27, 2012, 4; Nate Shapiro, "Jews and Guns," *The Jewish Press*, March 31, 2017, 7.

6 Samantha Melamed, "For Some Jews, Guns Now Option for Safety: Discussions Take On New Meaning since Pittsburgh," *Philadelphia Inquirer*, November 6, 2018, A1.

7 Gregg Carlstrom, "Give Guns to Europe's Jews, Rabbi Urges," *The Times*, January 14, 2015, 32.

8 Tanya Basu, "Ben Carson Defends Claim that Holocaust Could Have Been Prevented if Jews Had Guns," *Time*, October 11, 2015, https://time.com/4069415/ben-carson-defends-holocaust-guns/.

aftermath of the Tree of Life synagogue attack, President Donald Trump suggested that an armed synagogue guard might have prevented the assault.[9]

In the United States, violent Jewish gangsters like Meyer Lansky and Benjamin "Bugsy" Siegel have played an ambiguous role in popular culture dating back to the early twentieth century. Urban newspapers labeled them murderers, drug dealers, and thieves, while older Jewish-Americans—immigrants and first-generation Americans—often reviled the links of their communities to criminal villainy. However, by the early 1930s, younger Jewish-Americans began to adopt the sensationalist Hollywood vision of a glamorous criminal lifestyle. There was a further celebration of Jewish gangsters after 1970 with the appearance of a string of best-selling books on organized crime—*The Godfather* (1969) by Mario Puzo, *Lansky* (1971) by Hank Messick, and *Honor Thy Father* (1971) by Gay Talese. These and other popular representations of organized crime focused less on the criminality and violence of Jewish and other "ethnic" mobsters and more on what Wendy H. Bergoffen describes as their roles as "ethnic heroes." The romantic, likable image of the Jewish gangster has been repeatedly bolstered by Hollywood, as in the case of Hesh Rabkin (Jerry Adler), the appealing, explicitly non-Italian, Jewish confidant of protagonist Tony Soprano (James Gandolfini) in the HBO series *The Sopranos* (1999–2007). Rabkin's Jewish identity was "affirmed" in his role as a shady (but good-hearted) moneylender and in his use of Yiddish terms, but complicated by his enthusiastic friendship with the violent Italian-American mobsters in his circle, whom he sometimes bankrolled.[10]

In his chapter, "American Jewish Gangsters," Larry Gragg revives an important theme of an earlier volume on Jews in sports in the Americas, edited by Raanan Rein and David M.K. Sheinin.[11] Jews became American gangsters for reasons not unrelated to their gravitating to boxing and other sports in significant numbers. As immigrants and the children of immigrants growing up in poverty in US cities, and facing the violence of the streets as well as anti-Semitism, Jewish-Americans *became* American by expanding the boundaries of ethnic identity into all realms of American urban life—including those many considered unsavory and brutal. In his novel, *Joshua Then and Now* (1980),

9 Kelsey Walsh, "On Trump's Saying Pittsburgh Needed Armed Guard, 'Where do we stop?': City Official," *ABC News*, October 28, 2018, https://abcnews.go.com/Politics/trumps-synagogue-needed-armed-guard-stop-arming-society/story?id=58803241.

10 Rachel Rubin, "Gangster Generation: Crime, Jews and the Problem of Assimilation," *Shofar* 20, no. 4 (2002): 1–20; Wendy H. Bergoffen, "Guardians, Millionaires, and Fearless Fighters: Transforming Jewish Gangsters into a Usable Past," *Shofar* 25, no. 3 (2007): 91–110.

11 Raanan Rein and David M.K. Sheinin eds., *Muscling in on New Worlds: Jews, Sports, and the Making of the Americas* (Boston: Brill, 2014).

Mordecai Richler played with the notion of the uneasy expansion of Jewish identity markers by characterizing the Jewish-Canadian protagonist's father as a working-class former boxer, armed gangster, and Old Testament savant. And as Gragg points out, violent mobsters may have pushed back the boundaries of Jewish identity, but were never outside it. He notes, for example, that Meyer Lansky, Mickey Cohen, and others put their armed muscle to work as Jews, breaking up pro-Nazi rallies in several American cities during the 1930s.

In 2009, Quentin Tarantino took the likability of armed Jews to a new level in the revenge, what-if narrative of the film *Inglorious Basterds*. Tarantino built on the tension in Stephen Spielberg's Jewish-American soldier Private Mellish (Adam Goldberg), in *Saving Private Ryan* (1998). In one scene, Mellish ominously taunts fearful captured German prisoners with the Star of David around his neck as well as his gun. Tarantino's film also drew on journalist John Sack's investigation into brutal Jewish prison guards who ran internment camps for German civilians in Poland after World War II. In 1993, the German publisher Piper Verlag had been scheduled to distribute a printed run of 6,000 copies of *An Eye for an Eye: The Untold Story of Jewish Revenge against Germans in 1945*. When Piper Verlag changed chief executive officers on the eve of the book's distribution, Viktor Niemann, the new press head was appalled by the optics of Jewish soldiers killing Germans in the aftermath of the Shoah. The magazine GQ also reversed a decision to print a shorter version of Sack's study, though the *Village Voice* ran the story and Basic Books published the book.[12] In Tarantino's fictional version of armed Jewish retribution, the burden of journalistic proof faced by Sack vanished. In *Inglorious Basterds*, a US special service force made up entirely of Jewish-American soldiers reaches Europe in 1944 to terrify German soldiers in the last stages of the war by killing them with all manner of weapons in the most brutal of ways. Tarantino not only normalizes armed Jewish violence against Nazi soldiers but renders the Jewish-American liberators of Europe immensely amiable despite the gore (or perhaps because of it).[13]

No set of institutional problems underscores more sharply the contradictions of Jews with guns than the question of Jewish police officers in the diaspora. In popular culture and in academic literature, the identity of Jewish police officers is often highlighted in various ways, while the ethnicity or

12 David Streitfeld, "Book on Jewish-Run Revenge Camps Is Pulled," *Washington Post*, February 15, 1995, https://www.washingtonpost.com/archive/lifestyle/1995/02/15/book-on-jewish-run-revenge-camps-is-pulled/17f4868b-403f-4083-897d-1edeace16e27/.

13 Eyal Peretz, "What Is a Cinema of Jewish Vengeance? Tarantino's *Inglorious Basterds*," *The Yearbook of Comparative Literature* 56 (2010): 64–74; A. Dana Weber, "From Glorious Nibelungs to Inglorious Basterds: Quentin Tarantino's Refractive Retelling of Fritz Lang's Epic," *German Studies Review* 42, no. 3 (2019): 537–60.

race of their non-Jewish colleagues does not feature prominently. It is their Jewishness that becomes an analytical focal point. Michael Chabon writes of a dystopian Jewish past in *The Yiddish Policemen's Union* (2007), where a Jewish society formed in World War II-era Alaska is concomitantly a safe space for Jewish refugees from the Shoah and a de facto prison from which there may be no escape. On one level, the Jewish police force and its detectives are anomalous for their "normalcy" as potboiler shamuses right out of the sort of 1940s' Hollywood film that would never have featured Jewish characters in essentially "American" roles. Atypically, they are not Jewish policemen to whom ethnicity matters; they are American policemen, except that the "American" frontier in which they are living is entirely Jewish.[14] By the latter part of the twentieth century, Jewish police officer characters abounded in Hollywood films—played by Jewish and non-Jewish actors alike. Even so, the cultural improbability of their Jewish identities (even in historically accurate biopics) often underscored other oddities in their characters. Examples include Joe Mantegna's Bobby Gold (*Homicide*, 1991), Liev Schreiber's Dovi (*Fading Gigolo*, 2013), Russell Crowe's Richie Roberts (*American Gangster*, 2007), and Jerry Orbach's Gus Levy (*Prince of the City*, 1981). In the latter, at a moment of high tension when Levy refuses to take a fall for police corruption, another character refers to him admiringly as "one tough Jew."[15]

In the scholarly literature, Jewish police officers sometimes appear almost as caricatures. They are Jews that made good and in turn, were able to offer support to members of the Jewish community they were policing. Norton B. Stern and William M. Kramer's 1973 study of late-nineteenth-century Los Angeles police chief Emil Harris is hagiography that underlines a flawless, improbably heroic career. His "methodology as a detective is basically unrecorded" except that the authors report that his ability "is written large across the pages of the Los Angeles press." The authors shape Harris' morality as early 1970s' (not 1870s') liberal. He was a defender of Chinese Angelinos, for example, in an era of intense discrimination against that community. There is no insight into what Harris' Jewish identity meant, either to the officer himself or in the larger context of ethnicity and race in Southern California. It is unclear how, but the authors argue that a "large number of Jewish merchants … found in their coreligionist Emil Harris, a protector, defender, and friend."[16]

---

14 Mike Witcombe, "A Comedy of Eruvs: (Re)Locating Jewish Identity in Michael Chabon's *The Yiddish Policemen's Union* and Howard Jacobson's *J*," *Shofar 34*, no. 3 (2016): 29–51.

15 Vincent Brook, "Sidney Lumet, Political Jew," *Jewish Film & New Media 1*, no. 1 (2013): 20–43.

16 Norton B. Stern and William M. Kramer, "Emil Harris: Los Angeles Jewish Police Chief," *Southern California Quarterly 55*, no. 2 (1973): 163–92.

As does Larry Gragg's essay on Jewish gangsters, David M.K. Sheinin's chapter considers Phil Givens' transformation into a cop as a reflection of changing Jewish identity in 1960s' and 1970s' Toronto. In "Phil Givens, the Police, and the Battle for Toronto's Yonge Street," Sheinin argues that in his journey from mayor of Toronto in the early 1960s to police commission chair fifteen years later, Givens *became* a police officer in ways that reflected his Jewish-Canadian identity in the fight for social order, in defense of civil rights, and against anti-Semitism and other forms of discrimination. By contrast, in fiction and in the media, Jewish police officers have often become a circus sideshow *because* of the seeming novelty of that combined identity. "I got sick of people telling me 'You're Jewish. You don't have to be a cop,'" said [Steve] Itzko on the stereotype of Jewish men flocking to work in media, medicine and law. Despite Itzko's gripe, the *Philadelphia Daily News* item that focused on the city's patrolman, while celebrating a rich tradition of Jewish policing there, highlighted the oddity of Jewish policemen.[17] A 2019 London *Daily Telegraph* article reported that a Sussex police force officer who allegedly held a pepperoni pizza above a Jewish officer's meal was facing disciplinary action. On March 8, 2019, the formal allegation read: "Knowing that another officer is Jewish, Sgt Jacobs deliberately dangled a pizza containing pepperoni over that officer's food." The police force was "committed to promoting equality and respect for diversity," the story went on, noting that "religiously observant Jews will only eat pepperoni if it is certified kosher as they do not eat pork."[18] Similar curiosities were reported by *Jet* in 1983 (on Charleston, South Carolina's first Black Jewish police chief) and by *The Jerusalem Post* in 2019 (on Baltimore's first Jewish and first woman police chief).[19]

Five chapters in this volume are devoted to the high representation of Jews in left-wing parties and movements, and their involvement in armed struggles. Such was the case in 1930s' Spain following the military uprising headed by General Francisco Franco, aided by Nazi Germany and Fascist Italy. Tens of thousands of young people around the world traveled to the Iberian Peninsula to defend the Republic in the Spanish trenches. At least 35,000 volunteers arrived in Spain from more than fifty countries. The number of Jewish volunteers who joined the so-called International Brigades (IB) was likely over

17 Simone Weichselbaum, "Number of Jewish Police Declining: But Traditions, Ethnic Pride As Strong As Ever," *Philadelphia Daily News*, December 18, 2004, 5.

18 "Pepperoni Prank on Jewish Police Officers Backfires," *Daily Telegraph*, November 14, 2019, 13.

19 "Charleston S.C. Gets Black Jewish Police Chief (Ruben Greenberg)," *Jet*, March 1, 1983, 44; "Baltimore County Police Swear in 1st female and 1st Jewish Police Chief," *Jerusalem Post*, June 18, 2019, 5.

5,000. It is difficult to form a more precise estimate since they were scattered among the various national units, often with last names that did not indicate their ethnic origins or that had been replaced by monikers. One thing is clear, however: in most cases, the number of Jews among the volunteers from each nation was greatly disproportionate to their representation in the general population of those countries.[20]

Out of 5,000 Polish volunteers, 2,250, or 45 percent, were Jewish. At the time, Jews represented only 10 percent of the Polish population. In the case of the United States, common estimates are of 1,250 Jewish-American volunteers, 38 percent of the total number of Americans enlisted in the IB, when Jews represented only 4 percent of the country's population. The 1,043 Jewish-French volunteers constituted 15 percent of all French volunteers, compared with 0.5 percent of their share in the French population. Between 200 and 400 Jews left Britain for Spain; that is, 11–22 percent of all British volunteers, when Jews made up only 0.5 percent of the population.[21]

Michael Petrou's chapter, "Jewish-Canadians in the Spanish Civil War," pays close attention to the forty-four Jewish-Canadians out of a total of 1,600 fellow countrymen who went to fight in Spain. Like many non-Jews, Jewish volunteers went to Spain to fight Fascism, but at least some of them had particular ethnic motivations for leaving their homes and families and risking their lives in battle, first and foremost, the belief that Fascism posed a particular threat to Jews, and that their participation was an opportunity to combat it. Jewish-Canadians were radicalized at the time by the general political climate in Canada during the Great Depression, which intensified their discrimination and marginalization. It also contributed to pushing many of them toward Communism and the Communist Party which, as it did elsewhere, led the support of the Spanish Republic campaign.

The most famous Argentine volunteer in the Spanish Civil War was Mika Feldman Etchebehere.[22] Gabriela Jonas Aharoni analyzes the documentary *Mika, My War in Spain*, and follows her trajectory from the Jewish agricultural colony Moisés Ville, in the province of Santa Fe, through her political militancy in Communist and Trotskyist groups, until her arrival in Spain and her designation as the captain of an Anarchist and Trotskyist unit. Although the

20 Gerben Zaagsma, *Jewish Volunteers in the International Brigades and the Spanish Civil War* (London: Bloomsbury, 2017).

21 Martin Sugarman, "Against Fascism—Jews Who Served in the International Brigade in the Spanish Civil War," https://www.jewishvirtuallibrary.org/jsource/History/spanjews.pdf.

22 See her autobiography, Mika Feldman, *Mi guerra de España* (Madrid: Plaza y Janés, 1987).

documentary rejects the analytical category of ethnicity, it illuminates the role of Jewish-Argentines in the Spanish Civil War.[23]

Motivated by other ideological imperatives, approximately 1,500 Jewish-American volunteers joined the Israel Defense Forces (IDF) during the 1948 Arab-Israeli War. The total number of international volunteers who came to defend the newly-born state was more than 3,500, with the largest groups coming from the United States, Canada, France, Britain, and Latin America.[24] Amy Weiss emphasizes in her chapter that the illegality of serving in a foreign country's military, as well as Jewish-Americans' anxiety about their social integration in United States society, muted the public discussion around the reasons that brought them to temporarily join the IDF. Their participation in the war, however, shaped Israel's nascent armed forces.

"You Jews have to start defending yourselves because others are not gonna do it for you … We can't be sheep at the slaughter." This was the message preached by Yonatan Stern to a crowd of 1,500 Jewish men and women in a synagogue in Monsey, NY, in early January 2020.[25] Stern is the director of Cherev Gidon (Hebrew for "Sword of Gideon"), a Pennsylvania-based training academy, where Israeli veterans teach local Jews how to use firearms. Demand for the course has risen, following a series of violent anti-Semitic attacks in the United States during the COVID-19 pandemic.

Cherev Gidon began operating in the early 2010s and established an IDF-like training template, which flies the Israeli flag, uses Israeli uniforms and weapons, and charges $500 a day. The call for arms is explicit: "Every Jew, especially Jews who live abroad, need to own a weapon, know how to use it, and be ready to defend himself." Born and raised in the West Bank Jewish settlement of Kiryat Arbah, Stern served in the IDF in the Netzah Yehuda Battalion, a counter-terrorism infantry unit, specifically created for religious men. In an interview with an Israeli journalist, he explained: "We are here to avoid another Jewish Holocaust."[26]

Jewish self-defense activities outside Israel have a long history, starting as early as 1903 in Eastern Europe in response to a wave of anti-Semitic pogroms. Born in Odessa in 1880, the Revisionist Zionist thinker Vladimir (Ze'ev)

23 Raanan Rein, "A Trans-national Struggle with National and Ethnic Goals: Jewish-Argentines and Solidarity with the Republicans during the Spanish Civil War," *Journal of Iberian and Latin American Studies* 20, no. 2 (2014): 171–82.

24 Nir Arielli, "When Are Foreign Volunteers Useful? Israel's Transnational Soldiers in the War of 1948 Re-examined," *Journal of Military History* 78, no. 2 (2014): 703–24.

25 "Israeli Military Veterans Give Self-Defense Classes to Monsey Residents after Machete Attack," *New York Post*, January 9, 2020.

26 Shomon Ifergan, "In Case of a Holocaust" [Hebrew], *Mako Magazine*, February 7, 2018.

Jabotinsky was one of the founders of such a Jewish self-defense organization. As the prospects of additional pogroms increased, the organization aimed to safeguard Jewish communities throughout Russia. During the first decades of the twentieth century, Jabotinsky warned of the dangers facing Jews in the Diaspora and called for the establishment of a sovereign Jewish state in Palestine. He envisioned a new type of Jew: proud, generous, and fierce.[27]

With Joseph Trumpeldor, Jabotinsky co-founded the Jewish Legion of the British army in World War I to fight against the Ottomans who then controlled Palestine. He later began training Jews in warfare and the use of small arms. Jabotinsky died of a heart attack in 1940 while visiting a Jewish self-defense camp in Hunter, New York.[28] His ideology influenced at least several of the Jewish volunteers who went to fight in the 1948 war but also many Jewish youngsters in 1960s' South America. Following the kidnapping of Nazi war criminal Adolf Eichmann in Buenos Aires in 1960 and the assassination of the Latvian war criminal Herberts Cukurs in Montevideo in 1965, a wave of anti-Semitic attacks swept South America. Thousands of Jewish-Argentines, Jewish-Uruguayans, and Jewish-Chileans were recruited for self-defense activities, sponsored by the local organized Jewish communities, Israel's embassies, and the Israeli Mossad. Raanan Rein looks in his chapter at the Uruguayan case of Jewish self-defense and its key players, many of whom were affiliated with the Revisionist Zionist movement Betar, established in Eastern Europe in the 1920s.

Several members of Argentina's Jewish self-defense organization ended up in one of a number of guerrilla movements that sprang up in the 1960s in the context of the Cold War, the growing influence of the Cuban Revolution, and the increased legitimation of the use of violence for political purposes. In his chapter "Misguided Liberation," Lukas Böckman examines the Ejército Guerrillero del Pueblo (EGP), which tried to emulate the Cuban example in Argentina. Headed by Ricardo Mascetti, a close friend of Ernesto (Che) Guevara, the EGP had a relatively large number of Jewish guerrillas. Böckman links the participation of Jewish militants in revolutionary activities to Jewish messianic ideas of *tikkun olam* (healing the world).

While Böckman looks at Jewish-Argentines who left the Communist Party to join the guerrillas, Adrián Krupnik turns our attention to Jewish-Argentines who had been active in Zionist movements prior to their joining the armed

27 Colin Shindler, *The Triumph of Military Zionism: Nationalism and the Origins of the Israeli Right* (London: I.B. Tauris, 2009), 116–32.

28 For a recent biography of Zabotinsky, see Hillel Halkin, *Jabotinsky: A Life* (New Haven: Yale University Press, 2014).

ranks of the Liga Comunista Revolucionaria and the Montoneros. In her essay Stephanie Pridegeon offers a different insight into the constant tension between particularist and internationalist, or between ethnic (Jewish) and national (Argentine), causes. Pridegeon looks at Jewish-Argentine fiction, especially Marcelo Birmajer's *Tres mosqueteros* and Manuela Fingueret's *Hija del silencio*. Participation in the Montoneros movement is presented as both contradictory to and in congruence with Jewishness. In one way or another, this kind of tension is a leitmotif that runs through all texts included in this volume, which highlights the multiple ties of Jews in the Americas to the manufacture, distribution, and use of arms for ethnic, political, and criminal purposes.

As if to demonstrate yet again the lasting puzzle of armed Jews, in his 2018 Showtime television series *Who is America?*, the British shock humorist Sacha Baron Cohen developed the character Erran Morad, an Israeli anti-terrorism expert. Heavily disguised and boasting an over-the-top Israeli accent, Baron Cohen's Morad teaches unusual (comical) methods to counter pedophiles, immigrants, terrorists, and others. His foils on the show are gullible Americans. Morad proposes arming three-year-olds with guns. What makes him plausible to an unsuspecting Philip Van Cleave, a gun rights activist and the head of the Virginia Citizens Defense League, is that still most "mysterious" of all armed Jews, an Israeli Mossad agent.

Bibliography

Bergoffen, W.H. (2007) "Guardians, Millionaires, and Fearless Fighters: Transforming Jewish Gangsters into a Usable Past," *Shofar* 25, no. 3: 91–110.

Brook, V. (2013) "Sidney Lumet, Political Jew," *Jewish Film & New Media* 1, no. 1: 20–43.

Halkin, H. (2014) *Jabotinsky: A Life*, New Haven: Yale University Press.

Jones, D.C. (2011) "'A Beastly Affair': Visual Representations of Animality and the Politics of the Dreyfus Affair," *Canadian Journal of History* 46, no. 1: 35–62.

Peretz, E. (2010) "What is a Cinema of Jewish Vengeance? Tarantino's *Inglorious Basterds*," *The Yearbook of Comparative Literature* 56: 64–74.

Rein, R. (2014) "A Trans-National Struggle with National and Ethnic Goals: Jewish-Argentines and Solidarity with the Republicans during the Spanish Civil War," *Journal of Iberian and Latin American Studies* 20, no. 2: 171–82.

Rein, R., and D.M.K. Sheinin eds. (2014) *Muscling in on New Worlds: Jews, Sports, and the Making of the Americas*, Boston: Brill.

Rubin, R. (2002) "Gangster Generation: Crime, Jews and the Problem of Assimilation," *Shofar* 20, no. 4: 1–20.

Shindler, C. (2009) *The Triumph of Military Zionism: Nationalism and the Origins of the Israeli Right*, London: I.B. Tauris.

Stern, N.B., and W.M. Kramer (1973) "Emil Harris: Los Angeles Jewish Police Chief," *Southern California Quarterly* 55, no. 2: 163–92.

Weber, A.D. (2019) "From Glorious Nibelungs to Inglorious Basterds: Quentin Tarantino's Refractive Retelling of Fritz Lang's Epic," *German Studies Review* 42, no. 3: 537–60.

Witcombe, M. (2016) "A Comedy of Eruvs: (Re)Locating Jewish Identity in Michael Chabon's *The Yiddish Policemen's Union* and Howard Jacobson's *J*." *Shofar* 34, no. 3: 29–51.

Zaagsma, G. (2017) *Jewish Volunteers in the International Brigades and the Spanish Civil War*, London: Bloomsbury.

CHAPTER 2

"If I Can See One … I Can Build Them": David Lopez Jr. and Weapons Manufacturing during the American Civil War

Barry L. Stiefel

This study of "armed Jews" in the Americas will look at an early example of weapons repair and manufacturing during the American Civil War (1861–65), and especially at David Lopez Jr. (1809–84) of Charleston, South Carolina, a prominent member of the Jewish community, who was involved in the Confederate firing on the Union's Fort Sumter which triggered the Civil War. Following those first shots Lopez went on to become the builder and first superintendent of South Carolina's armory in Greenville, where weapons were both made and repaired.[1] Therefore, not only did Lopez produce arms, but he enabled several of South Carolina's fighting units during the Civil War—most of whom were obviously not Jewish conscripts—to have their own guns in their struggle for state sovereignty (and slavery) separate from the federalism of the United States. Prior to the Arab-Israeli War (1947–49) and the establishment of the defense industry in Israel, Jews as manufacturers of armaments on an industrial scale were practically unheard of. While Lopez was not the first person of Jewish origin to make a gun or repair a cannon, the others who preceded him—as well as many who came after him—were small-time tradesmen and smiths. Therefore, what do the arms manufacturing achievements of Lopez reveal about the changing position of Jews in society and in armed conflict, or at least in regard to the American South of the 1860s? Lopez's work in weapons manufacturing in Civil War—era South Carolina provides a foundational starting point for understanding Jews who actually made firearms in the Americas.

1 Much of the history on David Lopez Jr. and his building construction career comes from Barry L. Stiefel, "David Lopez, Jr. (1809–84): Builder, Defender of the Confederacy, and Industrialist," *American Jewish Archives Journal* 64, no. 1 (2012): 53–81.

 | DOI:10.1163/9789004462540_003

FIGURE 2.1 Portrait of David Lopez Jr. (1809–84)
COURTESY OF SPECIAL COLLECTIONS, COLLEGE OF CHARLESTON LIBRARY

1 Contextualizing David Lopez Jr.

Gunsmithing was the vocation of a small number of Jews in Early Modern Europe and the Ottoman Empire; Daniel Jütte and Willem M. Floor mention a few examples in passing.[2] However, like most other gunsmiths during that

2 Daniel Jütte, *The Age of Secrecy: Jews, Christians, and the Economy of Secrets, 1400–1800* (New Haven: Yale University Press, 2015), 70–71; Willem M. Floor, *Guilds, Merchants, & Ulama in Nineteenth-Century Iran* (Washington, DC: Mage Publishers, 2010), 117.

period, the work was primarily a cottage industry whereby craftsmen produced arms in small orders and carried out repairs for loyal customers. Jews were by no means making firearms in significant enough numbers to outfit an army, militia, or other paramilitary group or subunit, although as merchants they sometimes served as suppliers.[3] Many cottage industry gunsmiths did not make or repair firearms exclusively, but engaged also in locksmithing, watchsmithing, jewelry-making, and other skilled trades that involved metal work with small, mechanically-functioning objects because of the transferable skills and equipment involved. Smiths often had to be versatile in their products and services in order to be economically competitive. In *This Happy Land: The Jews of Colonial and Antebellum Charleston*, James W. Hagy found in municipal records another Jewish gunsmith who operated in Lopez's native city between the 1800s and 1820s, although he does not identify this person by name.[4] So, there was at least one other Jewish gunsmith in Charleston during Lopez's youth, although we are unable to determine whether they knew one another.

David Lopez Jr. was born January 16, 1809, as the tenth of twelve children in his immediate family. Lopez Jr. scarcely knew his merchant father and namesake, who died in 1812 at the age of sixty-two (1750–1812), a few years after Lopez Jr.'s birth. Thus, he was nurtured by his mother Priscilla Moses Lopez (1775–1866), older siblings, and enslaved African servants. Lopez Sr. had been a merchant and respected member of the community who had served as president of the local synagogue, Beth Elohim, and the founding president of the Hebrew Orphan Society in Charleston. Shortly after his father's death, Lopez Jr. also became a slave owner by inheriting the “Negro Boy Named Mathew (the Child of Nancy) and also two Union Bank Shares.”[5] Utilization of enslaved Africans is a phenomenon that recurred in the Lopez family until emancipation came about in 1865 with the conclusion of the Civil War.

During Lopez Jr.'s youth the family was religiously observant, practicing the Sephardic customs of their forefathers. With each rift of the Beth Elohim congregation that occurred in 1824 and 1840, between Orthodox and Reform adherents, the Lopezes sided with the former. The Lopez family bible, *Seder Parashiyyot ve-haftarot kol ha-shanah/Quinque Libri Mosis*, published in Amsterdam circa 1726, may be found in the Jewish Heritage Collection at the

3 Barry L. Stiefel, “Jews as Middlemen in the Neutral XVIIIth Century Atlantic World,” in Éric Schnakenbourg ed., *Neutrals and Neutrality in the Atlantic World in the 18th Century (1700–1820): A Global Approach* (Mordelles, France: Editions Les Perséides), 159–84.

4 James W. Hagy, *This Happy Land: The Jews of Colonial and Antebellum Charleston* (Tuscaloosa: University of Alabama Press, 1993), 192.

5 Will of David Lopez [Sr.], January 27, 1812, Charleston County Wills, Vol. 32, CCPLSCR, Charleston, SC, 521–22.

College of Charleston. It consists of six volumes, including all the books of Moses, as well as the *haftarot*. It is a *chumash* set that was given to the College of Charleston by Thomas J. Tobias (1906–70), a local Jewish historian and Lopez relative.[6] A bookplate has been pasted by David Lopez on the back cover of each volume, suggesting that he was familiar with Hebrew and Spanish, in addition to English.

Until Lopez Jr. became an independent adult at the age of twenty-one, there is little else that we know about him. In 1832, he married his first wife, Catherine D. Hinton (1814–43) of Edgefield, South Carolina, with whom he had half a dozen children, two of whom did not survive childhood. By 1834, the family had returned to Charleston. Hinton was not Jewish nor did she ever convert to Judaism, based on her burial place, which is located outside of what was originally the Jewish cemetery in a plot purchased in advance for her by Lopez. However, since the children of this intermarriage participated in Beth Elohim's congregational functions decades later, they appear to have been raised in a setting where Judaism was part of the home environment.[7] How, when, and why Lopez was in Edgefield during the early 1830s (or earlier) is unknown. Based on surviving information, he may have been working—possibly even apprenticing—with a master builder-carpenter there in order to learn a vocational trade. His employer may even have been his father in-law-to-be, John Hinton. What is clear is that Lopez did not continue the family businesses of mercantile trading or shopkeeping, and that he took up a profession that was new not only to the Lopezes but to Jews in general during the early nineteenth century.[8]

On March 31, 1834, Lopez placed an advertisement in the *Southern Patriot*, a local newspaper:

> Carpenter & Joiner: Subscriber having returned to Charleston, will carry on the CARPENTER and JOINER'S business in all their branches and hopes by close personal attention, neatness, dispatch and Punctuality, to merit a share of pub. patronage.[9]

Within a matter of months his business must have met with great success because by August 1834 Lopez was not only advertising his services but also

6 The *haftarah* is a selection of readings from the books of the Hebrew Prophets that follow the Sabbath Torah. A *chumash* is a Torah in bound printed form.

7 *Board of Trustee Minute Book (1866–1875)*, May 5, 26, 1867, and August 2, 1868, KKBE Collection, Box 3, Folder 1, SCCC, Charleston, SC, 93–94 and 154.

8 Stiefel, "David Lopez, Jr.": 53–81.

9 Classified Ad [no title], *Southern Patriot XXXI*, no. 5047, March 31, 1834, 3.

looking to hire "five or 6 good CARPENTERS, for whom good wages will be punctually paid and constant employment given."[10] Three years later Lopez purchased his first enslaved African, Christopher, sometimes called Kit, who was trained specifically as a carpenter on Isaac Fripp's plantation on St. Helena Island, South Carolina.[11] More enslaved help would be purchased over the following years. In 1837 Lopez completed the construction of his first building, 153–155 Queen Street, a tenement house used as an income-generating property.[12] Until his building-contracting services were utilized in the late 1830s and afterward, Lopez also made ends meet as a supplier of building materials for other contractors.[13]

In 1838, Charleston experienced a devastating fire that destroyed a significant portion of the city, including the synagogue that had been standing since 1794. This proved to be a boom opportunity for Lopez as the city sought to restore it from the ashes. In 1839, he obtained the contract to build a new synagogue on the same site, completing it the following year. The construction of the second Beth Elohim was actually the first instance in the Americas in which the builder was also Jewish. All previous synagogue contractors in that hemisphere were non-Jews. However, Lopez did not use the house of worship he built for very long. After its completion, a second rift between Orthodox and Reform emerged within the congregation, with the Reformers being awarded the building in 1841 at the end of a legal battle for title ownership. Lopez did not stay long with the new Orthodox congregation, Shearit Israel, leaving in frustration in 1843 because of their refusal to allow his non-Jewish wife burial in the Jewish cemetery. A few years later, the widower Lopez remarried Rebecca Moïse (1814–58), and together they had another six children. The Moïse family belonged to the now-Reform Beth Elohim congregation that used the synagogue Lopez had constructed, but he did not rejoin any Jewish congregation until after the Civil War.[14]

In addition to numerous residences and commercial buildings, Lopez's building accomplishments also included the Old Farmer's and Exchange Bank (1853–54), Institute Hall (1854), and the renovation of the county courthouse,

10 Classified Ad [no title], *Southern Patriot XXXII*, no. 5166, August 18, 1834, 3.

11 "Isaac Fripp of St. Helena Parish, District of Beaufort to David Lopez, Bill of Sale for a Slave Named Kit or Christopher, a Carpenter by Trade," June 30, 1837, Series: S213003, Vol.: 005T, South Carolina Department of Archives and History (henceforth, SCDAH), Columbia, SC, 200.

12 Jonathan H. Poston, *The Buildings of Charleston: A Guide to the City's Architecture* (Columbia, SC: University of South Carolina Press, 1997), 409.

13 Classified Ad 9 [no title], *The Charleston Mercury*, October 14, 1847, 3.

14 Stiefel, "David Lopez, Jr.": 53–81.

which was his first major government contract. Institute Hall was used to host fairs and conventions that could hold as many as 3,000 attendees, most significantly the Democratic Party convention of 1860, which led to the Ordinance of Secession declaring the withdrawal of South Carolina from the union of the United States of America. As a result of this event, the building acquired the name Secession Hall, since it served as a form of independence hall for what would become the Confederate States of America. In 1846, Lopez also served as the chairman of Charleston's City Board of Fire Masters, set up in order to prevent fires like the one that had occurred in 1838. But as experienced as Lopez was in carpentry, masonry, and metalwork construction from his building projects, as well as in fire control management, there is no evidence by 1860 that he knew anything about firearms manufacturing or repair. By the outbreak of the Civil War, however, his expertise and capabilities had changed profoundly.

2 David Lopez Jr. and the Repair and Manufacturing of Firearms

There is no surviving firsthand testimony from Lopez about how he began repairing and manufacturing firearms, though years after his passing he was immortalized by an observer to the pivotal event at Fort Sumter: William A. Courtenay (1831–1909), who enlisted in the Confederate Army during the Civil War and later became mayor of Charleston (1879–87). On February 24, 1897, the *Weekly News & Courier* reported Courtenay's eyewitness account of "Charlestonians in the War," beginning with the preparations to fire the first shots at the Battle of Fort Sumter on April 12–13, 1861:

> When Major [Robert] Anderson burned the gun carriage at Fort Moultrie [December 26, 1860] a new trouble was presented. Governor [Francis W.] Pickens sent in haste for Mr. David Lopez, the leading carpenter contractor of the city. New heavy gun carriages had to be constructed forthwith. No such thing had ever been built in Charleston. Mr. Lopez replied: "If I can see one, or the remnants of one, I can build them." Seasoned timber had to be found, iron machinery improvised and the work started so that Moultrie's guns could be restored to their positions. The writer later placed a copy of the latest edition of the United States Ordnance Manual in Lopez's hands soon after the work begun, which, of course, expedited the work.[15]

---

15 William A. Courtenay, "Charlestonians in the War," *Weekly News & Courier*, February 24, 1897, 8.

Shortly after Lopez repaired the cannons at Fort Moultrie, Governor Pickens and Brigadier General Pierre G.T. Beauregard (1818–93) ordered that additional defensive batteries be constructed around Charleston harbor, especially on Sullivan's Island near Fort Moultrie, Mount Pleasant, and Morris Island, overlooking the harbor's entrance.[16] To Lopez's credit, the repaired cannons at Fort Moultrie and the surrounding batteries he built around Charleston harbor resulted in a quick and sound victory for the Confederate attackers at the Battle of Fort Sumter, which lasted for little more than a day. Lopez had rendered himself indispensable in repairing cannons and constructing defensive works for rebelling South Carolina at a moment when the seceding state was in dire need of people with his talent and skills. As part of the family construction business, Lopez's sons John H. and Moses E., and son in-law Daniel S. Hart (1839–77), joined their father in this endeavor and fought for South Carolina and the Confederacy.[17] All three younger men served in the Palmetto Guards, and Moses E. was stationed at Morris Island's Ironclad Battery for the bombardment on Fort Sumter, which his father may literally have had a hand in building.[18]

Following the formal adoption, on February 22, 1862, of the Constitution of the Confederate States of America, the confederate and state governments undertook the objectives of both ruling and fighting for separation from the United States. Building a robust military was therefore critical. In South Carolina, with the support of Governor Pickens, and because of the reputation he had acquired in Charleston for preparing the city's successful attack on Fort Sumter, in early March 1962 Lopez was officially appointed Superintendent of State Works, with the task of "making cannon and so forth."[19] At the time, a makeshift armory for weapons repair had already been created on the grounds of the state capitol in Columbia. Soon after his arrival there, having brought equipment and supplies with him from Charleston, Lopez found that the

16 Russell A. Alger et al. eds., *The War of the Rebellion: A Compilation of the Official Records of the Union and Confederate Armies* (Washington, DC: GPO, 1898), 131–33. Also see Charles E. Cauthen ed., *The State Records of South Carolina: Journals of the South Carolina Executive Councils of 1861 and 1862* (Columbia, SC: SCDAH, 1956), 14, in respect to Governor Picken's orders; and To David Lopez, Comptroller General (S126175), Records of the State Auditor, Board of Ordinance Receipts, April 16, 1861, SCDAH.

17 Daniel S. Hart married Priscilla Lopez (1839–82) on December 3, 1860.

18 Barnett A. Elzas, *The Jews of South Carolina: From the Earliest Times to the Present Day* (Philadelphia, PA: J.B. Lippincott Company, 1905), 229–32.

19 Cauthen, *State Records of South Carolina*: 104.

FIGURE 2.2 Photograph showing an interior view of Fort Moultrie a few days after the April 12–13, 1861, bombardment, along with the cannons that David Lopez Jr. repaired weeks earlier
PHOTOGRAPH BY OSBORN & DURBEC, LIBRARY OF CONGRESS PRINTS AND PHOTOGRAPHS DIVISION

armory site would not do and he straightaway set about creating a better facility and seeking a more favorable location.[20]

By March 12, Lopez had found an ideal location for an armory at Greenville, South Carolina, a quiet southern village in the Piedmont region. In Lopez's opinion, which was shared by South Carolina's Department of the Military, Greenville was preferable to Charleston or Columbia because its remoteness and anonymity would not only keep the armory workers attentive to their tasks, since there was no city close by where they could waste their pay on gambling

20 Robert S. Seigler, *The Best Gun in the World: George Woodward Morse and the South Carolina State Military Works* (Columbia: University of South Carolina Press, 2017), 106.

and other unrefined leisure behaviors, but would make it less prone to attract the attention of the Union army.[21] The South Carolina Railroad Company also had a depot in the town, on the Greenville-Columbia line, which could connect the armory to the rest of the state and other areas of the Confederacy. Workers, supplies, arms, and laborers could be conveyed to where they were required by rail. Moreover, iron ore could be found in the mountains around Greenville, particularly Kings Mountain, as it was an essential material in the repair and manufacturing of arms, in addition to plentiful supplies of lumber and constant access to freshwater from valley streams. Situated not far from the armory's proposed location were the Nesbit Iron Manufacturing Company and Kings Mountain Iron Company, where iron ore could be refined and smelted in preparation for arms manufacturing. This was the strategic genius of Lopez as a contractor and project manager as applied to weapons development. William H. Gist (1807–74), chief of the South Carolina Department of Construction and Manufacture, under whom Lopez served, also concurred with this assessment.[22] A successful local business and landowner in the area, Vardry McBee (1775–1864), contributed twenty acres to the state government on which Lopez constructed an industrial compound of more than fifteen buildings. These included tool houses, stables, kitchens, a hospital, barracks, storehouses, a smokehouse, office, foundry, blacksmith shop, machinist shop, small arms shop, and a carpentry shop.[23] Arms repair operations were gradually transferred from Columbia to Greenville as the armory upstate was developed so as to mitigate the disruption of work and produce additional supplies.[24]

The State Armory at Greenville, better known as the State Works, was small compared with others constructed under the Confederacy, and was part of a larger effort to develop arms manufacturing facilities for the war effort. Prior to the outbreak of the Civil War, the only foundry in what became the Confederacy was in Richmond, Virginia, which remained the largest and most important in the South. Another rifle works existed in Fayetteville, North Carolina. However, the Greenville site's output of repaired and manufactured arms was among the longest running in the Confederacy because of Lopez's intuition. In April 1862, the cities of Nashville, Tennessee, and New Orleans, Louisiana, along with their very important armories were captured by the

21 David Lopez to J. Chestnut, General Correspondence, Executive Councils of 1861 and 1862 (S160003), March 12, 1862, SCDAH.

22 Seigler, *The Best Gun in the World*: 108.

23 Francis W. Pickens, *Convention Documents: Report of the Special Committee of Twenty-One, On the Communication of His Excellency Governor Pickens, together with the Reports of Heads of Departments, and Other Papers* (Columbia, SC: R.W. Gibbes, 1862), 171–74.

24 Seigler, *The Best Gun in the World*: 107.

Union Army, depriving the Confederates of both vital arms and ammunition. As a result, fighting a rebellion became significantly more challenging. A month later Baton Rouge, Louisiana, and its armory were overcome, followed by Memphis, Tennessee (June 1862). The Confederacy was able to manage it loses for a while thanks to its cities with armories, but then in May 1863 Jackson, Mississippi, fell, followed by Little Rock, Arkansas (September 1863), Atlanta, Georgia (September 1864), Savannah, Georgia (December 1864), and Fayetteville, North Carolina (March 1865). With the defeat of Richmond, the national capital of the Confederate States of America, in early April 1865, the Civil War ended on April 9, 1865 when General Robert E. Lee surrendered after the Battle of Appomattox Court House. Greenville and the State Works built there by Lopez were not captured by the Union Army until May 2, 1865—nearly a month after the "official" termination of the Civil War. Even though weapons production and repair were relatively minimal at the State Works in 1865 due to financial and material shortages, Lopez's foresight regarding the location in Greenville was successful since it was bypassed by the Union Army.[25] The one drawback to this site, which neither Lopez nor Gist had anticipated, was its heavy dependence upon steam engines and coal in order to connect the armory to the rest of the state and the South, which became an increasing challenge as the war progressed, supplies became limited, and operating expenses grew.[26]

With the fall of Nashville to the Union Army in April 1862, Lopez was in Atlanta and telegraphed the Tennessee governor, Isham Harris (1818–97), asking him to grant to the South Carolina state government the machinery from Nashville's armory which was being stored in Georgia for a fee that would be deferred until after the end of the Civil War.[27] Harris and Pickens, governors of Tennessee and South Carolina, respectively, each consented to the arrangement and Lopez was acclaimed a hero for his brilliant idea. Thus, Tennessee transferred to the armory at Greenville, South Carolina, much of its machinery for making and repairing arms, in addition to the materiel that Lopez had

25 Stiefel, "David Lopez, Jr. (1809–84)": 53–81.

26 Seigler, *The Best Gun in the World*: 115.

27 David Lopez to Governor Isham Harris, General Correspondence, April 26, 1862, Governor Isham Harri Papers (2857–1862), Box 2, Folder 8; Tennessee State Library and Archives; and Francis W. Pickens, *Convention Documents: Report of the Special Committee of Twenty-One, On the Communication of His Excellency Governor Pickens, Together with the Reports of Heads of Departments, and Other Papers* (Columbia, SC: R.W. Gibbes, 1862), 171–74.

brought with him from Columbia and Charleston.[28] During Lopez's business trip to Atlanta in April 1862, he also reached out to George W. Morse (1812–88), who had designed a very advanced breech-loading carbine. With this type of gun, the user loads the cartridge or shell through the rear of the barrel, called a breech, instead of in the front through the muzzle. It is generally a much faster and safer way of loading a gun, especially in the midst of battle. Prior to the Civil War, many military firearms were muzzleloaders, which required the use of a ramrod to load the powder and shot, taking some time before being ready to fire again. A carbine is a long firearm, but with a shorter barrel than a standard rifle or musket, making it lighter and easier to carry, as well as preferred by soldiers mounted on horseback. In Nashville, Morse had had the equivalent position of General Superintendent of the armory there prior to the Union invasion, and was offered a position under Lopez to develop his breech-loading carbine in Greenville, as well as other small arms, along with some of his former employees.[29]

As General Superintendent himself, Lopez's duties not only entailed constructing the State Works but also supervising the repair and manufacture of arms. The highest priority weaponry to be made were cannons, and Lopez and Gist worked closely with James K. Campbell of the Tredegar Ironworks in Richmond to determine whether South Carolina iron ore was suitable for casting these sorts of heavy arms.[30] Ten tons of Kings Mountain iron was procured and shipped to the Tredegar Ironworks in Richmond with which one twenty-four pounder and one twelve pounder cannon were made and successfully tested. Lopez then contracted with the Kings Mountain Iron Company for fourteen and a half tons of pig iron and an additional 1,000 pounds of iron bar for the State Works. However, while firearms, shell, and shot were made at the State Works, he was unable to acquire the additional equipment needed to cast cannonry because of the rush to start producing arms, of which smaller guns like the Morse breech-loading carbine were more feasible.[31]

By September 1862, the State Works in Greenville were mostly developed and workers were busy building Morse breech-loading carbines.[32] However,

---

28 William H. Gist, *Report of W. H. Gist, Chief of the Department of Construction and Manufacture, to this Excellency, Governor Pickens* (Columbia, SC: R.W. Gibbes, 1862), 164.

29 Seigler, *The Best Gun in the World*: 97.

30 David Lopez to Col. James Chestnut, Chief of Department of Military, General Correspondence, March 12, 1862, Executive Councils of 1861 and 1862 (S160003), SCDAH.

31 Seigler, *The Best Gun in the World*: 109; and see correspondence of James K. Campbell with Archer Anderson, Tredegar Iron Works Records, 1801–1957, Library of Virginia, Richmond.

32 *Reports and Resolutions of the General Assembly of the State of South Carolina: Passed at the Annual Session of 1862* (Columbia, SC: Charles P. Pelham, 1862), 327–18.

Lopez soon encountered a labor shortage. Both white and black, as well as enslaved and free black workers, were recruited through newspaper advertisements across the state. For skilled white laborers he offered:

> To Machinists, Finishers and Gunsmiths
>
> Good employment and wages will be given at the state works, at Greenville, for machinists and gunsmiths. The same exemption from Conscription and Military Duty apply to men engaged in "these works" as in other Government Works. Provisions and living is much cheaper than in other sections of the state [*sic*].
>
> March 16 David Lopez, General Superintendent[33]

And

> To Owners of Negro Mechanics
>
> Wanted at the state work in the interior of this State, in a healthy and safe location, carpenters, blacksmiths, machinists, and wheelwrights. They will be taken care of and good wages paid. Apply to David Lopez, Spartanburg or M[oses] H. Nathan, corner of Wentworth and Meetings Streets [Charleston], March 3.[34]

Moses H. Nathan (1821–75) was a fellow Jewish congregant, colleague, and friend of Lopez from Charleston who, as chief of the Fire Department from 1859 to 1873, assisted with recruitment for the State Works.[35] From outside of the Greenville area, enslaved workers from across South Carolina who had been drafted would have a railroad pass purchased on their behalf so that they could be sent to the State Works.[36] By 1863, Lopez had developed a

33 Classified Ad 2 [no title], *The Charleston Mercury*, March 21, 1863, 2.

34 Classified Ad 1 [no title], *The Charleston Mercury*, March 3, 1862, 2.

35 Marie Ferrara, "Moses Henry Nathan and the Great Charleston Fire of 1861," *South Carolina Historical Magazine 104*, no. 4 (October 2003): 258–80.

36 See David Lopez, "Two Railroad Passes for Blacks Signed by Lopez as Superintendent of State Works for South Carolina," March 25, 1862, David Lopez Collection, 1862–1905, Local Call #P-458, American Jewish Historical Society, New York.

work force of 144 employees, of whom eighty—well more than half—were enslaved Africans.[37]

On the sixth day of Hanukkah (December 22) 1862, Lopez wrote to the newly elected governor of South Carolina, Milledge L. Bonham (1813–90, served 1862–64), presenting him with an early Christmas gift:

> I manufactured a Breech Loading Carbine of Morse's patent … I took it before the Military Committee and they tested the weapon with the performance of which they were much pleased … I beg to submit the Gun to you for inspection. This [is] a very cheap and efficient arm and can be easily made. Shall I proceed in the matter or await the action of the Legislature?[38]

At the time, the General Assembly was on its winter recess so the order for one thousand Morse breech-loading carbines, requested by the South Carolina House of Representatives Military Committee, could not go through its general approval process. Thus, Governor Bonham responded to Lopez on January 3, 1863, that "you should proceed with the manufacture of the arms,"[39] and forthwith production began on the order. Bonham must have had great faith in Lopez's abilities to fulfill the order of a thousand firearms since the first prototype sent to him did not arrive. Lopez corrected this loss by sending a second Morse breech-loading carbine on January 20, 1863, along with a Union preferred Burnside breech-loading carbine confiscated at the Battle of Harpers Ferry (a Confederate victory on September 12–15, 1862), to demonstrate the superiority of the Morse over the Burnside gun, the former being easier to reload and more water-resistant.[40]

---

37 *Report to the Committee on the Military (Senate)*, General Assembly (S165005), 1863, No. 363, SCDAH.

38 David Lopez. Correspondence to Milledge L. Bonham, December 22, 1862, Letters and Telegrams (S512001), Gov. Lopez to Bonham, Milledge L. Bonham Papers, Box 1, Folder 1, SCDAH.

39 Milledge L. Bonham, Correspondence to David Lopez, January 3, 1863, Letters and Telegrams (S512001), Gov. Lopez to Bonham, Milledge L. Bonham Papers, Box 1, Folder 1, SCDAH.

40 David Lopez. Correspondence to Milledge L. Bonham, January 20, 1863, Letters and Telegrams (S512001), Gov. Lopez to Bonham, Milledge L. Bonham Papers, Box 1, Folder 6, SCDAH.

3 Resignation and Dismissal from the State Works in Greenville

Progress on the production of the thousand Morse breech-loading carbines was made as 1863 unfolded; however, the work did not evolve smoothly. In December 1862, James Tupper was appointed to the position of State Auditor, and the two disagreed immensely over best accounting practices.[41] Governor Bonham also made policy changes that created unnecessary challenges from Lopez's perspective. Lastly, George W. Morse, who was jealous of his superior for getting credit for a weapons product that he had originally designed, did not respect Lopez's administration of the State Works. By July 1863, Lopez had given notice of his resignation although the actual document cannot be found; however, he did continue in the position provisionally until a replacement could be appointed.[42]

In August 1863, before Lopez could leave on his own accord as General Superintendent of the State Works, it appears that he was dismissed due to an altercation between two workers, his son Moses E. Lopez and E.A. LeBlanc. Several testimonies have survived the incident, including those of G. Ralph Smith (who would replace Lopez as General Superintendent), Lopez's son in-law Daniel S. Hart, William W. Smith, Phillip A. Mullani, and E.A. LeBlanc. All witnesses, except LeBlanc, closely concur with each another, though there is no written deposition by David Lopez Jr. himself. Hart's observations best describe what occurred that day:

> On the Afternoon of August 14th LeBlanc entered Mr. [David] Lopez's [Jr.] office and in a very abrupt manner asks Mr. Lopez if he intended charging him with a piece of bacon [purchased for $15.00 on July 25th] he had refused to keep [due to spoilage]. Mr. Lopez replied certainly, whereupon LeBlanc called him a swindler and made some other remarks which I cannot call to memory. Mr. Lopez ordered him out of his office, and threatened to discharge him if he did not return to his work immediately. LeBlanc then used the most indelicate language, which caused Mr. Lopez to curse him, and proceed to put him out of the office by force. LeBlanc appeared as if he intended to strike Mr. Lopez as he advanced to put him out of the office, and then Mr. Lopez's son [Moses E.] said,

41 Charles E. Cauthen and J.T. Power, *South Carolina Goes to War, 1860–1865* (Columbia: University of South Carolina Press, 2005), 198.

42 Seigler, *The Best Gun in the World*: 145.

> placing his hand on a gun that was by him "If you do, I'll blow your brains out." Some few words passed between them and the affair ended.
>
> Very respectfully, Daniel S. Hart[43]

The incident is odd. There may have been a record of ill feeling between LeBlanc and Superintendent Lopez, which we know nothing about. Bacon was a very significant meat staple at the State Works in Greenville, which the employees took seriously. For example, from April 8, 1862, there is a record of a line amount for reimbursement that Lopez submitted to his superiors in Columbia for $3,500 worth of bacon from Ashville, North Carolina, the equivalent of approximately $89,000 today.[44] G. Ralph Smith followed with his own inquiry of LeBlanc's bacon in question, stating that,

> I examined this (rotten) bacon, and found two-thirds perfectly sound, and the balance preferred (by the negroes) to beef … I will add, that the bacon when given to LeB[lanc] was pronounced sound and [?] by several of the best men here, some of whom were anxious to get it.[45]

Although Lopez was vindicated over the LeBlanc bacon affair, a General Superintendent with a son employed at the works who was hasty to pull a firearm in a quarrel, particularly at an armory where weapons were being made, was a liability.

As of August 1863 a total of three Morse breech-loading carbine prototypes had been made at the State Works in Greenville, with hundreds of partials near completion. The first batch was only finished the following month, and the Edward M. Boykin Mounted Rifles picked up the completed Morse breech-loading carbines in November 1863. Half of the Mounted Rifles Squadron were later recommissioned into Company K, Seventh Regiment of the South Carolina Cavalry, but their Morse breech-loading carbines were replaced with Enfield rifles by the Confederate States government. As more carbines were completed, additional South Carolina military units received the firearms, including, in 1864, Percival's Company of Mounted Infantry which used them

---

43 Daniel S. Hart, Correspondence from Report of G. Ralph Smith, August 21, 1863, Comptroller General, State Auditor, SCDAH.

44 David Lopez, Correspondence to Lieutenant Governor W.W. Harlee, April 8, 1862, General Correspondence, Executive Council (S160003), SCDAH. No evidence exists of whether the Lopezes kept kosher while in Greenville.

45 G. Ralph Smith, Correspondence and Report of G. Ralph Smith to Mr. James Tupper, State Auditor, August 21, 1863, Comptroller General, State Auditor, SCDAH.

at the Battle of Aiken, South Carolina, in February 1865. Other units included Frederick's Company of State Cavalry, the Pendleton Mounted Infantry, and Brabham's Independent Calvary Company. When the State Works were closed officially in May 1865, an estimated 1,178 Morse breech-loading carbines had been made at the facility built under Lopez's supervision, although most of the arms were not completed until after he had ceased being the General Superintendent.[46]

Lopez remained a short while longer in Greenville before leaving. In May 1864, Lopez reported from Columbia that his enslaved worker Jim had run away. Lopez's other enslaved laborers, Kit, George, Barber, Marcus, Jacob, and John, continued at the State Works, with Lopez collecting their income. His son Moses E. Lopez returned to military service, enlisting with the Palmetto Guard Artillery.[47] Around the time of the Civil War's conclusion Lopez Jr. was in Charleston and during Reconstruction he helped to rebuild the damaged Beth Elohim synagogue he had originally constructed in 1839–40. Due to his experience as General Superintendent of State Works in Greenville, Lopez was later hired as the General Superintendent of the Coosaw Mining Company near Beaufort, South Carolina; a position he would hold for the remainder of his life. Specializing in phosphate extraction, Lopez's development and management of the Coosaw mines made the company very successful, and it became the largest producer in the state, with a net worth of above one million dollars. Moses E. Lopez was hired as an assistant and succeeded in the position of General Superintendent after his father's passing. Never again did David Lopez Jr., Moses E. Lopez, or any other member of his immediate family work with or take up arms again following the Civil War.[48]

4 Conclusion

In August 1960, in preparation for events commemorating the centennial anniversary of the Civil War, W.W. Luckett, superintended of what is now the Fort Sumter and Fort Moultrie National Historic Park, had displayed at the visitor center museum an unassuming United States Ordnance Manual from 1861. According to news coverage in *The Evening Post*, Luckett knew little about Lopez. Within the booklet was a handwritten note signed by W.A.C. (William A. Courtenay) and explaining why this particular manual mattered.

---

46 Seigler, *The Best Gun in the World*: 183–84, 240–41.

47 Ibid.: 151.

48 Stiefel, "David Lopez, Jr.": 53–81.

It opened an important window to a very significant place in American history that Luckett was charged with maintaining:

> A day or two after Major Anderson removed to Fort Sumter (from Fort Moultrie) with his Garrison, as I was entering the governor [Picken]'s office, which was then in Meeting Street, one door north of Horlbeck Alley, I met Mr. David Lopez. He asked "Can you tell me how I can get a book on the design and dimensions of carriages for heavy guns?"
>
> I had collected a few military books and had a copy of the U.S. Ordnance Manual just off the press. This I loaned to Mr. Lopez. It enabled him to make carriages for all the guns being mounted in fortifications [in Charleston]. The book was returned after the war and this note is to show its connection to the war.[49]

Based on this message from Courtenay, Luckett came to better understand the key events that occurred between December 1860 and April 1861 and which resulted in the Battle of Fort Sumter. While Courtenay's quick thinking on December 27 or 28, 1860, was essential when he lent Lopez the manual for rebuilding the cannon carriages at Fort Moultrie, in retrospect, his efforts had greater foresight since he documented that moment in the manual for posterity upon its return. We can infer from this testimony that Lopez, one of the most respected contractors in Charleston, as well as a prominent member of the local Jewish community, was not just the first Jewish person to produce arms during the Civil War in support of South Carolina's withdrawal from the Union, but the very first person to do so for the war as a whole. While Major Anderson may have attempted to quickly beat metaphorical swords into plowshares when he sabotaged the cannons as he and his soldiers withdrew from Fort Moultrie, Lopez was no biblical Isaiah and personally saw to it that the swords were made ready again for the first battle.

While there is no evidence to suggest that Lopez actually fired a gun or cannon in a combat situation, his skills pertaining to arms was to enable others to do the actual shooting. And just as one deed (good or bad) leads to another, so too did the making of arms for Lopez. His crucial work in Charleston, combined with his years of experience and reputation as a contractor, resulted in him becoming the General Superintendent of the State Works in Greenville, where he both built the facility and made and repaired firearms there. His efforts in the Greenville area brought about the casting of two cannons and

49 Belvin Horres, "Rare Book to be Shown in Fort Sumter Museum," *The Evening Post* (Charleston, SC), August 15, 1960.

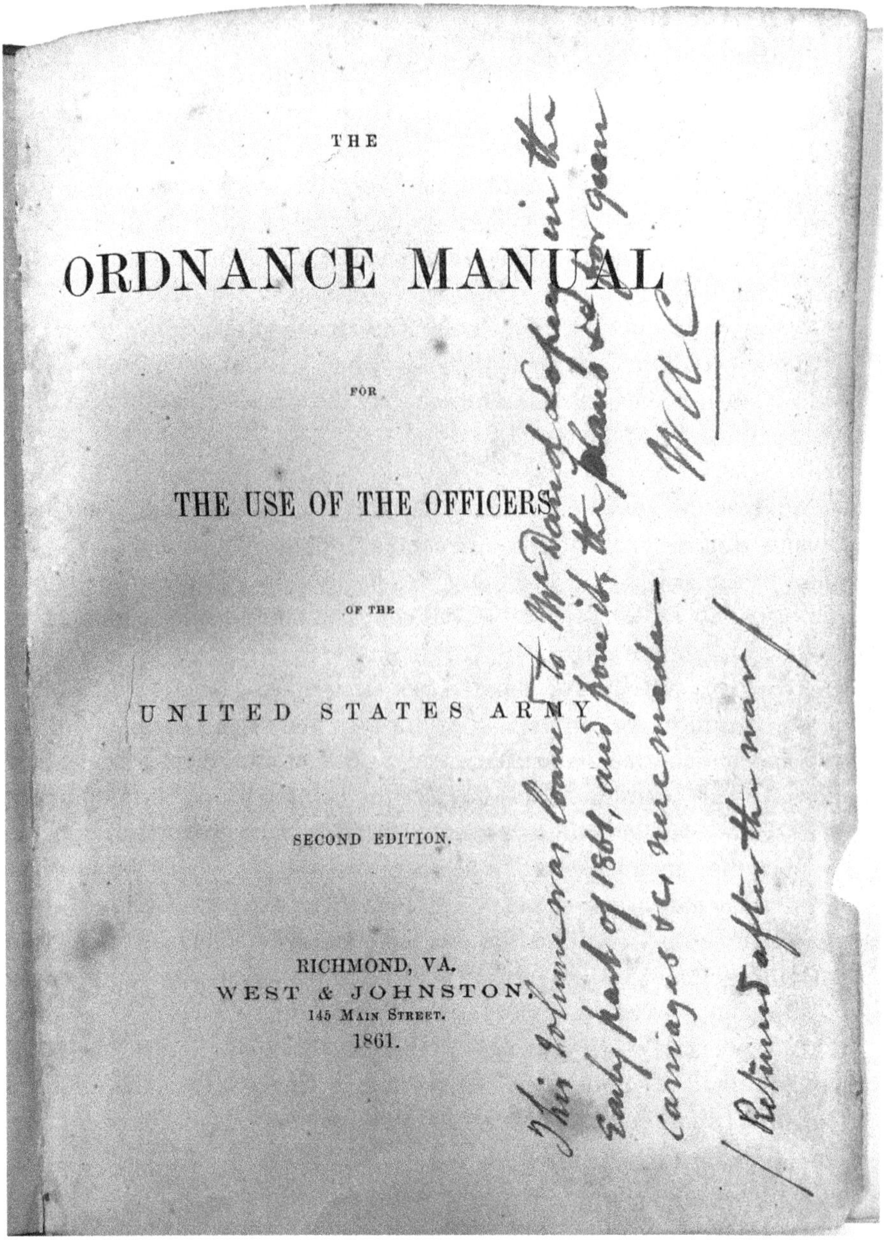

FIGURE 2.3 The cover page of the U.S. Ordinance Manual lent by William A. Courtenay with a second inscribed hand note: "This volume was loaned to Mr. David Lopez in the early part of 1861, and proves it, the plans and for gun carriages etc., were made. Returned after the war. WAC."
COURTESY OF U.S. DEPARTMENT OF THE INTERIOR, NATIONAL PARK SERVICE, FORT SUMTER NATIONAL MONUMENT

the manufacture of more than 1,100 Morse breech-loading carbines, a rifle considered to be one of the best small arms in the world at the time.[50] While the cause he chose, the Confederate States of America, lost the war and in recent years has been regarded as being on the wrong side of history due to its insistence on maintaining the institution of African chattel slavery (of which Lopez also engaged in heavily, even in weapons manufacturing), Lopez probably achieved more than any other Jew in the Americas regarding the production of arms until that point. Lopez's achievements demonstrate that Jews in contemporary society, or at least in North America, were becoming increasingly more accepted as equals and that they could participate as such on both the frontlines and the home front when it came to the making and handling of arms.

Bibliography

Alger, R.A. et al., eds. (1898) *The War of the Rebellion: A Compilation of the Official Records of the Union and Confederate Armies*, Washington, DC: GPO.

Cauthen, C.E., ed. (1956) *The State Records of South Carolina: Journals of the South Carolina Executive Councils of 1861 and 1862*, Columbia, SC: SCDAH.

Cauthen, C.E., and J.T. Power (2005) *South Carolina Goes to War, 1860–1865*, Columbia: University of South Carolina Press.

Courtenay, W.A. (1897) "Charlestonians in the War," *Weekly News & Courier*, February 24, 8.

Elzas, B.A. (1905) *The Jews of South Carolina: From the Earliest Times to the Present Day*, Philadelphia, PA: J.B. Lippincott Company.

Ferrara, M. (2003) "Moses Henry Nathan and the Great Charleston Fire of 1861," *South Carolina Historical Magazine 104*, no. 4 (October): 258–80.

Floor, W.M. (2010) *Guilds, Merchants, & Ulama in Nineteenth-Century Iran*, Washington, DC: Mage.

Gist, W.H. (1862) *Report of W.H. Gist, Chief of the Department of Construction and Manufacture, to His Excellency, Governor Pickens*, Columbia, SC: R.W. Gibbes.

Hagy, J.W. (1993) *This Happy Land: The Jews of Colonial and Antebellum Charleston*, Tuscaloosa: University of Alabama Press.

Horres, B. (1960) "Rare Book to be Shown in Fort Sumter Museum," *The Evening Post* [Charleston, SC], August 15.

Jütte, D. (2015) *The Age of Secrecy: Jews, Christians, and the Economy of Secrets, 1400–1800*, New Haven: Yale University Press.

50 Seigler, *The Best Gun in the World*: 51.

Pickens, F.W. (1862) *Convention Documents: Report of the Special Committee of Twenty-One, On the Communication of His Excellency Governor Pickens, Together with the Reports of Heads of Departments, and Other Papers*, Columbia, SC: R.W. Gibbes.

Poston, J.H. (1997) *The Buildings of Charleston: A Guide to the City's Architecture*, Columbia, SC: University of South Carolina Press.

Reports and Resolutions of the General Assembly of the State of South Carolina: Passed at the Annual Session of 1862 (1862), Columbia, SC: Charles P. Pelham.

Seigler, R.S. (2017) *The Best Gun in the World: George Woodward Morse and the South Carolina State Military Works*. Columbia: University of South Carolina Press.

Stiefel, B.L. (2012) "David Lopez, Jr. (1809–84): Builder, Defender of the Confederacy, and Industrialist," *American Jewish Archives Journal 64*, no. 1: 53–81.

Stiefel, B.L. (2015) "Jews as Middlemen in the Neutral XVIIIth Century Atlantic World," in Éric Schnakenbourg ed., *Neutrals and Neutrality in the Atlantic World in the 18th Century (1700–1820): A Global Approach*, Mordelles, France: Editions Les Perséides, 159–84.

CHAPTER 3

American Jewish Gangsters

Larry D. Gragg

"By the 1950s," states Rachel Rubin, author of *Jewish Gangsters of Modern Literature*,

> Jewish gangsters had disappeared from the newspaper headlines and political discussions so thoroughly that nearly every historical work I have read on Jewish gangsters written after 1960 opens with some kind of expectation that readers will find it hard to believe that there ever was such a thing as a powerful Jewish mob in the first place.[1]

Yet, the Prohibition era produced many Jewish gangsters. The heaviest concentration was in New York, but they could also be found in most major cities, from Chicago and Philadelphia to Detroit and Minneapolis. The successful ones transitioned from the riches made in bootlegging liquor to labor racketeering, prostitution, contract killings, the narcotics trade, and gambling. A few found legitimacy in the rapidly growing post-World War II tourist center of Las Vegas. Regardless of their paths, most Prohibition-era Jewish gangsters represented a one-generational phenomenon because they distanced their families from their illegal, often violent endeavors and because most Jews enjoyed substantial upward mobility in the postwar economy that made the gangster pathway much less attractive.

Prior to the large immigration of Jews to America beginning in the late nineteenth century, most Americans saw them as a peaceful people. A Washington DC newspaper, for example, noted that Jews had "fewer criminals among them than any other race," and *Century* magazine claimed "that only two murderers have sprung from their ranks in two hundred and fifty years."[2] Moreover, as

1 Rachel Rubin, "Gangster Generation: Crime, Jews and the Problem of Assimilation," *Shofar: An Interdisciplinary Journal of Jewish Studies* 20, no. 4 (Summer 2002): 1–2.

2 Quoted in Jenna Weissman Joselit, *Our Gang: Jewish Crime and the New York Jewish Community, 1900–1940* (Bloomington, IN: Indiana University Press, 1983), 1.

 | DOI:10.1163/9789004462540_004

historian Gil Ribak discovered, "many Jews" had come to see "themselves as a nation that loathed and shunned violence."[3]

Yet, by the first decade of the twentieth century, a radically different view of Jews had emerged in the national press. In 1908, New York City Police Commissioner Theodore A. Bingham charged in the *North American Review* that "the Hebrews" were responsible for half of all crimes committed in his city. Specifically, he characterized them as "burglars, firebugs, pickpockets and highway robbers."[4] A 1909 article in *McClure's*, the popular muckraking magazine, claimed that Jews not only controlled prostitution in New York City but were also responsible as procurers and pimps for spreading it across the country.[5] There was also a rapidly rising tide of anti-Semitism in that era. As Leonard Dinnerstein has pointed out, "Jews, by definition, were considered inferior because they were not Christians" and they often "became scapegoats for the ills in society."[6] These attitudes led to barriers for Jews in housing, jobs, and education.

Those changing perceptions were due, in large part, to the arrival of over two million Jews in America between 1880 and 1915, 600,000 of them between 1903 and 1907 alone. The majority came from Russia and Poland.[7] Once they cleared Ellis Island, most settled on Manhattan's Lower East Side, or in Brooklyn. Of the Lower East Side, Cornell economics professor Walter E. Lagerquist wrote in the *New York Times* in 1910: "In no equal area can we find so many people as are crowded together in this little quarter."[8]

Most of the Jewish immigrants had some "industrial skills," and the majority worked in the clothing trades. Those without skills found themselves on the "lowest rung of the entrepreneurial ladder as vendors and peddlers."[9] Decent

3 Gil Ribak, "'The Jew usually left those crimes to Esau': The Jewish Responses to Accusations about Jewish Criminality in New York, 1908–1913," *Association for Jewish Studies Review 39*, No. 1 (April 2014): 2.

4 Theodore A. Bingham, "Foreign Criminals in New York," *North American Review 188*, no. 634 (September 1908): 384–85.

5 George Kibbe Turner, "The Daughters of the Poor: A Plain Story of the Development of New York as a Leading Centre of the White Slave Trade of the World, under Tammany Hall," *McClure's Magazine 34* (November 1909): 45–61.

6 Leonard Dinnerstein, *Anti-Semitism in America*, paperback ed. (New York: Oxford University Press, 1995), 246.

7 Samuel Joseph, *Jewish Immigration to the United States From 1881 to 1910*, reprint (New York: Arno Press, 1969), pp. 162–74, and Stephen Birmingham, *"The rest of us": The Rise of America's Eastern European Jews* (Boston: Little, Brown and Co., 1983), 15.

8 Walter E. Lagerquist, "Social Geography of the East Side," *New York Times*, April 3, 1910, SM1.

9 Moses Rischin, *The Promised City: New York's Jews, 1870–1914* (Cambridge: Harvard University Press, 1962), 59, 67; Thomas Kessner, *The Golden Door: Italian and Jewish Immigrant Mobility in New York City, 1880–1915* (New York: Oxford University Press, 1977), 19.

affordable housing was beyond the means of most of these immigrants. Overcrowded apartments with meager furnishings were the rule, a challenge exacerbated by the necessity to take in boarders to help make ends meet.[10] When not in school or at work, boys and young men spent much of their time in the streets. A contemporary journalist argued that this milieu provided a "nursery in crime."[11] "A most important factor in street life, and one that easily escapes the chance observer," according to a 1900 University Settlement Society Report, was "the 'tough' gang. These gangs, made up of fellows from six or eight to twenty years of age, are popularly known as 'grafters', or pickpockets." The "hangouts" were "on street corners, in alleyways, and in poolrooms frequented only by boys and young men, where gambling at cards goes on openly."[12]

Ethnicity also played a role. Jewish gangs, as well as Irish and Italian gangs, defended neighborhoods and families. Meyer Lansky recalled being "beaten up daily by the Irish boys." He and his friends "could run away, or we could fight back—and fighting back meant everything connected with that."[13] He would have agreed with gang leader Max "Kid Twist" Zweifach, who contended, in the dismissive language of the era, that Jewish gangs would let "no 'wop' and no 'mick'" take over their neighborhoods.[14]

Beyond defending their people and their turf, gang activity often led to arrests, largely for committing property crimes. In New York City between 1900 and 1915, only about 10 percent of arrests of Jews were for violent crimes.[15] This fit the pattern of criminal behavior that prevailed among Jews in Eastern Europe, where "Jewish malefactors were usually involved in property-related crimes and vice, and stayed away from violent offenses."[16]

Still, prior to 1920, a few powerful gangsters that were willing to use violence to achieve their ends emerged on the Lower East Side. Monk Eastman assembled a large gang in the early 1890s which provided muscle in elections for political boss Tim Sullivan, extorted money from stores, restaurants, and street vendors, and committed numerous robberies while organizing gambling houses and prostitution. According to his biographer, by 1900 "Monk's gang had grown into a twelve-hundred-strong army of thieves, pickpockets,

10 Kessner, *Golden Door*: 131; Birmingham, *Rest of Us*: 23.
11 Quoted in Joselit, *Our Gang*: 24.
12 Quoted in Irving Howe and Kenneth Libo eds., *How We Lived: A Documentary History of Immigrant Jews in America, 1880–1930* (New York: Richard Marek, 1979), 61.
13 Quoted in Robert Lacey, *Little Man: Meyer Lansky and the Gangster Life* (Boston: Little, Brown and Co., 1991), 37.
14 Ribak, "Jewish Responses to Accusations": 8.
15 Joselit, *Our Gang*: 33.
16 Ribak, "Jewish Responses to Accusations": 5.

safecrackers, prostitutes, and thugs." Their various enterprises made Eastman and his men rich, and they knew how to spend the money. They "wore expensive clothes, drank good liquor, squired the best-looking prostitutes, and had money to burn."[17] Monk's reign as a gangster boss ended in 1904 when he was dispatched to Sing Sing Prison after a conviction for robbery and felonious assault. Others like Zweifach and Big Jack Zelig followed Eastman.[18] Their notoriety appealed to some Jewish teens. "Our heroes were great fighters, soldiers or strong arm hoodlums who were top gangsters," Samuel Goldberg recalled. "Wrongly, we tried to emulate them. Police and law-enforcement officers were our enemies. We were continually at war between ourselves or with gangs from other districts that were of different races and religions."[19]

Yet, the vast majority of young Jewish men did not opt for a life of crime. Scholars have puzzled over the key motives of some among that generation of Jews who chose crime as a way of life. There is little doubt that Goldberg's admission of hero worship helps explain the appeal, but historian Robert Rockaway has a more persuasive answer. He asserted that first and second generation immigrants faced a truly challenging economic environment and that anti-Semitism was widespread. Yet, he pointed out, those who chose crime as a way of life "were no more deprived or suffering than their peers who grew up in the same slum or overcrowded immigrant quarters, but went legitimate." For example, gangster Louis "Lepke" Buchalter's brothers and half-brother became a dentist, a pharmacist, and a rabbi; his half-sister became a schoolteacher.[20] "Most likely," Rockaway concluded, "these men selected careers in crime because they wanted money, power, recognition and status, and they wanted it fast. Crime offered them a quick way to realize their dreams."[21] Political scientist Peter Lupsha agreed. "Their choice was an individual decision," he argued, "a self-choice, reinforced by peers, experience, and a talent for violence."[22] Theirs was "a chosen career path, a rational choice,

---

17 Neil Hanson, *Monk Eastman: The Gangster Who Became a War Hero* (New York: Alfred A. Knopf, 2010), 16–57; Joselit, *Our Gang*, 25–7; Herbert Asbury, *The Gangs of New York: An Informal History of the Underworld* (Garden City, NY: Garden City, 1928), 272–98.

18 Robert A. Rockaway, *But He Was Good to His Mother: The Lives and Crimes of Jewish Gangsters*, paperback ed. (Jerusalem: Gefen Publishing House, 2000), 106–12.

19 Quoted in Howe and Libo, *How We Lived*: 60.

20 Paul R. Kavieff, *The Life and Times of Lepke Buchalter: America's Most Ruthless Labor Racketeer* (Fort Lee, NJ: Barricade Books, 2006), 9.

21 Rockaway, *Good to His Mother*: 59, 61.

22 Peter A. Lupsha, "Individual Choice, Material Culture, and Organized Crime," *Criminology: An Interdisciplinary Journal* 19, no. 1 (May 1981): 15–16.

rooted in one perverse aspect of our values: namely, that only 'suckers' work, and that in our society, one is at liberty to take 'suckers' and seek easy money."[23]

The ratification of the Eighteenth Amendment to the Constitution, which prohibited the manufacture, sale, and distribution of "intoxicating liquors," and the passage of the Volstead Act, which implemented the amendment, both approved in 1919, began the Prohibition era and made the gangster lifestyle even more attractive.[24] Italian, Irish, and Jewish gangs saw the limitation on Americans' personal liberty as an opportunity to make enormous profits through illegal liquor trafficking. While it is impossible to determine precisely how much gangsters made through bootlegging, Emory Buckner, the US attorney for New York's Southern District, claimed that illegal liquor sales were about $3.6 billion a year.[25] It was a lure few criminals could eschew. As author Paul J. Vanderwood explained, "Bootlegging encouraged petty criminals to organize in mobs and then syndicates." The results were often stunning, as "uneducated tough guys became millionaires: brutal killers morphed into Robin Hoods and heroes."[26]

Arnold Rothstein emerged as the most consequential of the New York City Jewish gangsters. Unlike other Jewish mobsters of that era, most of whom were first or second generation immigrants born into poverty, Rothstein, who was born in 1882, grew up in a middle-class family. However, his father's successful cotton goods business had no appeal for the young man, who loved gambling. Rothstein established successful gambling halls both upstate at Saratoga Springs and in Manhattan. He fixed horse races and invested in nightclubs, Broadway plays, and illegal narcotics, while becoming a big player in labor racketeering. By most accounts, he also made thousands by helping fix the 1919 World Series.[27]

By 1920, Rothstein had become a respected powerbroker. As a recent biographer described his range of influence:

---

23 Ibid.: 22.

24 There are many good studies of Prohibition. See, for example, Daniel Okrent, *Last Call: The Rise and Fall of Prohibition*, paperback ed. (New York: Scribner, 2011); Michael A. Lerner, *Dry Manhattan: Prohibition in New York City*, paperback ed. (Cambridge, MA: Harvard University Press, 2008).

25 Okrent, *Last Call*: 274.

26 Paul J. Vanderwood, *Satan's Playground: Mobsters and Movie Stars at America's Greatest Gaming Resort* (Durham, NC: Duke University Press, 2010): 104.

27 David Pietrusza, *Rothstein: The Life, Times, and Murder of the Criminal Genius Who Fixed the 1919 World Series*, paperback ed. (New York: Basic Books, 2011), 1–105, 147–92, 316–29.

> Arnold Rothstein became not just the Great Bankroll, but the great go-between. If politicians wanted something from gamblers and vice lords, they approached A.R. If the underworld sought protection from Tammany's judges and prosecutors and pliant police officers, it, too, approached Rothstein. He made things happen, quietly and without fuss. More importantly, he left no trail of evidence. Everything ran smoothly and profitably.[28]

Rothstein's biggest impact was his role as a key figure in developing true organized crime. As Jenna Weissman Joselit has argued, "Rothstein transformed criminal activity from a haphazard, often spontaneous, endeavor into one whose hallmarks—specialized expertise, administrative hierarchy, and organizational procedure—corresponded to the classic sociological model of a bureaucracy."[29] Because of this, contemporaries simply called him "the Brain." Before his murder in 1928, Rothstein had provided the funding for several other gangster enterprises, always looking for new protégés for whom he would be the behind-the-scenes investor. According to Meyer Lansky's biographer Robert Lacey, these were "self-contained deals that were sealed with a handshake and recorded, on both sides, in the head."[30]

Two Jewish gangsters in particular benefitted from Rothstein's largesse. Irwin Wexler, better known as "Waxey" Gordon, was born in 1888. The son of immigrant parents on the Lower East Side, Gordon took to the streets at a young age and joined a gang led by Dopey Benny Fein. He became a skilled pickpocket, known as the young man with "waxey" hands. Gordon moved on to bookmaking, strikebreaking, and burglary. In the early 1920s, with funding from Rothstein, Gordon developed a lucrative trade in imported liquor from England, the West Indies, and Canada. By 1925, Gordon was banking about $200,000 a month. He used these profits to invest heavily in illegal distilleries and breweries in the United States.[31] According to historian Stephen Fox, Gordon "probably made and distributed more illegal beer than anyone else on the East Coast."[32]

Meyer Lansky, who had many conflicts with Gordon in the liquor trade, was Rothstein's leading protégé. Lansky later recalled his first encounter with

28 Ibid.: 92.

29 Joselit, *Our Gang*: 143.

30 Lacey, *Little Man*: 50.

31 Rockaway, *Good to His Mother*: 15–16; Albert Fried, *The Rise and Fall of the Jewish Gangster in America* (New York: Holt, Rinehart, and Winston, 1980), 94; Joselit, *Our Gang*: 95–96.

32 Stephen Fox, *Blood and Power: Organized Crime in Twentieth-Century America*, paperback ed. (New York: Penguin Books, 1990), 30.

Rothstein, who told him that "he had picked me because I was ambitious and 'hungry'."[33] Following his mentor's death, Lansky became New York's key banker and "financial manipulator" in the illegal liquor trade.[34] The son of immigrant parents, Lansky left school at fifteen to begin work in a tool-and-die shop where the foreman told him by the time he was thirty-five he could be making $8 a day. While he was good at his job, "for better or worse," Lansky "decided that would not be my path in life."[35] Instead, a year after passage of the Volstead Act, and with funding from Rothstein, Lansky opened a garage on the Lower East Side with friends Benjamin "Bugsy" Siegel and Moe Sedway. They rented trucks to bootleggers, and Lansky and Siegel also hired on as guards for men like Frank Costello. When they began their own liquor trafficking, Lansky and Siegel invested in a couple of speakeasies, which expanded rapidly thanks to liquor imported from Great Britain. Lansky's network extended as far as Cleveland, Ohio, where he did business with Jewish gangster Moe Dalitz, the most powerful bootlegger in that city.[36]

Historian Mark Haller has shown that Jews like Lansky and Gordon were more important than any other ethnic group among Prohibition-era gangsters. "Data on those who were leading bootleggers by the late 1920's," Haller wrote, "indicate that some 50 percent were of Jewish background, some 25 percent of Italian background, and the rest primarily Irish and Polish in background."[37] Since nearly half of American Jews lived in New York City, it is not surprising that "Jewish bootlegging in the United States was centered" there.[38]

Some Jewish gangsters were not involved in bootlegging, or it was only one of many illegal activities in which they engaged. For example, "Lepke" Buchalter and his partner Jacob "Gurrah" Shapiro focused upon labor racketeering in the garment, food, and trucking industries. Utilizing as many as 250 enforcers, Buchalter and Shapiro made millions in fees and payoffs.[39]

There were notable Jewish gangsters in many cities, including Abner "Longie" Zwillman and Joseph "Doc" Stacher in Newark, Charles "King" Solomon in

33 Lacey, *Little Man*: 49.

34 Fox, *Blood and Power*: 58.

35 Dennis Eisenberg, Uri Dan, and Eli Landau, *Meyer Lansky: Mogul of the Mob* (New York: Paddington Press, 1979), 59–60.

36 Lacey, *Little Man*: 51–5.

37 Mark H. Haller, "Bootleggers and American Gambling, 1920–1950," *Gambling in America*, Appendix 1 (Washington: Commission on the Review of the National Policy toward Gambling, 1976), 109.

38 Fox, *Blood and Power*: 25.

39 Kavieff, *Lepke*: 29–68; Joselit, *Our Gang*: 122–27.

Boston, Max "Boo Boo" Hoff in Philadelphia, Moe Dalitz in Cleveland, Jack "Greasy Thumb" Guzik in Chicago, Abe Bernstein in Detroit, and Isadore "Kid Cann" Blumenfeld and Dave Berman in Minneapolis. However, the largest concentration was in New York City. In addition to Rothstein, Lansky, Gordon, Buchalter, Shapiro, Siegel, and Sedway, Arthur "Dutch Schultz" Flegenheimer, Harry Greenberg, and Abraham Reles were among the most notable New York Jewish gangsters.[40]

When Prohibition ended in 1933, Jewish gangsters focused upon other endeavors, many of which they had pursued in the 1920s—investing in nightclubs, prostitution, narcotics trafficking, racketeering, and gambling. It was the latter that often proved to be the most lucrative. Knowing that most Americans liked to gamble, gangsters exploited the numbers racket—an illegal game in which the player places a bet with a bookmaker on three numbers—provided slot machines, fought to gain control of the race wire business, and opened illegal casinos. Meyer Lansky was the most successful in making the transition to gambling.

With the repeal of the Eighteenth Amendment to the Constitution, Lansky opted not to invest in the narcotics trade or prostitution. Rather, he focused almost all of his energy on gambling. He pursued opportunities wherever they emerged, from Upstate New York and the Deep South, to Cuba and Las Vegas, Nevada. In the process, Lansky contributed to the development of the "carpet joint" rather than the "sawdust joint" style of casinos. Besides table games and slot machines, a "carpet joint" usually had a hotel and a stylish nightclub with dinner shows featuring well-known entertainers. Lansky had actually explored opportunities to expand gambling in Saratoga Springs, New York, before the end of Prohibition. The city boasted the nation's first horse-racing track, elegant hotels, mineral springs, and casinos established by men like John Morrissey and Richard Canfield in the nineteenth century. These features drew Lansky like a magnet, and he transformed a restaurant into the lavish Piping Rock casino. He catered to the wealthy and famous who came for the horse racing each August. In 1938, as Lansky developed one of the premier carpet joints in Saratoga, Cuban dictator Fulgencio Batista invited the gambler known for running honest and successful casinos to clean up gambling operations in Havana. Lansky not only did so, but in 1957 also opened the Riviera, the largest Havana casino. In order to attract gamblers from Miami in the 1930s and the 1940s, Lansky opened casinos in Hallandale just north of the city. In 1941, Lansky and

40 Allan R. May includes biographies of several of the New York gangsters in *Gangland Gotham: New York's Notorious Mob Bosses* (Santa Barbara, CA: Greenwood Press, 2009).

a partner opened a dog-racing track in Council Bluffs, Iowa, and five years later he had a large interest in the Beverly Club outside New Orleans.[41]

Lansky almost always had partners in these ventures, some of whom were Italian gangsters he had known since the early days of Prohibition, among them Frank Costello and Joe Adonis who collaborated with him in the Piping Rock casino and Costello in the Beverly Club. Vincent "Jimmy Blue Eyes" Alo joined him in developing casinos in Florida. These alliances illustrate how Jewish gangsters, and not just Lansky who had cooperated often with Lucky Luciano, prospered with Italian counterparts in organized crime. They had learned that it was essential to overcome decades-long ethnic hostilities in order to succeed in the illegal liquor trade. As historian Marni Davis explained, "Such cooperation was required by bootlegging's division of labor into manufacturing, accounting, armed protection of stock, and transport of goods to speakeasies and warehouses."[42] While he may have overstated the case, George Wolf, a Jewish lawyer from that era who represented both Italian and Jewish gangsters, elaborated: "The two groups have always worked in surprisingly good harmony. The Italians respecting the Jews for their financial brains, and the Jews preferring to stay quietly behind the scenes and let the Italians use the muscle needed."[43] Jewish domination of contract murders in the 1930s through an organization journalist Harry Feeney called "Murder, Inc." was an obvious exception to Wolf's conclusion.

Throughout the 1930s, "Murder, Inc." carried out numerous contract hits. Headed by "Lepke" Buchalter and Italian gangster Albert Anastasia, the nearly two dozen contract killers responsible for a large number of murders were mainly Jewish gangsters, many of whom came from the Brownsville section of Brooklyn. Abe Reles, Martin "Buggsy" Goldstein, Harry "Pittsburgh Phil" Strauss, Allie "Tick Tock" Tannenbaum, Harry Greenberg, and Abraham "Pretty" Levine were the most prominent hitmen. Reles became an informer of the organization in 1940 and gave evidence regarding about 70 murders. Reles, who took the nickname of "Kid Twist" from his hero Max Zweifach, died from a fall, or a push, from a hotel room on Coney Island while under police protection. His demise led many to label him the "canary that could sing, but could not fly." His

---

41 Lacey, *Little Man*: 79–111, 149.

42 Marni Davis, *Jews and Booze: Becoming American in the Age of Prohibition* (New York: New York University Press, 2012), 149.

43 Quoted in Selwyn Raab, *Five Families: The Rise, Decline, and Resurgence of America's Most Powerful Mafia Empires* (New York: Thomas Dunne Books, 2005), 40. See also Fox, *Power and Blood*: 74–75.

evidence, however, led to the arrest of many, including "Lepke" Buchalter who became the only major gangster to die in the electric chair.[44]

When Las Vegas began to develop a significant resort-casino business in the 1940s, predictably Lansky was part of that initiative. He invested in the El Cortez Hotel in 1945, the Flamingo Hotel in 1946, and the Thunderbird Hotel in 1948.[45] Because Nevada was the only state with legal gambling and legal off-track betting on horseraces, many Jewish gangsters joined Lansky's pilgrimage to the emerging resort city. Dave and Charles "Chickie" Berman, along with Willie "Ice Pick" Alderman, from Minneapolis, partnered with Lansky, Moe Sedway, "Bugsy" Siegel, and Phoenix gangster Gus Greenbaum in the purchase of El Cortez. Siegel, Sedway, and New York gangster Morris Rosen collaborated with Lansky in raising funds for the Flamingo. Moe Dalitz and the "Mayfield Road Gang" from Cleveland provided most of the funding for the Desert Inn which opened in 1950. And that was just the beginning. As historian Michael Green has shown: "In the 1950s came the Sahara, Sands, Royal Nevada, Riviera, New Frontier, Dunes, Fremont, Tropicana, and Stardust, mostly run by Jews tied to Lansky."[46] Besides the opportunity to make profits comparable with, if not greater than, those gained during Prohibition, some of those Jewish gangsters saw a chance to gain a sense of legitimacy in Las Vegas because gambling was not only legal but also increasingly important to the Nevada economy. Moe Sedway, for example, became "a hardworking fundraiser for nonprofit causes both local and national" and was "a candidate for Las Vegas City Commission."[47] Moe Dalitz, in partnership with developers like Merv Adelson and Irwin Molasky, built Sunrise Hospital and a number of commercial developments in Las Vegas. He also became a philanthropist, contributing to a Jewish school, the local YMCA, and the University of Nevada, Las Vegas. Indeed, journalist John L. Smith concluded that Dalitz "was responsible for

---

44 There are several books on Murder, Inc., including Robert Weldon Whalen, *Murder, Inc., and the Moral Life: Gangsters and Gangbusters in La Guardia's New York* (New York: Fordham University Press, 2016); Edmund Elmaleh, *The Canary Sang but Couldn't Fly: The Fatal Fall of Abe Reles, the Mobster Who Shattered Murder Inc.'s Code of Silence* (New York: Union Square Press, 2009); Joselit, *Our Gang*.

45 Lacey, *Little Man*: 151–55, 218–19.

46 Michael Green, "The Jews," in Jerry L. Simich and Thomas C. Wright eds., *The Peoples of Las Vegas: One City, Many Faces* (Reno: University of Nevada Press, 2005), 166. See also Alan Balboni, "Southern Italians and Eastern European Jews: Cautious Cooperation in Las Vegas Casinos, 1940–1967," *Nevada Historical Society Quarterly 38*, no. 3 (Fall 1995): 153–73.

47 Jessalynn Strauss, "Becoming Respectable: A History of Early Social Responsibility in the Las Vegas Casino Industry," *UNLV Gaming Research and Review Journal* 19, no. 2 (2015): 13.

much of the growth and prosperity the Las Vegas Valley experienced from the 1960s through the early 1980s."[48]

The quest for legitimacy was something that also drove "Bugsy" Siegel, the most notable Jewish gangster involved in the development of Las Vegas in the 1940s. Indeed, his life offers a useful case study, one that illustrates the collective experiences of Jewish gangsters in the first half of the twentieth century.

Siegel was born on the Lower East Side in 1906. His parents had immigrated to the United States three years earlier from Galicia in the Austro-Hungarian Empire. Siegel's father Max had found work in the needles trade, but he and his wife Jennie had to take in boarders to support their growing family. To help his parents and four siblings, Ben quit school when he was eleven and went to work for a trucking company.[49] Barely in his teens, Siegel embraced the gang culture of the Lower East Side, committing his first crime with two other boys who robbed a loan company office.[50] He gained the attention of Meyer Lansky, who became a life-long friend, in a street fight, one in which Siegel fought with such ferocity that Lansky nicknamed him "Bugsy." Although he always hated the label, the name stuck to the fearless Siegel.[51] Unlike his younger brother Maurice who became a doctor, Ben opted for a life of crime when he joined Lansky in what became known in about 1921 as the Bug and Meyer Gang, which launched a successful bootlegging enterprise.[52] Between 1926 and 1931, authorities arrested Siegel six times for rape, carrying a concealed weapon, narcotics possession, gambling, vagrancy, and "consorting" with criminals. The last arrest at the Franconia Hotel in New York in 1931 was the most significant. Police also apprehended Ben "Lepke" Buchalter, "Gurrah" Shapiro, Harry Greenberg, and five other men in an effort to disrupt Buchalter's labor racketeering ring, but all were released except Greenberg who was deported.[53] Despite the arrests, Siegel prospered throughout the 1920s. While he also invested in a garage company, speakeasies, a dog track in Atlantic City, and Broadway shows, and was an active trader on the stock market, Siegel accumulated most of his wealth

48 John L. Smith, *Sharks in the Desert: The Founding Fathers and Current Kings of Las Vegas* (Fort Lee, NJ: Barricade Books, 2005), 49.

49 Larry D. Gragg, *Benjamin "Bugsy" Siegel: The Gangster, the Flamingo, and the Making of Modern Las Vegas* (Santa Barbara, CA: Praeger, 2015), 1–3.

50 Dean Jennings, *We Only Kill Each Other*, paperback ed. (New York: Pocket Books, 1992), 22–23.

51 Eisenberg, *Mogul of the Mob*, 55–57.

52 Lacey, *Little Man*: 51; Gragg, *Bugsy*: 8–9.

53 Gragg, *Bugsy*: 12.

from bootlegging, as much as $800,000 (nearly $12 million in 2020), according to an internal revenue agent.[54]

Whatever he made, Siegel spent, always indulging in conspicuous consumption. In this, he was following the lead of Lucky Luciano and Al Capone, two of the most visible gangsters of the era who validated their rapid success through their public appearance and image.[55] Outfitted with expensive suits and hats, Siegel regularly made the night club circuit, always with beautiful women. A contemporary recalled that Siegel "was a legend with those girls. They were dancers and actresses and models and girls like that."[56] After he married Esther Krakauer in 1929, the couple moved into the Waldorf-Astoria before relocating to the wealthy suburb of Scarsdale in Westchester County.

With the end of Prohibition, Siegel's life trajectory changed dramatically. In 1935, the wealthy bootlegger moved his family (including two daughters) to southern California. There are no shortages of explanations for the move. Did Lucky Luciano, Joe Adonis, or his close friend Meyer Lansky send him to handle an expansion of "Syndicate" business on the West Coast? Did Siegel relocate because he feared prosecution by the crusading New York special prosecutor Thomas E. Dewey? Was he simply seeking new economic opportunities in the wake of the legalization of the manufacture and distribution of alcoholic beverages? Siegel later told police he had been visiting southern California since the mid-1920s and liked his prospects there better than in New York.[57]

Regardless of the reason for the move, Siegel rapidly engaged in a number of legal and illegal ventures. He remained an active investor in stocks and bonds, but he also held stakes in a variety of enterprises: a dog track in Culver City, California, a racetrack in Tijuana, property in Redondo Beach, the popular Clover Club nightclub off Sunset Boulevard, and bootlegger Tony Cornero's popular s.s. *Rex* gambling ship. Like Buchalter and Shapiro back in New York, Siegel was involved in labor racketeering, gaining control of the movie extras union. He used that leverage to solicit payoffs from film stars. Most importantly, Siegel took over the lucrative race-wire business in the American Southwest, working with Mickey Cohen in southern California, Moe Sedway in Las Vegas, and Gus Greenbaum in Phoenix.[58]

---

54 Ibid.: 9–17; Lacey, *Little Man*: 52.

55 Laura Beshears, "Honorable Style in Dishonorable Times: American Gangsters of the 1920s and 1930s," *The Journal of American Culture* 33, no. 3 (September 2010): 197.

56 Quoted in Eisenberg, *Mogul of the Mob*: 96.

57 Gragg, *Bugsy*: 19.

58 Ibid.: 23–27, 69–88.

The success of these ventures is difficult to gauge, although in 1941 Siegel told reporters that he was "worth $500,000."[59] He certainly made enough to live in luxury, exceeding that of his New York days. He had a thirty-five-room mansion with a large swimming pool, always had sporty new cars, and spent lavishly on monogrammed silk shirts, elegant suits, and sports jackets. Even when arrested in 1940 under suspicion of the murder of Harry Greenberg, a reporter wrote that Siegel looked "like a picture out of a man's fashion magazine," wearing a "fancy brown sports coat, brown slacks with a razor-edge pair of creases, and brown shirt (open at the throat) which he later changed to a white one, and cream-colored socks."[60] He also traveled widely and often. New York and Chicago were his most frequent destinations, but Siegel also loved visiting Mexico, Cuba, Italy, Florida, Minneapolis, and Hot Springs, Arkansas.[61] When at home he spent much of the day gambling rather than managing his numerous enterprises. In addition to frequent appearances at the track at Santa Anita, Siegel admitted that he bet between $2,000 and $5,000 a day on "football, baseball, hockey, any sort all over the United States," as well as political contests.[62]

Hollywood was also an allure Siegel could not resist. Bedazzled by movie making, he often accompanied his pal actor George Raft to the set when the latter was filming. He had affairs with a number of actresses, including Jean Harlow, Ketti Gallian, Marie "the Body" McDonald, and the woman his daughter Millicent called the "love of his life," Wendy Barrie. He also became involved with the wealthy Countess Dorothy di Frasso and mob moll Virginia Hill.[63]

In both 1940 and 1942, Siegel faced trials in Los Angeles for the murder of Harry Greenberg. On November 22, 1939, someone had murdered Greenberg, evidently for trying to shake down "Lepke" Buchalter for whom he had supplied muscle in the latter's labor rackets in New York. While there was sufficient evidence to demonstrate that Siegel likely helped plan the hit, and that he was probably at the murder scene in Hollywood, there was not enough for a conviction in either case.[64] By the time the judge had dismissed the charges in his second trial in February 1942, Siegel had already turned his attention to the emerging resort town of Las Vegas.

59 "Final Pleas in Siegel Hearing," *Los Angeles Examiner*, May 27, 1941, 2.

60 Gragg, *Bugsy*: 28–9, 55; "'Bugsy' a Fashion Plate, Puts up No Battle," *Los Angeles Examiner*, August 17, 1940, 6.

61 Gragg, *Bugsy*: 34–38; Larry Gragg, "The Anti-Nazi Gangster," *History Today* 65, no. 6 (June 2015): 49–54.

62 Gragg, *Bugsy*: 27; Jennings, *We Only Kill*: 106–7.

63 Gragg, *Bugsy*: 30–34, 85–88.

64 Ibid.: 47–67.

Established in 1905 as a convenient locale for repairs on the newly established San Pedro, Los Angeles, and Salt Lake Railroad because it was roughly about halfway between Los Angeles and Salt Lake City, Las Vegas had benefitted from the building of the nearby Boulder Dam and the tourists that this civil engineering marvel attracted. In addition, in 1931, the state of Nevada passed an open-gambling law, as well as the shortest residence requirement for divorce in the country. The number of tourists had reached about 800,000 by 1939. Two years later, state legislators also made Nevada the only state to allow off-track betting on horse races. By 1942, there were several downtown casinos and four popular hotel-casinos. Two, the El Cortez and Nevada Biltmore, were downtown, and the El Rancho Vegas and Hotel Last Frontier were on Highway 91.[65]

Eager to take advantage of the gaming business Las Vegas offered, in 1941 Siegel sent his old friend Moe Sedway to the gambling center to act on his behalf. Over the next five years, Sedway gained a monopoly on the race wire service to seven casinos, as well as a share of those establishments' race book revenues, a move that earned Siegel as much as $25,000 a month. Siegel also had substantial stakes in the downtown Northern Club, Frontier Club, and Las Vegas Club. Two years after failing to acquire the El Rancho Vegas in 1943, he joined with Lansky and several others in the purchase of the profitable El Cortez. In early 1946, he took the lead among a syndicate of investors in taking control of the construction of the first truly luxurious hotel-casino in Las Vegas. Billy Wilkerson, the owner of the influential *Hollywood Reporter* magazine and popular southern California restaurants, had the inspiration for the Flamingo, but could not raise the required funding to complete the project. Incorporated as the Nevada Project Corporation, the Siegel-led syndicate finished the casino for a December 26, 1946, opening. After a spectacular first weekend, business flagged, and in February Siegel closed the casino until the hotel re-opened on March 1 the following year.[66]

Besides Siegel's love of and fascination with the glittering world of nightclubs acquired while living in New York and Los Angeles, his embrace of the Flamingo reveals that he, like so many Jewish gangsters, saw that developing or investing in hotel-casinos in Las Vegas was a way to gain a measure of legitimacy. His daughter Millicent recalled her last conversation with her father in

65 Good accounts of the city's development in this era can be found in Eugene P. Moehring and Michael S. Green, *Las Vegas: A Centennial History* (Reno: University of Nevada Press, 2005), and David G. Schwartz, *Roll the Bones: The History of Gambling* (New York: Gotham Press, 2006).

66 Gragg, *Bugsy*: 89–131; W.R. Wilkerson, III, *The Man Who Invented Las Vegas* (Beverly Hills, CA: Ciro's Books, 2000).

October 1946. Although he did not acknowledge being a killer, Siegel did admit to her that he had been a bootlegger, but the Flamingo was opening a new door, he claimed. "Now," he told her, "I am building a hotel and am 100 percent legitimate."[67]

It was not to be. Despite bringing in well-known entertainers such as Jimmy Durante, Lena Horne, the Andrews Sisters, Abbot and Costello, Joe E. Lewis, and Pearl Bailey, attracting many Hollywood celebrities to the showroom, and holding weekly afternoon bingo sessions and drawings for new automobiles, Siegel was unable to turn the Flamingo's fortunes around. Contractors filed liens on the property in March, and his checks to them began to bounce in May.

As Siegel struggled to make the Flamingo a success, rumors developed about threats to eliminate him, largely because part of the funding for the project came from figures in organized crime. In January 1947, an FBI informant reported that Siegel had told him that Frank Costello, Joe Adonis, and Meyer Lansky "may want to kill him for losing their money." In early June, a local FBI agent warned Siegel of rumors that had reached his office about an imminent "hit," and on June 19 there were long distance calls to the Flamingo with the message: "Bugsy—you've had it." Efforts to trace the calls failed.[68] Siegel brushed aside the rumors.

In the early hours of June 20, Siegel flew to Los Angeles to conduct some business and pick up his two daughters who were traveling by train to be with him for a summer vacation. However, late that evening, a gunman murdered Siegel while he sat on a sofa in his girlfriend Virginia Hill's rented mansion in Beverly Hills. Although there was an investigation in southern California as well as in Las Vegas, the identity of the assailant remains unknown.[69]

In all of his ventures, legitimate and illegal, Siegel sought to distance his family from his underworld connections. When he was away, his wife would tell his daughters Millicent and Barbara that he was on a business trip. Moreover, Millicent recalled that her father was "always there for anybody" in his family. In the 1930s he became his parents' sole source of support, his father-in-law moved into their mansion in the late 1930s, and he gave his sister Bessie's husband a job in Las Vegas in the 1940s.[70] Siegel's actions mirrored

67 Gragg, *Bugsy*: 146.

68 Ibid.: 124; Curtis O. Lynum, *The FBI and I: One Family's Life in the FBI during the Hoover Years* (Bryn Mawr, PA: Dorrance, 1988), 86; Jennings, *We Only Kill*: 193–94.

69 Gragg, *Bugsy*: 133–41.

70 Ibid.: 44–5.

those of several Jewish gangsters. As far as the FBI could determine, murdering gangster "Lepke" Buchalter appeared "to have been genuinely devoted" to his family.[71] Susan Berman, whose father Dave had been a powerful gangster in Minneapolis before becoming one of Siegel's business partners in Las Vegas, revealed that her father did all he could to make sure she would "never know the secrets of his past because the knowledge might destroy me."[72] Meyer Lansky promised his wife "he wouldn't let the children get involved" in his business.[73] As California thug Mickey Cohen explained, "The people in my walk of life" had "the highest regard and respect for their family."[74]

Beyond protecting family, Siegel, as did other Jewish gangsters, engaged in philanthropy. In May, Siegel hosted a benefit concert at the Flamingo to support the new Damon Runyon fund. A famed newspaperman, Runyon died of cancer in 1946, and his friend, broadcaster Walter Winchell, encouraged the entertainment industry to hold benefit shows in order to help fund cancer research. Entertainers from the Nevada Biltmore, El Cortez, El Rancho Vegas, Last Frontier, and Flamingo performed. Ticket sales and an auction of luxury items raised over $8,000. Earlier in the spring, Siegel had contributed to an Eddie Cantor United Jewish Appeal fund raiser in Las Vegas, and in June he made a contribution to Lou Costello's Youth Foundation in Los Angeles.[75] On the surface, as Wendy Bergoffen has explained, Siegel's actions reflected a tradition among Jews to "give as a way to guarantee just and rightful treatment for those in need."[76] More importantly, as Robert Rockaway has noted, the philanthropy of Jake Guzik and Longie Zwillman represented the effort these gangsters made to become "acceptable" in the Jewish community.[77]

Rich Cohen has argued that gangsters like Lansky "made sure their kids lived normal lives, stayed in school, studied, got into college."[78] In emphasizing that their children should seek a better life through a more acceptable way, Jewish gangsters succeeded in making the gangster life a largely one generation experience. They joined other Jewish parents in investing heavily in their children's

---

71 Quoted in Rockaway, *Good to His Mother*: 206.

72 Susan Berman, *Easy Street* (New York: Dial Press, 1981), 46, 48.

73 *Rockaway, Good to His Mother: 202.*

74 Mickey Cohen, as told to John Peer Nugent, *In My Own Words: The Underworld Autobiography of Michael Mickey Cohen* (Englewood Cliffs, NJ: Prentice-Hall, Inc., 1975), 182.

75 Gragg, *Bugsy*: 130.

76 Wendy Bergoffen, "Guardians, Millionaires, and Fearless Fighters: Transforming Jewish Gangsters into a Usable Past," *Shofar: An Interdisciplinary Journal of Jewish Studies* 25, no. 3 (Spring 2007): 95.

77 Rockaway, *Good to His Mother*: 241–44.

78 Rich Cohen, *Tough Jews: Fathers, Sons, and Gangster Dreams*, paperback ed. (New York: Vintage Books, 1999), 152.

education and "as school teachers, social workers, civil servants, doctors, lawyers, dentists, accountants, and business owners, Jews occupied a decidedly middle class position in the nation's economy." Indeed, by mid-century, Hasia Diner has written, "American Jews made up the richest and most powerful Jewish community in the world, and they ranked among the most affluent ethnic groups in America."[79] To be sure, some barriers remained. Few Jews, for example, had been able by mid-century to scale the heights of the corporate world, to serve as chief operating officers or chairmen of companies.[80] But that did not stop the accomplished and ambitious. For example, when "'white-shoe' law firms in New York City were closed to Jews," they "created their own or became partners in law firms" in other cities, as shown by Marc Lee Raphael.[81]

One of the more intriguing debates about Jewish gangsters concerns the significance of accepting their resort to violence. First, there had to be an acknowledgment of their reality. According to Joe Kraus, a student of Jewish gangsters in Chicago, in the 1980s he "had only the vaguest idea that there had ever been anything like Jewish gangsters. Jews weren't tough; we were cerebral. We didn't own guns we lobbied our congressmen for gun control."[82] However, through a number of movies, notably *Once Upon a Time in America* (1984), *Billy Bathgate* (1991), and *Bugsy* (1991), and books like Robert Lacey's *Little Man: Meyer Lansky and the Gangster Life* (1991), the image of violent Jewish gangsters became familiar. The key question was how to view that reality.

The bravado and ruthless actions of men like Buchalter and Reles had an appeal for some, such as author Rich Cohen. In a widely read and reviewed book called *Tough Jews*, Cohen portrayed Jewish gangsters in a positive light. He argued:

> If Jewish gangsters still thrived today, if they hadn't gone legit, if Jews of my generation didn't regard them as figments, creatures to be classed with Big Foot and the Loch Ness monster, I think the Jewish community might be better off. After all, everyone needs someone who gives them the illusion of strength.[83]

---

79 Hasia R. Diner, *The Jews of the United States, 1654–2000* (Berkeley: University of California Press, 2004), 230.

80 Howard M. Sachar, *A History of Jews in America* (New York: Alfred A. Knopf, 1992), 655.

81 Marc Lee Raphael, "Rethinking the American Jewish Historical Experience," *Religion and American Culture: A Journal of Interpretation* 17, no. 2 (Summer 2007): 252.

82 Joe Kraus, "The Jewish Gangster: A Conversation across Generations," *American Scholar* 64, no. 1 (Winter 1995), 53.

83 Cohen, *Tough Jews*: 20.

Cohen embraced the gangsters because they demonstrated a now-lost Jewish manliness.

It is true that many who grew up on the Lower East Side in the early twentieth century acknowledged the attraction of strength offered by those gangs. Singer and comedian Eddie Cantor revealed in his autobiography that being in a gang gave him his "first urge to power." He appeared invincible as he pulled his hat over his eyes, and "donned a big red sweater and flourished a bat, looking for all the world like the chief of a gas-house gang."[84] Cantor, like most young Jewish men, ultimately declined to pursue a violent life, but to those like Abe Reles "the gangster life meant violence, which equaled masculinity, which meant power, which meant the ability to have the sorts of things the working stiffs in the neighborhoods could only dream of, which meant status and respect."[85]

It was Cohen's admiration for those like Reles and his associates in Murder Inc. that prompted widespread criticism of his book. Jewish gangsters, Adam J. Levitin wrote in *Commentary*, "were equal opportunity brutes, happily extorting, robbing, and murdering their fellow Jews."[86] In his *New Republic* review of *Tough Jews*, Leon Wieseltier asserted that the men Cohen considered as demonstrating a model of strength "believed in nothing, except money and attitude." Getting straight to the point, Wieseltier argued that Jewish gangsters were no heroes, they were just "dumb fucks."[87]

Robert Rockaway, who devoted a chapter titled "Defenders of Their People" in his popular anecdotal history of Jewish gangsters *But He Was Good to His Mother*, as well as a journal article titled "Hoodlum Hero: The Jewish Gangster as Defender of His People, 1919–1949," also faced criticism from scholars like Rachel Rubin and Wendy Bergoffen. Beyond protecting neighborhoods from Italian and Irish gangs in the early decades of the century, Rockaway described how Meyer Lansky, Longie Zwillman, Dave Berman, and Mickey Cohen were among the many Jews who broke up pro-Nazi rallies in several American cities as the threat to Jews in Europe grew in the late 1930s. He also detailed how some Jewish gangsters aided in the post-World War II arms-smuggling effort in support of the drive to create a Jewish state.[88] Rockaway was careful to note that gangster defenders of their communities commonly "got their start in crime by

84 Eddie Cantor, *My Life Is in Your Hands & Take My Life: The Autobiographies of Eddie Cantor*, reprint (New York: Cooper Square Press, 2000), 40.

85 Whalen, *Murder, Inc.*: 128.

86 Adam J. Levitin, "The Old Gang," *Commentary 106*, no. 92 (August 1998): 69.

87 Leon Wieseltier, "Yidfellas," *The New Republic 218*, no. 21 (May 25, 1998): 42.

88 Robert A. Rockaway, "Hoodlum Hero: The Jewish Gangster as Defender of His People, 1919–1949," *American Jewish History* 82, nos. 1–4 (1994): 221–26, 229–33.

preying on fellow Jews," and that "these men were not latter-day Robin Hoods and should not be glorified as such."[89] Still, Bergoffen, argued that Rockaway championed "gangsters as guardians" by presenting "a highly selective portrait of the historical record … one that enables Jews today to accept physical violence as one way Jewish men took care of their own."[90]

The criticism of Cohen's and Rockaway's works reflects the debate among American Jews on how best to come to terms with Jewish gangsters, a struggle that has lasted over a century. "In the past," Wendy Bergoffen has noted, "respectable and assimilated members of the Jewish community did everything in their power to disavow kinship with criminals."[91] Now, there are contrary voices. The divide, often along gender lines, is clear. Should Jewish gangsters be dismissed as a one-generation anomaly in the chronicle of a historically peaceful people, or should they be included as one piece in the complex tapestry of the Jewish experience in the United States? The debate will continue.

Bibliography

Books

Asbury, H. (1928) *The Gangs of New York: An Informal History of the Underworld*, Garden City, NY: Garden City Publishing, Co.

Berman, S. (1981) *Easy Street*, New York: Dial Press.

Birmingham, S. (1983) *"The Rest of Us": The Rise of America's Eastern European Jews*, Boston: Little, Brown, and Co.

Cantor, E. (2000) *My Life Is in Your Hands & Take My Life: The Autobiographies of Eddie Cantor*, reprint, New York: Cooper Square Press.

Cohen, M., as told to John Peer Nugent (1975) *In My Own Words: The Underworld Autobiography of Michael Mickey Cohen*, Englewood Cliffs, NJ: Prentice-Hall, Inc.

Cohen, R. (1999) *Tough Jews: Fathers, Sons, and Gangster Dreams*, paperback ed., New York: Vintage Books.

Davis, M. (2012) *Jews and Booze: Becoming American in the Age of Prohibition*, New York: New York University Press.

Diner, H. (2004) *The Jews of the United States, 1654–2000*, Berkeley: University of California Press.

---

89 Rockaway, *Good to His Mother*: 253–54.

90 Bergoffen, "Guardians, Millionaires, and Fearless Fighters": 97. See also Rubin, "Gangster Generation": 4, 9–11.

91 Bergoffen, "Guardians, Millionaires, and Fearless Fighters": 92.

Dinnerstein, L. (1995) *Anti-Semitism in America*, paperback ed., New York: Oxford University Press.

Eisenberg, D., U. Dan, and E. Landau (1979) *Meyer Lansky: Mogul of the Mob*, New York: Paddington Press.

Elmaleh, E. (2009) *The Canary Sang but Couldn't Fly: The Fatal Fall of Abe Reles, the Mobster Who Shattered Murder Inc.'s Code of Silence*, New York: Union Square Press.

Fox, S. (1990) *Blood and Power: Organized Crime in Twentieth-Century America*, paperback ed., New York: Penguin Books.

Fried, A. (1980) *The Rise and Fall of the Jewish Gangster in America*, New York: Holt, Rinehart, and Winston.

Gragg, L. (2015) *Benjamin "Bugsy" Siegel: The Gangster, the Flamingo, and the Making of Modern Las Vegas*, Santa Barbara, CA: Praeger.

Green, M. (2005) "The Jews," in Jerry L. Simich and Thomas C. Wright eds., *The Peoples of Las Vegas: One City, Many Faces*, Reno: University of Nevada Press.

Haller, M. (1976) "Bootleggers and American Gambling, 1920–1950," *Gambling in America*, Appendix 1, Washington: Commission on the Review of the National Policy Toward Gambling.

Hanson, N. (2010) *Monk Eastman: The Gangster Who Became a War Hero*, New York: Alfred A. Knopf.

Howe, I., and K. Libo eds. (1979) *How We Lived: A Documentary History of Immigrant Jews in America, 1880–1930*, New York: Richard Marek.

Jennings, D. (1992) *We Only Kill Each Other*, paperback ed., New York: Pocket Books.

Joselit, J. (1983) *Our Gang: Jewish Crime and the New York Jewish Community, 1900–1940*, Bloomington: Indiana University Press.

Joseph, S. (1969) *Jewish Immigration to the United States From 1881 to 1910*, reprint, New York: Arno Press.

Kavieff, P. (2006) *The Life and Times of Lepke Buchalter: America's Most Ruthless Labor Racketeer*, Fort Lee, NJ: Barricade Books.

Kessner, T. (1977) *The Golden Door: Italian and Jewish Immigrant Mobility in New York City, 1880–1915*, New York: Oxford University Press.

Lacey, R. (1991) *Little Man: Meyer Lansky and the Gangster Life*, Boston: Little, Brown and Co.

Lerner, M. (2008) *Dry Manhattan: Prohibition in New York City*, paperback ed., Cambridge, MA: Harvard University Press.

Lynum, C. (1988) *The FBI and I: One Family's Life in the FBI during the Hoover Years*, Bryn Mawr, PA: Dorrance.

May, A. (2009) *Gangland Gotham: New York's Notorious Mob Bosses*, Santa Barbara, CA: Greenwood Press.

Moehring, E., and M.S. Green (2005) *Las Vegas: A Centennial History*, Reno, NV: University of Nevada Press.
Okrent, D. (2011) *Last Call: The Rise and Fall of Prohibition*, paperback ed., New York: Scribner.
Pietrusza, D. (2011) *Rothstein: The Life, Times, and Murder of the Criminal Genius Who Fixed the 1919 World Series*, paperback ed., New York: Basic Books.
Raab, S. (2005) *Five Families: The Rise, Decline, and Resurgence of America's Most Powerful Mafia Empires*, New York: Thomas Dunne Books.
Rischin, M. (1962) *The Promised City: New York's Jews, 1870–1914*, Cambridge, MA: Harvard University Press.
Rockaway, R. (2000) *But He Was Good to His Mother: The Lives and Crimes of Jewish Gangsters*, paperback ed., Jerusalem: Gefen Publishing House.
Sachar, H. (1992) *A History of Jews in America*, New York: Alfred A. Knopf.
Schwartz, D. (2006) *Roll the Bones: The History of Gambling*, New York: Gotham Press.
Smith, J. (2005) *Sharks in the Desert: The Founding Fathers and Current Kings of Las Vegas*, Fort Lee, NJ: Barricade Books.
Vanderwood, P. (2010) *Satan's Playground: Mobsters and Movie Stars at America's Greatest Gaming Resort*, Durham, NC: Duke University Press.
Whalen, R. (2016) *Murder, Inc., and the Moral Life: Gangsters and Gangbusters in La Guardia's New York*, New York: Fordham University Press.
Wilkerson, W. III (2000) *The Man Who Invented Las Vegas*, Beverly Hills, CA: Ciro.

Articles

Balboni, A. (1995) "Southern Italians and Eastern European Jews: Cautious Cooperation in Las Vegas Casinos, 1940–1967," *Nevada Historical Society Quarterly 38*: 153–73.
Bergoffen, W. (2007) "Guardians, Millionaires, and Fearless Fighters: Transforming Jewish Gangsters into a Usable Past," *Shofar: An Interdisciplinary Journal of Jewish Studies 25*: 91–110.
Beshears, L. (2010) "Honorable Style in Dishonorable Times: American Gangsters of the 1920s and 1930s," *The Journal of American Culture 33*: 197–206.
Bingham, T. (1908) "Foreign Criminals in New York," *North American Review 188*: 383–94.
Gragg, L. (2015) "The Anti-Nazi Gangster," *History Today 65*: 49–54.
Kraus, J. (1995) "The Jewish Gangster: A Conversation across Generations," *American Scholar 64*: 53–65.
Levitin, A. (1998) "The Old Gang," *Commentary 106*: 68–70.
Lupsha, P. (1981) "Individual Choice, Material Culture, and Organized Crime," *Criminology: An Interdisciplinary Journal 19*: 3–24.
Raphael, M. (2007) "Rethinking the American Jewish Historical Experience," *Religion and American Culture: A Journal of Interpretation 17*: 247–60.

Ribak, G. (2014) "'The Jew usually left those crimes to Esau': The Jewish Responses to Accusations about Jewish Criminality in New York, 1908–1913," *Association for Jewish Studies Review* 39: 1–28.
Rockaway, R. (1994) "Hoodlum Hero: The Jewish Gangster as Defender of His People, 1919–1949," *American Jewish History* 82: 215–35.
Rubin, R. (2002) "Gangster Generation: Crime, Jews and the Problem of Assimilation," *Shofar: An Interdisciplinary Journal of Jewish Studies* 20: 1–17.
Strauss, J. (2015) "Becoming Respectable: A History of Early Social Responsibility in the Las Vegas Casino Industry," *UNLV Gaming Research and Review Journal* 19: 11–19.
Turner, G. (1909) "The Daughters of the Poor: A Plain Story of the Development of New York as a Leading Centre of the White Slave Trade of the World, under Tammany Hall," *McClure's Magazine* 34: 45–61.
Wieseltier, L. (1998) "Yidfellas," *The New Republic* 218: 42.

Newspaper Articles

"'Bugsy' a Fashion Plate, Puts up No Battle," *Los Angeles Examiner*, August 17, 1940.
"Final Pleas in Siegel Hearing," *Los Angeles Examiner*, May 27, 1941.
"Social Geography of the East Side," *New York Times*, April 3, 1910.

CHAPTER 4

"Our Fight Is Not Yet Over": Jewish-Canadians in the Spanish Civil War

Michael Petrou

In July 1937, while crossing the Atlantic Ocean on the Cunard-White Star liner *Alaunia*, Maurice Constant, a former student at the University of Toronto, wrote a letter to Helen Dunsay, a woman he had met while working at a Hashomer Hatzair Zionist youth camp in upstate New York, with an elaborate plan.

The Civil War in Spain had been raging for the past year, pitting a left-wing Republican government against an insurrection led by General Francisco Franco, and backed with men and weapons by Fascist Italy and Nazi Germany. When André Malraux, the French author and an organizer of the republic's small air force, came to speak at the University of Toronto's Hart House Theatre in March that year, Constant was captivated by his lecture in the cavernous, smoke-filled hall, and by his cool nonchalance as afterward he stood outside, smoking in the university grounds with a beautiful blonde French woman by his side. "He had a certain character," Constant recalled years later. "The reason I admired this guy, that he was my idea of a model, he was an intellectual and a man of action. And that was my model."[1]

Constant decided he had to get to Spain. But his goal, he told Dunsay, was not to fight, but to report on the war covertly, so as not to provoke deadly retaliation from government spies. "I have contracted to smuggle that most precious of commodities, uncensored stories, out of the Iberian Peninsula," he wrote. Constant planned to disguise stories in his letters to her that she would then be tasked with assembling, typing, and forwarding to a Toronto newspaper. He would indicate which parts of his letter were meant for publication by coded signals, such as how he dotted his 'i's and 'j's. It would be a lot of work, he said, but he hoped she would feel rewarded by its unusual nature. "I don't know whether the observation will flatter you, but one reason I am telling you all this is because I judged you as unromantic enough (in the better sense of the word) not to make a fuss over the apparent mystery [of] it all."[2]

1 Author's interview with Maurice Constant, August 2002.
2 Letter from Constant to Dunsay, July 17, 1937, in possession of the author.

 | DOI:10.1163/9789004462540_005

Constant made it to Spain and there abandoned his plans to merely observe the war. "I soon found out that the only thing worth writing about is the struggle, that the only point from which the struggle may be properly viewed is the front line, and that the only real understanding of the front line is obtained by the soldier," he wrote in a September letter to Dunsay in which he explained that he had joined the International Brigades.

By taking that step, Constant became one of some 1,600 Canadians, including at least 44 Jews, to fight in Spain.[3] Jews were over-represented among volunteers from many of the contributing countries. Estimates vary, but out of 35,000 to 40,000 volunteers who joined the International Brigades or other militias and units that fought on the anti-Fascist side of the war, Jews accounted for at least 3,500, and perhaps many more. They included about 20 percent of Polish volunteers, 20 percent of British volunteers, and one-third of American ones.[4] The Jewish-Canadian contingent was much smaller, but it was still slightly disproportionate. Jews made up about 1.5 percent of Canada's population in 1931; they comprised about 3 percent of all Canadian volunteers.[5]

It is tempting to incorporate the story of Jewish-Canadians in Spain within a larger narrative positing that so many Jews volunteered to fight Fascism in Spain because, as Jews, they were uniquely aware of its danger and uniquely motivated to confront it. Everything that came after the Spanish Civil War, culminating in the gas chambers of Auschwitz and the killing pits of Ukraine,

3 Working from a database of Canadian volunteers I built while researching a book on Canadians in the Spanish Civil War (*Renegades: Canadians in the Spanish Civil War*, University of British Columbia Press, 2008), I have assembled a smaller database of 44 Canadian Jews who fought or served in a non-military capacity in Spain during the Civil War. The figure is likely not exact. Jewish volunteers often identified themselves as such in biographical questionnaires completed in Spain or were identified as Jewish in reports completed by their commanders and others there. In other cases, such details might have been unearthed elsewhere—including from postwar sources—but some surely declined to provide this information, and in several cases the biographical information available is limited. I have counted as Canadian any volunteer who was born in Canada or lived there for a significant length of time, including more than ten immigrants from Europe. These criteria are subjective. Several volunteers whom I have counted among the Canadians also lived for significant amounts of time in the United States and have been considered Americans in other compilations.

4 For a discussion of numbers, see Gerben Zaagsma, *Jewish Volunteers in the International Brigades and the Spanish Civil War* (London: Bloomsbury, 2017), 22–23. The figure for Jews in the British volunteer contingent is from Richard Baxell, *British Volunteers in the Spanish Civil War: The British Battalion in the International Brigades, 1936–1939* (London: Routledge, 2004), 18. The figure for Jewish volunteers from America is from Peter Carroll, *The Odyssey of the Abraham Lincoln Brigades* (Stanford: Stanford University Press, 1994), 18.

5 Bernard L. Vigod, *The Jews in Canada* (Ottawa: Canadian Historical Association, 1984), 9.

confirms the clarity of such foresight. And given the awesome scale of the tragedy Jews suffered at the hands of Fascists during World War II, their resistance to Fascism in Spain was important to their collective identity in the years that followed.

This idea was championed by Albert Prago, an American academic and Jewish veteran of the Spanish Civil War, in 1979. He wrote:

> The International Brigades became the vehicle through which Jews could offer the first *organized armed resistance* to European fascism. Their combatant role in Spain proved they could fight well, and that as early as 1936 they were actively resisting fascism. Not all went passively to the concentration camps and crematoria.[6]

"Should not the full story be told of the heroism of the Jews of the International Brigades?" Prago asked in a subsequent article. "Should not historians examine the ample evidence which refutes the peculiar interpretations of the Arendts and Bettelheims?"[7] Prago was referring to Hannah Arendt and Bruno Bettelheim, writers who alleged that Jews responded to their own slaughter at the hands of the Nazis during the Holocaust with passivity and cooperation.[8] Prago's interpretation was echoed by none other than Chaim Herzog, then president of Israel, during a 1986 ceremony in the country to mark the fiftieth anniversary of the outbreak of the Spanish Civil War. He spoke of "history's warning trumpets" in Spain and noted that during the Civil War "there were … six million of our brethren still alive in Europe who did not yet realize that a sword was poised over their necks." But some did see the sword, he continued. "There were people who realized just what a fascist victory in Spain would mean … They united in a common front and fought against the perdition and the Holocaust that was threatening the world."[9]

But the story of Jews in the Spanish Civil War—and Jewish-Canadians among them—is more nuanced than one of farsighted resistance to the coming Holocaust as described by Herzog and Prago. And it has been shaped by hindsight that was not available to young Jews contemplating joining the International Brigades in 1936 or 1937. Even Maurice Constant's personal story is less straightforward than it appears initially, including his identity as a Jew.

---

6 Albert Prago, "Jews in the International Brigades," *Jewish Currents* (February 1979): 16 [my emphasis].
7 Prago, "Jews in the International Brigades: 2," *Jewish Currents* (March 1979): 27.
8 Zaagsma, *Jewish Volunteers*: 122–23.
9 Ibid.: 145.

"It also occurs to me that you may not recognize the name," Constant wrote toward the end of his letter from the ocean liner, "it's a far cry from Maurice Constant to Moshe Cohen. Just now it seems to symbolize a very distant period, full of incomprehensibilities."[10] It is unclear whether Moshe Cohen was the name Constant was given at birth, or whether it was simply how he was known among campers at Hashomer Hatzair. Constant's children, who were unaware of his Jewish heritage until many years later, only heard of Moshe Cohen when they were given copies of Constant's letters to Dunsay, discovered following her death more than seventy years after they were written. Despite Constant's youthful involvement with Hashomer Hatzair, he had rejected religion as a boy and refused to have a bar mitzvah.[11]

Other volunteers were religious or more steeped in their Jewish identity. But that does not mean they went to Spain primarily because they were Jews. The decision made by Jewish-Canadians to fight in the Spanish Civil War was influenced by many factors, including a belief that Fascism posed a particular threat to Jews, and that their participation was an opportunity to combat it. There is little evidence however, that this was the dominant motive for most of them. Jewish-Canadians who fought in Spain were also radicalized by the general political climate in Canada during the Great Depression, which intensified the discrimination and marginalization they faced as Jews in Canada. It also contributed to pushing many Canadian Jews toward Communism and the Communist Party of Canada, which championed the Spanish cause. The past, however, never stands still in the minds of the present. And if being Jewish had only a little to do with why many Jewish-Canadians fought in Spain, it would come to matter more for those who looked back on the participation of Jewish-Canadians in the war years later.

The most high-profile Canadians who set off for Spain in summer 1936 never got there. The boxers Sammy Luftspring and Norman "Baby" Yack were darlings of Toronto's Jewish community in the 1930s. Both were contenders to compete for Canada in the Olympic Games that were to be held that summer in Berlin. Luftspring said his parents convinced him not to go because of the anti-Semitism of Germany's Nazi government, and he came to agree with them. "[T]here was enough Jew in me that I knew I had to make a protest, and the only protest I could make was to declare my refusal to go on political grounds," Luftspring later wrote in his memoir. In a letter to the *Globe* newspaper, Luftspring and Yack explained:

10 Letter from Constant to Dunsay, July 17, 1937.

11 Author's personal communication with David Constant, Maurice's son.

> We are making a personal sacrifice in refusing the chance and we are sure that all true Canadian sportsmen will appreciate that we would have been very low to hurt the feelings of our fellow-Jews by going to a land that would exterminate them if it could.[12]

In fact, the letter was likely written for the two by an official of the Canadian Jewish Congress, which had been pressuring them not to travel to Berlin. Moreover, the boxers only agreed not to go after the Canadian Jewish Congress promised it would send them instead to Barcelona, where Spain's Republican government was hosting a "People's Olympiad" as an alternative to the official Olympics hosted by the Nazis.[13] The anti-Fascist symbolism of the People's Olympiad does not appear to have been a major factor in Luftspring's calculations. "I must say I have never been much of a student of politics," he wrote.[14] The Congress held a fundraiser and produced the money need to finance the voyage for Luftspring, Yack, and their trainer, Harry Sniderman. Luftspring recalled "the whole ghetto" turning out to march the boxers to Union Station on their way to Spain.[15] For the Canadian Jewish Congress, which was campaigning for a boycott of the Berlin Olympics, the boxers' departure was not so much about their going to Spain as it was about their not going to Germany. Paying for the boxers to go to Spain was simply the price of keeping them from Berlin.[16]

Republican Spain, in other words, was not a beacon for Jewish-Canadians that summer. But the Civil War, which began with an attempted coup on the night of July 17, 1936—causing Luftspring, Yack, and Sniderman to turn around in Toulouse and head back to Canada—had the potential to change this. Nazi Germany was a major ally of the rebel Nationalists. It ferried Spanish and colonial troops from Spanish Morocco to mainland Spain in the early days of the war and sent its own Condor Legion of air and ground units to aid the rebel cause. This support was decisive. The Nazis were openly and aggressively anti-Semitic when the Civil War began. By 1936, Germany had enacted legislation known as the Nuremberg Laws which, among other things, excluded Jews from citizenship rights and forbade marriage or extra-marital sex between Jews and non-Jewish German citizens.

---

12 Sammy Luftspring with Brian Swarbrick, *Call Me Sammy* (Scarborough: Prentice-Hall, 1975), 85.

13 Richard Menkis and Harold Troper, *More Than Just Games: Canada and the 1936 Olympics* (Toronto: University of Toronto Press, 2015), 165–66.

14 Luftspring and Swarbrick, *Call Me Sammy*: 86.

15 Ibid.: 88.

16 Menkis and Troper, *More Than Just Games*: 167.

Spain's Nationalists were also anti-Semitic. Historian Gerben Zaagsma describes their propaganda as "a mix of traditional Catholic anti-Semitism and contemporary theories about a world Judeo-Freemason-Bolshevik conspiracy."[17] General Gonzalo Queipo de Llano who, in lurid radio broadcasts, promised that the wives of "red cowards" would be raped by the real men in his armies—"kicking their legs about and squealing won't save them"—publicly described the Nationalist struggle as "not a civil war, but a war for Western civilization against the Jews."[18] This sort of rhetoric tapped into a deeper tradition of Spanish Catholic and right-wing anti-Semitism, argues historian Paul Preston.[19] Even with the war won, Franco upheld the specter of a Jewish-Bolshevik-Masonic conspiracy as justification for continued vigilance (and repression) by the victors. "Let us not deceive ourselves: the Jewish spirt, which permitted the alliance of big capital with Marxism and which was behind so many pacts with the anti-Spanish revolution, cannot be extirpated in a day and still beats in the hearts of many," Franco declared in a speech on May 19, 1939, when a massive victory parade was held in Madrid.[20]

And yet, says Zaagsma, anti-Semitism was only one element of Nationalist propaganda and not its central one. Franco even told the Jews of Tetuan to ignore the anti-Semitic broadcasts of Radio Seville. Few Jews lived in areas under Nationalist control, and systemic persecution of them did not occur. Some were extorted, jailed, and murdered in Spanish Morocco. But, Zaagsma adds, Nationalists tried to limit such outrages, in part because of fear of what they might do to the image of the Nationalist cause outside Spain.[21] Franco's forces killed tens of thousands of non-combatants during and after the war. Compared with those massacres, which Preston describes as a "holocaust," the abuses visited on Spain's Jewish community were mild.[22]

At the same time, Jews around the world were facing graver threats—in Nazi Germany, but also in Poland, where violence and pogroms boiled—and in Palestine, then a British mandate, where local Arabs were revolting against British control and its plans to partition the territory in order to create a Jewish "national home," and where Jewish militias fought Arab ones. These threats

---

17 Zaagsma, *Jewish Volunteers*: 4.

18 On "red cowards," see Paul Preston, *The Spanish Holocaust: Inquisition and Extermination in Twentieth-Century Spain* (London: W.W. Norton, 2012), 149; on "not a civil war," see Zaagsma, *Jewish Volunteers*: 148.

19 Preston, *The Spanish Holocaust*: 4–5.

20 Ibid.: 471.

21 Zaagsma, *Jewish Volunteers*: 4–5.

22 Preston explains his reasoning for using this term in the prologue of *Spanish Holocaust*.

featured much more prominently in mainstream Jewish publications in Europe and America.[23]

The Spanish Civil War did not dominate public debates among Jewish-Canadians either. But many supported the Spanish Republic and those who fought for it. In 1938, the poet A.M. Klein, for example, published the following tribute to his childhood friend Samuel Abramson, then a volunteer in Spain:

To One Gone to the Wars
  For S.H.A.

Unworthiest crone of my grammar days
  Expectorator in learning's cuspidor
Forsaking the scholar's for the gamin's ways,
  The gates of knowledge for the cubicular door.

How you have shamed me, me the nobler talker,
  The polisher of phrases, stainer of verbs,
Who daily for a price serve hind and hawker
  Earning my Sabbath meat, my daily herb

'Tis you who do confound the lupine jaw
  And stand protective of my days and works,
As in the street-fight you maintain the law
  And I in an armchair weigh and measure Marx.

Alas, that fettered and bound by virtues long
since rusty,
I must, for spouse and son,
Withhold, as is befitting any prison trusty,
My personal succour and my uniformed aid,
And from the barracks watch the barricade—
Offering you, meek sacrifice, unvaliant gift,
My non-liturgic prayer:
For that your aim be sure,
Your bullet swift,
Unperilous your air, your trenches dry,

23 Zaagsma, *Jewish Volunteers*: 4–5.

> Your courage untainted by defeat,
> Your courage high.[24]

Such support was also reflected by Rabbi Harry Stern of Temple Emanu-el in uptown Montreal who, in 1937, printed a letter from Abramson in his weekly *Canadian Jewish Review* column. He introduced the letter with his own expressed concern:

> We are not so sure about all the issues involved in the fratricide war raging in Spain ... We know there are those who look upon General Franco as ineffable. Then there are the uncompromising "left wingers" who are convinced that Barcelona's conduct is perfect and that by all means, fair or foul, the Loyalists are justified in uprooting the vested interests in Iberia.[25]

It is reasonable to detect in such a statement the author's fear, as a leader of a minority community, of straying too far from accepted public opinion, especially in Quebec, where French Catholics generally backed Franco's rebels. Jewish-Canadians in the 1930s had to contend with strong and broad currents of anti-Semitism in both French and English Canada, and with more specific anti-Semitic aspersions that conflated Jewishness with Communism. This was a global phenomenon—reflected in Nazi and Spanish Nationalist propaganda—but also, in Quebec, in the mind of Catholic priest and well-known Quebec nationalist Abbé Lionel Groulx, and among broader sections of the populace. Responding to a rally in support of the Spanish Republic in 1937, Université de Montréal students marched, chanting: "À bas, à bas, à bas les Communistes. À bas les juifs!" (Down with the Communists. Down with the Jews!)[26]

Jewish-Canadians in the 1930s had good reason to be attracted to the Communist Party. Many were immigrants or the children of immigrants from Central and Eastern Europe and were marginalized within mainstream Canadian society as a result. Underlying this was the everyday anti-Semitism they faced on the street and from the country's political establishment. Gordan Lunan, a non-Jewish immigrant from Scotland who was jailed after World War II on charges of spying for the Soviets, later recalled:

---

24 Reprinted in Gerald Tulchinsky, *Branching Out: The Transformation of the Canadian Jewish Community* (Toronto: Stoddart, 1998), 139.

25 Quoted in ibid.: 137.

26 Ibid.: 141.

> A sizeable section of the movement … was essentially [a] middle class Jewish association of well-informed, culture-oriented, freethinking people who did not or would not fit into conventional Montreal society … For many, the odour of persecution and pogroms was still in their nostrils.[27]

Communist candidates from mainly Jewish constituencies were elected in provincial and federal elections in 1943. Fred Rose (also later convicted for spying) won in a federal by-election in Montreal, and Joseph B. Salsberg, who campaigned in Toronto, won a seat in Ontario.

Despite such visible successes, Jews made up only about 10 percent of the Communist Party of Canada's membership in 1943, figures which were likely similar during the Spanish Civil War.[28] But the presumed association had wide currency, including among the Canada's national police force, the Royal Canadian Mounted Police (RCMP). Reporting on a crowd waiting to welcome Canadians returning from Spain in 1939, a RCMP observer claimed that of those in attendance approximately 85 percent were foreigners and "Jews as usual were well represented."[29]

It was mostly among the minority of Jews who were Communists, however, that the Spanish cause mattered enough to fight for. About 75 percent of all Canadian volunteers were members of the Communist Party by the time they had arrived in Spain. (Several joined the party for the first time there.) Among Jewish-Canadian volunteers, the figures were almost the same.[30] Membership in the Communist Party was surely a major factor propelling Canadians—both Jewish and not—to Spain, although a far lower percentage of Jews fought in Spain than joined the Communist Party.

Those who were not party members were unlikely to be hostile to it and might have identified their political affiliation as "anti-Fascist" when asked on questionnaires in Spain. Party leaders in the Brigades were keen to bring promising anti-Fascists into the fold and took special note of those among returning vets who might be useful to the party. "Good. Should be recruited into C.P.," reads a summary of Sidney Cohen's characteristics on a list of survivors

27 Tulchinsky, *Canada's Jews: A People's Journey* (Toronto: University of Toronto Press, 2008), 282.

28 Ibid.: 281.

29 Canadian Security Intelligence Service (CSIS) files from access to information request 86-A-57, RCMP report, February 5, 1939.

30 Of 38 Canadian Jews in Spain for whom there is information available, twenty-nine had enrolled in the Communist Party or the Young Communist League before leaving for Spain.

returning to Manitoba. "Jewish. Good. Politically confused," is how Gerald "Jerry" Glow is described in the same document.[31]

Samuel Abramson, the subject of A.M. Klein's poetic tribute, was singled out among vets returning to Quebec. Abramson joined the party for the first time in Spain, and his evaluator suggested it "may be advisable that he does not work openly as CP member, being in a position to do work amongst a certain circle of people."[32] Abramson belonged to several Jewish community organizations in his native Montreal, and party members might have hoped that he could develop some influence there. Abramson had also worked as a journalist prior to the war, and the party leadership may have wanted him to cultivate more connections among other journalists and writers.

One Jewish-Canadian volunteer belonged to another far-left party that was stridently opposed to the pro-Soviet Communism of the Communist Party of Canada. William Krehm, a graduate of the University of Toronto, was a member of the League for a Revolutionary Workers' Party (LRWP), a tiny collection of Marxists who supported many of the principles espoused by Leon Trotsky, the boogeyman of Stalin-supporting Communists, but which considered itself independent. The LRWP did not send any volunteers to fight in Spain, where anarchists and other non-Stalinist militias had also taken up arms, but in October 1936 they dispatched a delegate—Krehm—to a conference of leftists in Brussels opposed to Soviet Communism. Krehm crossed the anarchist-controlled border for a short visit to Barcelona prior to the conference and returned to Spain afterward. He spent the next eight months working informally for the Partido Obrero de Unificacíon Marxista (POUM), an anti-Stalinist political party and militia—though Krehm's work consisted of propaganda rather than fighting.

During this time, the Communist Party of Spain and its Soviet backers gradually increased their control over the Spanish Republican government. The government outlawed POUM in June 1937 and POUM's fighters were disarmed. Its leader, Andreu Nin, was tortured to death. Many of its members were jailed. Krehm was caught up in the resulting sweep and spent more than two months in jail before he was released; he returned to Canada, his Spanish adventure at an end.[33]

---

31 RGASPI, 545 6 537, "Characterizations of Returned Men. District Seven." Cohen is identified as a non-party member elsewhere, too. However, an evaluation of Cohen by the Brigade Party Committee of the Communist Party of Spain's Central Committee says Cohen was a party member in Toronto "but has more or less refused to do any political work" in Spain. See RGASPI 545 6 549.

32 RGASPI 545 6 537, "Characterizations of Returned Men. District Two."

33 For details on Krehm's time in Spain, see Petrou, *Renegades*: 148–57.

Whether Krehm's Jewish background had anything to do with that adventure is unclear. His family's history is in many ways typical of early twentieth century Jewish-Canadian experience. His parents, Hyman and Sarah, emigrated separately from parts of the Russian Empire that are now Ukraine and Belarus, leaving regions soaked in anti-Semitism and upended by deadly anti-Jewish pogroms. Their families settled for a time among other Jews in the St. John's Ward neighborhood of Toronto, where they met and married. Here, Hyman, who had been a furrier in Ukraine, partnered with his brother to open a fur-selling business—joining thousands of other Jews in the garment trade.[34] Toronto was very much a Protestant- and Anglo-Saxon-dominated city at the time, and the Krehms felt the sting of anti-Semitism and exclusion in their new home, too. William, one of the few Jews to attend Parkdale Collegiate high school, faced taunts and fists because he was a Jew. A teacher there would wait until Jewish holidays, when most Jews stayed home from school, to warn other students of the threat that Jews posed to their British identity. William, however, did not stay home and organized a walk-out strike among students in protest.[35]

Krehm's earliest activism, which may have steered him toward a path that led to Spain, was therefore tied to his Jewish identity. But it seems likely he was motivated more by the injustice of his teacher's bigotry than by a personal affiliation with Judaism. Although Krehm had a bar mitzvah, he stopped keeping kosher when quite young and religion held little interest for him.[36] "At a young age he became a Marxist and thumbed his nose at everything," is how his son, Jonathan, described his father's approach to Judaism. "The sense of Jewish identity, it wasn't about religion. It was about tribal, ethnic identity."[37]

Parsing the biographies of Jewish-Canadian volunteers reveals additional similarities and differences with their non-Jewish comrades.

Jewish volunteers, like the other Canadians, were mostly workers. At least four were furriers or garment workers prior to their departure for Spain; five were drivers; four were mechanics, construction engineers, metal workers, or farm workers; three were sailors; another three were salesman. Others included a barber, a glass worker, a musician, a social worker and an upholsterer. These working-class volunteers were often members of unions or other

34 A comprehensive exploration of Krehm's life is provided in Tyler Wentzell's forthcoming *Comrades and Scoundrels: William Krehm, the International Anti-Stalinist Left, and the Spanish Revolution*.

35 Ibid.

36 Ibid.

37 Author's conversation with Jonathan Krehm, May 21, 2020.

organizations advocating on behalf of workers and the unemployed. Israel "José" Goldenberg, for example, belonged to the Jewish Labour League and Local 100 of the International Fur Workers Union in Toronto where, he said in a questionnaire he completed for the War Commissariat of the International Brigades, every member of the union knew him.[38] Goldenberg, who traveled to Spain from New York using the passport of an American named Joseph Budish, wrote in another biographical questionnaire, completed as part of an application for membership in the Communist Party of Spain, that he had been involved in many furriers' strikes in Toronto and was arrested twice during a strike in 1926 "for beating up scabs" and twice in 1935 for picketing during another strike.[39] A note in his personnel file says he had been a member of a secret party cell. He was described as a good comrade in Spain, but one who, like so many Canadians, had too little discipline.[40]

Goldenberg's activism was typical of that of future Canadian volunteers during the Great Depression. Many belonged to unions and Communist-affiliated organizations such as the Canadian Labour Defence League. At least 100 had been arrested—usually for offenses associated with labor militancy or simply being unemployed, vagrancy, or illegal assembly.

The Canadian contingent in Spain was also shaped by the experiences of many volunteers in unemployment relief camps established by Canadian prime minister Richard Bennett in the fall of 1932 to house single, unemployed men and provide them with meagre financial compensation in exchange for labor on public works projects such as in national parks and on road crews. The camps became hotbeds of radicalization, thanks in part to the work of the Communist Party of Canada within them. For men in the camps, the war in Spain was part of a much larger clash between the Left and the Right which included their own struggles in Canada. Ron Liversedge, a leading Communist and non-Jewish volunteer in the International Brigades, said Spain was a chance for Canada's poor "to actually fight against the people, against the same kind of people, whom they thought were responsible for their condition, and who were oppressing them here in Canada."[41]

Activism among camp workers boiled over in the summer of 1935. Some 2,000 men from British Columbia relief camps went on strike, protesting in the streets of Vancouver before climbing aboard boxcars heading east with the

38 RGASPI 545 6 549, Israel Goldenberg's personnel file, War Commissariat of the International Brigades questionnaire.

39 RGASPI 545 6 549, Communist Party of Spain questionnaire, completed May 23, 1938.

40 Ibid.

41 Petrou, *Renegades*: 31.

goal of reaching Ottawa. The RCMP stopped the "On to Ottawa Trek" in Regina. Mounted police charged protesters with clubs and shot into the crowd in the ensuing riot. Protesters beat a plainclothes police officer to death, and another demonstrator who was beaten by police died, apparently of injuries suffered in the attack, three months later.

These camps, and the dissent they bred, had a significant impact on the character of the Canadian volunteers. Liversedge estimated that 500 men who took part in the "On to Ottawa Trek" later fought in Spain.[42] Edward Cecil Smith, the leading Canadian commander in Spain, tried to explain the irreverence of his countrymen to party officials in the International Brigades by noting that the largest single group of Canadians came from British Columbia's "slave camps." In another report to party officials in Canada, Cecil-Smith blamed the relief camps for the indiscipline and "anarchist tendencies" of the "Vancouver volunteers." [43]

Few Jewish-Canadian volunteers, however, shared this experience. There is evidence that Harvey Rockman was involved in relief camp activism and the riot in Regina. He noted in the biography he completed for an application to the Communist Party of Spain that he had been indicted in Regina in July 1935 for "Class Struggle Propaganda Illegal." He was jailed for two months and freed that September.[44] But, in contrast to the hundreds of non-Jewish Canadian volunteers who were relief camp workers, miners, and lumberjacks, the Jewish-Canadians appear on average to have led more urban lives, mostly in Montreal, Toronto, and Winnipeg. They also included a higher percentage of volunteers whose jobs required more schooling or were creative in nature. Five identified themselves as writers or journalists in various questionnaires—including Maurice Constant, for whom such a title was at the time aspirational. Two were pharmacists, two were nurses, and two were medical doctors. Approximately twelve Jewish-Canadian volunteers had received some form of higher education. This represents about one-third of all Canadians in Spain who studied beyond high school.

The biggest difference between the Jewish and non-Jewish volunteers relates to their place of birth. Almost 80 percent of all Canadian volunteers were immigrants, and 80 percent of these immigrants had arrived in Canada

---

42 Ibid.: 31.

43 Ibid.: 43. Note that Cecil-Smith's reference to "slave camps" is from a report to officials in Spain. His quote about Canadians' anarchist tendencies is from a report on British Columbia volunteers sent to the Vancouver Committee of the Communist Party of Canada.

44 RGASPI 545 6 566, Harvey Rockman's personnel file, Communist Party of Spain questionnaire, June 23, 1938.

between 1925 and 1930. Of Jewish-Canadian volunteers, more than 60 percent were born in Canada. The foreign-born Jewish-Canadians in Spain came mostly from Eastern Europe, from places that were part of the Russian Empire when they were born, part of the Soviet Union during the Spanish Civil War, and are now part of Ukraine, Poland, Belarus, and Lithuania. These were contested areas where borders changed during the first half of the twentieth century because of war and diplomacy. Benjamin Mezei, for example, was born in the town of Berehove, in 1904, when it was part of Hungary. His birthplace was part of Czechoslovakia for most of the Spanish Civil War before reverting to Hungarian control in 1938. Muni Erlick, known in Spain as Jack Taylor, was born in a Ukrainian town that was later occupied by Romania. One Jewish-Canadian volunteer was born in Denmark, another in Finland, and a third in the United States.

The national origins of Jewish-Canadian volunteers broadly mirrored patterns of Jewish immigration to Canada in the early twentieth century, and did not differ greatly from those of their non-Jewish colleagues. The largest contingent of foreign-born Canadian volunteers—about 240—were ethnically Ukrainian, many of whom came from areas under Polish control at the time of the Spanish Civil War. Hungarians numbered over one hundred, as did Finns. About sixty were Polish.[45]

In *Renegades: Canadians in the Spanish Civil War*, I note that many of the Canadian volunteers emigrated from places that had been shaken by the upheavals of World War I, as well as by the Russian Revolution and the various uprisings and civil wars that had erupted in its wake. For some of these immigrants, such as Ukrainians from towns and cities that came under Polish control after World War I, or Ukrainians from Bukovina who resented Romania's occupation, the Soviet Union was seen as a potential liberator. Hungarian Communists declared a short-lived Soviet republic in 1919, which was backed by the new Communist government of the Soviet Union. In 1918, Finland had its own civil war between "Reds" supported by the Soviet Union and "Whites" backed by the German Empire. Finns who immigrated to Canada brought these divisions with them.

This was a significant reason why Communism, and the Soviet Union, appealed to numerous immigrants to Canada during the 1920s. But, for many immigrant volunteers, Republican Spain was far more than a Communist, or a Soviet-backed cause. Indeed, among criticism leveled at Canadians in Spain, by far the most common was that they were "politically undeveloped" or possessed "anarchist tendencies." This did not mean they were insufficiently

45 Petrou, *Renegades*: 22–23.

opposed to Fascism but that they lacked commitment to the party. To the frustration of their American and European political bosses, too many of the Canadians were irreverent and indifferent Communists.[46]

For such men, Spain mattered in part because it was a malleable cause, a screen on which they could project their own dreams and frustrations over failed revolutions and liberation struggles in their home countries. In the case of conflicts tied to the Russian Revolution, a perceived link to the Spanish Civil War was fairly easy to make given the prominent role of both the Soviet Union and international Communism in Spain. In a letter to a Ukrainian-language newspaper in Canada, William Krysa, a Ukrainian immigrant to Canada, described a visit to the volunteers in Spain by a Ukrainian who told them about the mistreatment of his compatriots living under Polish rule in Europe. The volunteers cheered their visitor. "When we have defeated the Fascists in Spain we promised to go to Poland and help the Ukrainians in their struggle against the Fascist regime," Krysa wrote.[47] Another Canadian volunteer, Hugh Lehtovirta, an immigrant from Finland, said he volunteered to fight in Spain to avenge "the murder of my sister and others" in the 1918 Finnish "revolution."[48]

Other Canadian volunteers saw in Spain a reflection of conflicts elsewhere that lacked this obvious, if indirect, connection. More than forty Canadian volunteers were Irish immigrants. Most Irish volunteers in the International Brigades had been involved with the anti-British Irish Republican Army, as were several Irish-Canadians in Spain. These included Patrick "Paddy" O'Daire who, before fighting in Spain, entertained striking workers in Canada by singing Irish rebel songs and ballads.[49] Many Croatian-Canadians and Americans—encouraged by Communist publications—also believed a link existed between the cause of the Spanish Republic and their own wish for an independent homeland. The US-based *Radnicki glasnik* told its readers in January 1937 that battles in Spain were fought "not only for the freedom of Catalonia and Euskadi [the Basque country] within Spain" but also for "Croatian national liberation."[50]

Some Jewish-Canadians in Spain were motivated by a conviction that the Spanish Civil War could be a proxy for other conflicts, or that it was at least part of the same constellation of nationalist and revolutionary struggles. Art Siven, a Finnish immigrant Jew, said in an interview years after the war that he

46 Ibid.: 38–45.
47 Ibid.: 37.
48 Ibid.: 37.
49 Ibid.: 35–6.
50 Quoted in ibid.: 35.

knew what Fascism was "because we had it in Finland."[51] Israel Goldenberg, who immigrated to Canada in 1929, said in his Communist Party of Spain questionnaire that he had been interested in the proletarian movement "since I was a child in the Ukraine in 1919, when the Bolsheviks defended the Jewish people against the white bandits."

But many Jews in the 1930s believed they had a national liberation struggle of their own to worry about in Palestine. Most Jews in Palestine supported the Spanish Republic, but they were reluctant to weaken the Zionist cause there by fighting for it. Almost "every group in the Yishuv, including the various gradations of the Zionist Left, opposed the idea of sending youths from Palestine to join the volunteers in the IB [International Brigades]," wrote historian Raanan Rein.[52] Ya'akov Riftin, a leading member of the Hashomer Hatzair youth movement said in September 1938:

> We have neither the time nor the ability to deal day in and day out with the events taking place on the Spanish fronts. We too are defending ourselves in the trenches for the third year now. We are losing people every day. And no one knows if we have reached the height of terror or if the worst is still ahead of us.[53]

This attitude was summed up in the slogan, "Hanita [a frontier kibbutz] before Madrid."[54] Between 180 and 250 Jews (and at least one Arab, a Yiddish-speaking baker) nevertheless left Palestine for Spain, out of a population of approximately 400,000 Jews in Palestine at the time.[55] This was a significant contribution, but the majority of Jews in Palestine would not have accepted the argument that these volunteers were fighting for Jewish national liberation in the way that, say, *Radnicki glasnik* claimed Croatian volunteers in Spain were fighting for Croatian independence, too. Rather, Palestinian Jews who fought in Spain were putting Madrid before Hanita.

Jews were therefore somewhat distinctive among the Canadian volunteers. To be sure, most were Communists, like Canada's non-Jewish volunteers, and the International Brigades were a Communist outfit. But the Jews still did not quite fit into the same mold as other Canadian volunteers. As noted, they were,

---

51 Quoted in ibid.: 37.

52 Raanan Rein, "Tikkun Olam and Transnational Solidarity: Jewish Volunteers in the Spanish Civil War," *Politics and Religion* 10, no. 2 (2016): 220.

53 Quoted in ibid.: 220.

54 Ibid.

55 Ibid.: 221.

on average, more educated and mostly lacked the galvanizing experience of the relief camps shared by so many non-Jews from Canada. Far fewer were immigrants, so Spain held little promise of revenge for lost European wars, or of second chances for unsuccessful uprisings. For them, Spain was above all else an ideological, anti-Fascist struggle—which brings us back to the original question of whether their Jewish faith and identity meant they felt the necessity of this struggle more keenly because they were Jews.

Peter Carroll makes this argument regarding Jewish-American volunteers in his book *The Odyssey of the Abraham Lincoln Brigade*. The fact that so many American volunteers were Jewish, he asserts, was in part because of the large number of Jews in the American Communist Party. "It also suggests that American Jews, more than other ethnic minorities, appreciated the dangers of fascism in Europe. To them, Hitler was no abstraction."[56] American Hyman Katz wrote to his mother from a hospital bed in Spain: "Don't you realize that we Jews will be the first to suffer if fascism comes?" To an aunt, acknowledging his mother's opposition to his fighting in Spain (where he would subsequently die), Katz said his mother was "partly to blame because she sacrificed to give me a good Jewish education, and the whole history of our people has taught me to admire the prophets and fighters who died for Liberty." He said he was a bad son "because I fight against the greatest oppressors of my people."[57]

That Jews fought in Spain because they believed they had a special stake in the outcome of the struggle against Fascism there was echoed in official International Brigade narratives during the war. In August 1937, the Brigades' Press Service put out its first edition of a Yiddish bulletin called *Freyhaytskemfer* (freedom fighter), with an editorial claiming: "Every victory over barbaric Fascism and inhuman anti-Semitism is a step toward our final goal: our social and national liberation." Another article declared: "The struggle in Spain will decide whether fascist barbarism, anti-Semitism, Hitlerite racism will govern Europe or if democracy and peace will be victorious."[58] The Brigades would later designate a Jewish unit within the Palafox Battalion of the mostly Polish 13th Dombrowski Brigade. Named the Naftali Botwin Company, after a Polish Jewish Communist executed in 1925 for killing a police informer, it was set up in part to recognize the large number of Jews in the International Brigades, many of whom wanted their own unit, and to stand as a rebuke to the idea that Jews were naturally submissive and unsuited to fighting.[59] Luigi Longo, inspector

56 Carroll, *The Odyssey*: 18.
57 Ibid.: 17–18.
58 Zaagsma, *Jewish Volunteers*: 37–8.
59 Ibid.: 45.

general of the International Brigades, highlighted this latter point in an introduction he wrote for a pamphlet on Jewish volunteers in Spain compiled by the Communist journalist Gina Medem. Longo, designated "Gallo" in Spain, described meeting a Jewish member of the International Brigades who asked for the formation of a Jewish unit. "We wish to prove what we Jews are capable of," the volunteer, later identified as Albert Nakhumi Weitz from Palestine and Poland, said, according to Longo. "The Fascists everywhere accuse us of being cowards. We will show to the world that we know how to conduct ourselves as heroes."[60]

The Botwin Company operated under Jewish military and political command, although many of its members were non-Jewish internationals and Spaniards. Two Jewish-Canadians fought in it: Benjamin Mezei, who survived; and Jozef Matlak, a Polish immigrant who died in Spain during the "Retreats" of 1938. The company's banner, in Yiddish and Polish, echoed the slogan of the Dombrowski Brigade: "For your freedom and ours."

The Canadian contingent in Spain had fewer Jews than the American one, so it is more difficult to make the case that their faith and Jewish identity was as consequential. However, the fact that their biographical profiles are a little different than those of non-Jewish Canadians suggest Jewish identity might have played at least a minor role. And there is some circumstantial evidence to suggest that, for some, it did. Many of the Jewish-Canadian volunteers were involved in Jewish and Zionist welfare and community organizations before traveling to Spain. When asked their "National Origin" on an August 1937 questionnaire for the Historical Commission of the International Brigades, Montreal garment workers Jack Pascal and Isaac Schatz answered "Jewish" rather than Canadian. Questioned why he had decided to travel to Spain, Schatz replied: "Being of Jewish origin, my hatred for fascism was so great as to compel me to leave a good home and fight these bastards."[61]

Not all Canadian volunteers embraced the fight against anti-Semitism as a component of the anti-Fascist cause they joined in Spain. At least three Canadians were judged in evaluations by their political bosses to have anti-Semitic tendencies.[62] According to the RCMP, on returning to Canada, Joseph Turnbull said he never would have enlisted if he had known he would be

60 Prago, "Jews in the International Brigades": 20. Longo's introduction is reprinted in its entirety, translated into English from Spanish by Prago. For the volunteer's identity, see Zaagsma, *Jewish Volunteers*: 41.

61 RGASPI 545 3 511 and 545 3 512, August 1937 questionnaires for the Historical Commission of the International Brigades.

62 RGASPI 545 6 537, evaluations of returning Canadian veterans.

fighting solely for the Jews.[63] And a note in John McGrandle's personnel file says he had complained that the battalion was "run by a clique of New York [Young Communist League] Jews."[64] It is true that that Americans held most leadership positions in the North American battalions, a fact that Canadian volunteers resented. In a February 1938 report, Jack Taylor described morale among Canadian and American volunteers as needing some work. Apparently one of the Canadians working on the kitchen staff of the Mackenzie-Papineau Battalion referred to his American comrades as "New York ice cream boys."[65]

While in Spain, Samuel Abramson worked to maintain a connection with Montreal's Jewish community. He wrote several letters to Hananiah Meir Caiserman, general secretary of the Canadian Jewish Congress. "Fascism is more than an academic problem to you these days. Don't give way an inch, for if you do, you'll get no thanks from the enemy," he said in one. "Treat the fascists like what they are. And when you strike, strike hard. I wish I were home to take a hand, but that must remain for the future."[66]

Abramson also passed on contact information for the Canadian Jewish Congress to the propaganda section of the International Health Service in Albacete, Spain, where he worked as an ambulance driver. The service duly sent Caiserman five copies of a pamphlet titled "We Fight to the Death," along with its newspaper. "We hope that you and your friends will be interested in our activity and make all good democrats sympathize with the doctors and nurses and first of all with the wounded antifascists in our hospitals," Gusti Jirku wrote in an accompanying letter. "There are many Jews among them who fought splendidly."[67] Caiserman replied to Jirku the following month, promising to do everything he could to bring Jirku's message to the larger Jewish community, adding:

> I was thrilled to learn that my dear friend, S.H. Abramson, is doing excellent work as an ambulance driver. If you have occasion, kindly tell him of my great admiration for his courage and his determination and that

63 LAC, RG 146, vol. 1880, CSIS files 95-A-00088, Friends of the Mackenzie-Papineau Battalion of Canada, part 3, box 59, RCMP report, February 20, 1939.

64 RGASPI 545 6 559, McGrandle's personnel file, undated report signed by Helge Meyer.

65 RGASPI 545 6 511, "Report by Jack Taylor to the Commission of Work among Internationals and the Communist Party of Spain on the situation in the XV Brigade," February 20, 1938.

66 Alex Dworkin Canadian Jewish Archives, Montreal, Canadian Jewish Congress organizational records, fond no. CJC0001, series no. ZB, Abramson to Caiserman, May 9, 1938.

67 Alex Dworkin Canadian Jewish Archives, fond no. CJC0001, series no. ZB, Gusti Jirku, Service de Propaganda du Service Sanitaire International, Albacete to Caiserman, September 26, 1937.

> I pray he keep well, so that he may return in time to our community, to the development of which he has given a good share.[68]

Abramson told Caiserman he was trying to return to Montreal to start a "Jewish Committee for Medical Aid to Spain."[69] In a June 1938 letter, he also explained to Jack Taylor what he had in mind, in a roundabout, passive-aggressive way. It seems Abramson's father had taken some steps—which were not fully described in Abramson's letter but may have involved public or political appeals—to get Abramson back to Montreal. Abramson began his letter by assuring Taylor that his father had done so without his knowledge or permission and that he still fully supported the cause of the Spanish Republic. Abramson then said he would not be applying for repatriation himself:

> There is no need to propagandize me or tell me "there can be no talk of able-bodied men returning home." I've been around and worked in Barcelona and know differently. Able-bodied men are going home for one reason or another … In most cases, I imagine, the reasons are important enough.

Abramson reminded Taylor that he had been active for many years in Jewish organizations in Canada. "Of course, I am no longer a Zionist today," he said, professing his support, perhaps insincerely, for Communist policy on Palestine. He then told Taylor about his desire to organize a Jewish aid committee. The committee he envisioned would not be confined to Jewish radicals, he said, but would seek support from bourgeois Jews he had gotten to know by working in Zionist organizations. Abramson said his Jewish colleagues in Canada insisted they could not start the work without him and that he needed to return to Canada to do it. He said he had brought up the matter with Bob Kerr, head of the Canadian cadre service in Spain, but had received no encouragement. Abramson said he would continue driving ambulances, important work. "But sometimes, in the quiet of the night, I think of the ambulances the Canadian Jewish Committee might have sent over. But then the dozens of 'political' workers who have gone home are probably doing good work too."[70] Party leadership in Spain had suspicions about Abramson's desire to launch an aid committee in

68 Alex Dworkin Canadian Jewish Archives, fond no. CJC0001, series no. ZB, Caiserman to Jirku, October 20, 1937.

69 Alex Dworkin Canadian Jewish Archives, fond no. CJC0001, series no. ZB, Abramson to Caiserman, March 10, 1938.

70 RGASPI, 545 6 541, Samuel Abramson's personnel file, Abramson to Taylor, June 25, 1938.

Canada. "Too much concerned about his own personal safety—egoist in political sense," concluded an evaluation in his personnel file, which also described him as an "able and good comrade."[71]

But if Abramson wanted to return to Montreal, he was genuinely sad to go when the time came to do so. "I leave Spain with deep regret. We have all come to love this fine people," he told Caiserman in what appears to have been his final letter from Spain, on November 7, 1938.[72] Ten days earlier, Spaniards in Barcelona held a parade to bid farewell to the International Brigades. "It was simply magnificent, almost beyond words," Abramson wrote in a letter to Caiserman the next day.

> Hundreds of thousands of people turned out, the streets were strewn with flowers, women kissed us as we walked along, everyone was quite overcome with emotion. I hope to be home about the end of the year, and back to work. There's plenty to be done. Our fight is not yet over.[73]

The Spanish Civil War has been continuously re-imagined in the more than eighty years since its conclusion. "We are fighting a jihad and this is the first Islamic international brigade in the modern era," Pakistan's spy chief Hamid Gul told journalist Ahmed Rashid in 1989, explaining why he was funneling Muslim volunteers from all over the world into Afghanistan to fight that country's Soviet-backed government. "The Communists have their international brigades, the West has NATO, why can't Muslims unite and form a common front?"[74] More recently, some leftist international volunteers who traveled to Syria to fight the so-called Islamic State have modeled themselves after the volunteers who came from all over the world to fight Fascism in Spain during its civil war.

The participation of Jewish-Canadians in the Spanish Civil War is no different. In an interview decades after the war, Ethel Magid, wife of Canadian doctor Aaron Magid, explained why she and her husband had gone to Spain: "Well, we were hearing all these things from Germany about what was happening

---

71 Ibid. The form on which this evaluation is written is faded, making it impossible to discern its exact providence.

72 Alex Dworkin Canadian Jewish Archives, fond no. CJC0001, series no. ZB, Abramson to Caiserman, November 7, 1938.

73 Alex Dworkin Canadian Jewish Archives, fond no. CJC0001, series no. ZB, Abramson to Caiserman, October 29, 1938.

74 Ahmed Rashid, *Taliban: Militant Islam, Oil, and Fundamentalism in Central Asia* (New Haven: Yale University Press, 2000), 129.

to the Jews, and we wanted to fight the fascists, so that is what we did."[75] That may well be true and, had she been asked the same question during the war, Ethel might have given the same response. But it is also true that, in 1981, Ethel, and the rest of the world, knew things they had been unaware of half a lifetime earlier, before the Holocaust, before a reckoning about how it happened and how Jews responded when it was underway. This surely shaped how she remembered the war and why she joined it, and it has shaped how others, including writers and poets, have understood the war, and the participation of Jewish-Canadians in it, since.

In her 1992 poem, "The Woman in the Hall," Miriam Waddington, who was born in 1917 and grew up speaking Yiddish in Winnipeg's North End, describes an enigmatic woman living alone "in some upstairs room/in a nebulous city":

> Whoever she is
> She is also Guernica
> and Madrid she is Moscow
> besieged and Dieppe
> betrayed …
>
> That woman in the hall
> is hunger and desolation,
> she is the addled brain
> of the crazy woman who
> survived the camps
> to sit out the days on the
> shores of the Dead Sea
> muttering curses and
> incantations …

Here, Waddington ties the Spanish Civil War ("Guernica") to the Holocaust ("the camps"). According to professor of English Emily Sharpe, Waddington's poetry goes a step further, presenting the Holocaust as an outgrowth of the Spanish conflict.[76]

That the two conflicts are linked is a widely held contention, especially among those who advocated opposing rather than appeasing Fascism in 1936—and a historically sound one. But there is a danger in drawing a causal

75 Quoted in Petrou, *Renegades*: 38.

76 Emily Robins Sharpe, "'The heart above the ruins': Miriam Waddington's Poetry, the Spanish Civil War, and Jewish Canadian Literature," *Canadian Jewish Studies* 26 (2018): 57.

connection between Spain and the Nazi death camps—as Sharpe does elsewhere when she claims that in World War II and the Holocaust, "the Nazis employed the techniques of warfare and mass killing they had practiced in Spain."[77] In the case of the Holocaust, this is inaccurate. The Germans bombed civilian towns in Spain, but they did not practice the methods of genocide they would later use against Europe's Jews.

And yet that connection between the Holocaust—in which Jews were slaughtered—and the Spanish Civil War—in which they pre-emptively fought back—is an appealing one. It is no coincidence that the title character in Mordecai Richler's novel *St. Urbain's Horseman* is a veteran of the Spanish Civil War who grew up on Montreal's heavily Jewish Saint Urbain Street and, decades later, the novel's protagonist imagines hunting the escaped Nazi doctor of Auschwitz, Josef Mengele, on horseback through South America.

The Spanish Civil War is a recurring topic in Richler's books. "Among the wonders of St. Urbain, our St. Urbain ... were two men who had served with the Mackenzie-Paps in the Spanish Civil War," he writes in his collection of memoirs and short stories, *The Street*.[78] Richler often blended truth and fiction in his writing, but the detail is exact. Saul Shapiro and Hyman Nelson, known in Spain as Kosmos Stavrous, lived on St. Urbain Street. Samuel Abramson attended St. Urbain's Baron Byng High School—renamed Fletcher's Field High School in Richler's stories about the neighborhood.

Maurice Constant, the Canadian who traveled to Spain hoping to be a war correspondent and then decided the cause was too important not to fight for, once speculated that Richler, who was born in 1931, felt a "romantic failure" at not having been old enough to join the war.[79] Canadian veterans of Spain discussed approaching Richler to ask him to write a book about them.[80] Nothing came of it, but the fictional protagonist in Richler's *Joshua Then and Now* does write such a book. Joshua Shapiro is fixated on the war. As a boy, he cuts photographs of Spain out of magazines and glues them into his scrapbook. He pins a map of wartime Spain to his bedroom wall. As a young man, in order to impress girls, he calls himself Robert Jordan, hero of Ernest Hemingway's Spanish Civil War novel. Kissing them, he cannot help but wonder if the "earth moved" for them the way it did Robert Jordan's love interest in Hemingway's book.[81]

77 Emily Robins Sharpe, *Mosaic Fictions: Writing Identity in the Spanish Civil War* (Toronto: University of Toronto Press, 2020), 27.

78 Mordecai Richler, *The Street* (Toronto: McClelland & Stewart, 1969), 19.

79 Author's interview with Maurice Constant, 2002.

80 Library and Archives Canada Mackenzie-Papineau Battalion collection, MG30-E173, R2609-0-0E, vol. 1, file 8, letter from William Beeching to Walter Dent.

81 Richler, *Joshua Then and Now* (Toronto: McClelland & Stewart, 1980), 63, 88–89, 98.

"I wish I had been old enough to fight in the Spanish Civil War," another character, Seymour Kaplan, drunk and melancholic, blurts out over dinner some four decades after the conflict ended.

> I wanted to fight on the Ebro. Come back with a wound, maybe. Nothing serious. I mean, not like Jake Barnes. But enough so that people would point me out even now. Sure he's in knitwear, but you know that limp, he got it in Spain. Do you understand?

Joshua does understand. He understands, too, why Seymour drinks.[82]

The Spanish Civil War, and Hemingway's writing about it, are how the book's Jewish characters gauge manliness, says Sharpe.[83] In this way, they are like the poet A.M. Klein, who wrote of his friend and volunteer Samuel Abramson: "How you have shamed me, me the noble talker/The polisher of phrases, stainer of verbs." For the characters in Richler's stories, the war also had a similar meaning for Albert Nakhumi Weitz, the Jewish fighter who asked for the formation of a Jewish unit in Spain: a chance to prove that Jews were not cowards, that they would stand against Fascism when it threatened them.

In the 1970s and 1980s, children at Camp Gesher, located near Cloyne, Ontario, and affiliated with the Habonim Dror Labour Zionist youth movement, gathered around campfires to sing. Guided by their counsellors, they reached back through time for music that might build fellowship and reinforce the camp's progressive values of solidarity and justice. Among the songs they sang, amid pine trees, crisp evening air, and fading dusk light, was "Si me quieres escribir" (If you want to write me), a song popular among Republican troops and International Brigade members during the Civil War.[84] A translation of one of its many versions begins: "If you want to write me, you know where I am posted/On the Gandesa front, first line of fire."

Several Jewish-Canadians died near Gandesa, the scene of difficult fighting for the International Brigades in the spring and summer of 1938. It is a catchy tune, and one perhaps should not read too much into its inclusion among skits and other campfire games. But Habonim Dror has roots, through its Dror component, that stretch to the Warsaw Ghetto uprising and other acts of Jewish resistance, and the song was probably not chosen for its melody. Singing it allowed young Jews at the camp to feel connected to a different generation of

82 Ibid.: 151–52.
83 Sharpe, *Mosaic Fictions*: 49.
84 Author's communication with a former camper.

Jewish-Canadians whose decision to fight Fascism in Spain would, in the years and decades that followed, matter more to Jews not yet born than they possibly could have imagined when their forefathers first took up arms.

Bibliography

Archives

Alex Dworkin Canadian Jewish Archives, Montreal.

Library and Archives Canada (LAC).

Russian State Archive of Socio-Political History (RGASPI). Comintern Archive. International Brigades Records (online at http://sovdoc.rusarchives.ru/sections/organizations/cards/94999).

Secondary Sources

Baxell, R. (2004) *British Volunteers in the Spanish Civil War: The British Battalion in the International Brigades, 1936–1939*, London: Routledge.

Carroll, P. (1994) *The Odyssey of the Abraham Lincoln Brigades*, Stanford: Stanford University Press.

Luftspring, S. with Brian Swarbrick (1975) *Call Me Sammy*, Scarborough: Prentice-Hall.

Menkis, R., and H. Troper (2015) *More Than Just Games: Canada and the 1936 Olympics*, Toronto: University of Toronto Press.

Petrou, M. (2008) *Renegades: Canadians in the Spanish Civil War*, Vancouver: University of British Columbia Press.

Prago, A. (1979) "Jews in the International Brigades," *Jewish Currents* (February 1979): 15–21.

Prago, A. (1979) "Jews in the International Brigades: 2," *Jewish Currents* (March 1979): 6–27.

Preston, P. (2012) *The Spanish Holocaust: Inquisition and Extermination in Twentieth-Century Spain*, London: W.W. Norton.

Rashid, A. (2000) *Taliban: Militant Islam, Oil, and Fundamentalism in Central Asia*, New Haven: Yale University Press.

Rein, R. (2016) "Tikkun Olan and Transnational Solidarity: Jewish Volunteers in the Spanish Civil War," *Politics and Religion 10*, no. 2: 207–30.

Richler, M. (1969) *The Street*, Toronto: McClelland and Stewart.

Richler, M. (1971) *St. Urbain's Horseman*, Toronto: McClelland & Stewart.

Richler, M. (1980) *Joshua Then and Now*, Toronto: McClelland & Stewart.

Sharpe, E.R. (2018) "'The heart above the ruins': Miriam Waddington's Poetry, the Spanish Civil War, and Jewish Canadian Literature," *Canadian Jewish Studies 26*: 56–74.

Sharpe, E.R. (2020) *Mosaic Fictions: Writing Identity in the Spanish Civil War*, Toronto: University of Toronto Press.

Tulchinsky, G. (1998) *Branching Out: The Transformation of the Canadian Jewish Community*, Toronto: Stoddart.

Tulchinsky, G. (2008) *Canada's Jews: A People's Journey*, Toronto: University of Toronto Press.

Vigod, B.L. (1984) *The Jews in Canada*, Ottawa: Canadian Historical Association.

Wentzell, T. (forthcoming) *Comrades and Scoundrels: William Krehm, the International Anti-Stalinist Left, and the Spanish Revolution.*

Zaagsma, G. (2017) *Jewish Volunteers in the International Brigades and the Spanish Civil War*, London: Bloomsbury.

CHAPTER 5

Mika, mi Guerra de España: Ideology and Commitment

Gabriela Jonas Aharoni

The documentary *Mika, mi guerra de España*, is based on the autobiography of Micaela (Mika) Feldman de Etchebéhère, published in 1976. Mika Feldman was born in 1902, in the Jewish colony of Moisés Ville, in the province of Santa Fe, Argentina. She was a unique woman who achieved the rank of captain in the Spanish Civil War.[1]

This chapter addresses the film as a documentary road movie in which mobility and movement play a central role. Visual figuration of spatial mobility as a sign of cultural, social, and political shifts, and especially of inquiry into identity and collective memory, is a salient feature of many documentaries produced in Latin America since the year 2000.

I argue that the documentary not only reconstructs the legendary figure of Mika Feldman but also draws a parallel between the Mika Feldman–Hipólito Etchebéhère couple traveling through different cities, regions, and continents, and their ideological and social commitment, which remained intact over the years in spite of changing international circumstances, and their critical approach and above all, combativeness.

Mika, mi Guerra de España highlights indirectly the role of Jews in the Spanish Civil War. The road movie reconstructs and animates historical facts out of the personal and the subjective. It provides an internal point of view that represents the historical world of Mika and Hipólito Etchebéhère, while creating the opportunity to discuss and analyze the role played by thousands of Jews from all over the world who voluntarily enlisted in the International Brigades (*Brigadas Internacionales*) in order to take an active part in the war. Thus, the film mingles discursive elements that may seem initially to be opposites: the general and the particular, the individual and the collective, and the political and the personal. Social and political issues are addressed from a

1 *Mika, mi guerra de España*, directed by Javier Olivera and Fito Pochat. Argentine, 2013.

 | DOI:10.1163/9789004462540_006

private and individual perspective, presenting a different (and subjective) version of a war that was decisive for European and world history.[2]

The International Brigades were a peculiarity of the Spanish Civil War. Between 1936 and 1937, foreign volunteers from fifty-four countries arrived in Spain with the aim of halting Fascism in Europe.[3] An estimated 4,000–8,000 Jews took part in the conflict. Although dispersed over different units, they represented the second largest foreign contingent in the International Brigades, a significant number considering the proportion of Jews in the general population of their native countries.[4] Rein holds that many Jewish-Argentines supported the Spanish Republic: not only did they identify ideologically with the left-wing parties there, but they were also concerned about the potentially dire consequences of the rise of Fascism in general and of Nazism in particular. In addition, their transnational solidarity with the Republican regime enabled them to promote and address social and ethnic causes in Argentina itself.[5] Gould considers the Jewish-Argentine experience a lens for discussing the relationship between transnational solidarity and the organization of social groups aimed at social and political change.[6]

1 The Role of Women in the Conflict

The Civil War was a turning point in terms of the presence of women in the Spanish public sphere. Women adopted new functions and responsibilities, responding out of instinct and shifting their traditional roles to the needs arising during the conflict. Once mobilized, they were exposed to transformative

2 Valeria Valenzuela, "Giro subjetivo en el documental latinoamericano: De la cámara-puño al sujeto-cámara," *La Fuga* 12 (2011): 1–7.

3 The estimated number of volunteers ranges from 35,000 to 40,000, meaning that the International Brigades were only a fraction of the People's Army. They were ideological volunteers who fought in units that were even less organized than the Spanish brigades of the Republican Army, which at least had some career soldiers among their ranks. Michael Alpert, "Una trompeta lejana—Las brigadas en la guerra de España: Una reconsideración sesenta años después," *Espacio, Tiempo y Forma* 12, (1999): 225–38.

4 Gerben Zaagsma, *Jewish Volunteers, the International Brigades and the Spanish Civil War* (London: Bloomsbury, 2017).

5 Raanan Rein, "A Trans-National Struggle with National and Ethnic Goals: Jewish-Argentines and Solidarity with the Republicans during the Spanish Civil War," *Journal of Iberian and Latin American Research* 20, no. 2 (2014): 171–82.

6 Carol C. Gould, "Transnational Solidarities," *Journal of Social Philosophy* 38, no. 1 (2007): 148–64; Jackie Smith and Hank Johnston eds., *Globalization and Resistance: Transnational Dimensions of Social Movements* (Lanham, MD: Rowman & Littlefield, 2002).

experiences. These processes accelerated and gained visibility in Republican Spain, with the result that many women began to question the gender roles established during the first stages of the war. The urgent situation and the fragmentation of the Republican state created new opportunities for the direct participation of women. Therefore, a significant number of women were actively involved at the front and even took on leadership positions.[7] Lugschitz documented the participation of 400 foreign women in the Spanish Civil War, but it is estimated that besides serving as combatants between 600 and 700 women worked as nurses, translators, doctors, drivers, and distributors of food and military supplies to the Republican troops.[8]

The role of women in the Civil War in general and in militias in particular has fallen through the cracks of historiography. The autobiographies of many female protagonists and the studies of feminist researchers have allowed for both a re-evaluation and a re-valuation of the protagonist roles of these women in the line of battle. Save for some exceptions, frontline militiawomen were members of mixed battalions. In the rearguard, militiawomen were organized in companies made up exclusively of their gender.[9] These processes accelerated and gained visibility especially but not only in Republican Spain. The Seccion Femenina de la Falange (SF; female section of the Spanish Fascist Party) promoted women's political and professional rights within the authoritarian Franco regime in Spain.[10]

The vast majority of militiawomen of POUM (Workers' Party of Marxist Unification), and even of the Republican Army, went into battle under the same conditions as men, and carried out equally important auxiliary roles. According to Lines, women bore a double burden: not only were they expected to fight but also to perform tasks traditionally considered "women's work," like cooking, cleaning, sewing, and laundry.[11]

---

7 Helen Graham, "Mujeres y cambio social en los años 30," *Historia del presente 1*, no. 2 (2003): 9–24; Ana Aguado, "Memorias de la Guerra Civil e identidades femeninas antifranquistas," *Amnis* 2 (2011).

8 Renee Lugschitz, *Spanienkämpferinnen: ausländische Frauen im spanischen Bürgerkrieg 1936–1939* (Vienna: Lit, 2012).

9 As explained by Ballesteros in her study of the life of "forgotten brigade members" like Mika Etchebéhère, among others. Rosa María Ballesteros Garcia, "El efecto de Cronos. Brigadistas olvidadas por la historia," *Aposta. Revista de Ciencias Sociales 38* (2008): 1–40, www.apostadigital.com.

10 See Inbal Ofer, *Señoritas in Blue—The Making of a Female Political Elite in Franco's Spain: The National Leadership of the Sección Femenina de la Falange (1936–1977)* (Eastbourne/Portland: Sussex University Press, 2009).

11 Lisa Lines, "Female Combatants in the Spanish Civil War: *Milicianas* on the Front Lines and in the Rearguard," *Journal of International Women's Studies 10*, no. 4 (2009): 168–87.

An important number among the hundreds of female volunteers who joined the International Brigades were Jewish-Argentine women. Sara Segal, Raquel Levenson, Fany Edelman, the Paulina and Adelina Abramson sisters, and the legendary Mika Etchebéhère are some of the Argentine-Jewish women who took part in the Spanish Civil War, performing various roles.[12]

2 Micaela Feldman/Mika Etchebéhère, the Captain

Micaela Feldman's story is akin to that of many Jewish immigrants who escaped in the late nineteenth century from the pogroms of Tsarist Russia. Her parents settled in the farming colony of Moisés Ville, in the province of Santa Fe, Argentina, the land that eventually became her birthplace. At first, the colonization experience was tough for the immigrants, not only because they found themselves in hostile territory, but also because the landowner, Pedro Palacios, failed to keep his promises. News of famine and disease among the colonists reached the ears of Baron Mauricio Hirsh, the creator of the Jewish Colonization Association (JCA), which founded over fifteen colonies in the provinces of Entre Ríos, Santa Fe, La Pampa, Buenos Aires, and Santiago del Estero.[13] Mika's father was a Yiddish language teacher and as an educator was considered an important figure in the Moisés Ville community.

After graduating from secondary school in the city of Rosario, Mika moved to Buenos Aires to study dentistry. That is where she met Hipólito Etchebéhère, an engineering student. Feldman-Etchebéhère had already expressed her political commitment in her adolescence, joining an anarchist group and co-founding a feminist-libertarian collective. Once in Buenos Aires, she participated actively in the events that led to the University Reform, and became a member of the Insurrexit group, a left-wing political and cultural movement which published an eponymous magazine. Together with Etchebéhère, Mika joined the Argentine Communist Party, from which they were eventually expelled due to their radicalization and Trotskyist inclinations.[14] During the mid-1930s, Feldman and Etchebéhère traveled to the old continent: first to Germany, then France, and finally Spain. When the war began, they joined

12 Raanan Rein, "Metas nacionales y étnicas en la lucha transnacional: argentinos-judíos en solidaridad con el bando republicano durante la Guerra Civil Española," *Investigaciones y Ensayos* (Academia Nacional de la Historia, Buenos Aires) *61* (2014): 405–25.

13 Ricardo Feierstein, *Historia de los judíos argentinos* (Buenos Aires: Galerna, 2006), 77–80.

14 Rein, "Metas nacionales": 442; Cynthia Gabay, "Identidad, género y prácticas anarquistas en las memorias de Micaela Feldman y Etchebéhère," *Revista Forma 14* (2016): 35–56.

the ranks of POUM.[15] It should be noted that Feldman-Etchébère's name does not appear in the list of brigade members because she enrolled, together with Hipólito Etchebere, in the militias of POUM, which had 700 international volunteers among their total of 10,000 members. When Hipólito died in battle, Mika took control of the regiment as captain.

Although Mika was the only woman who held military rank during the Civil War, and the only Argentine woman who fought on the battlefield, her story, as well as those of many other women who played an active role in the war, resurfaced in a more prominent way only in the early 2000s.[16] Her autobiography, *Mi guerra de España*, published in 1976, and a subsequent edition in 1987, inspired two of her grandnephews-in-law, Fito Pochat and Javier Olivera, to make a documentary film about her life; they also published a new edition of the book in 2013. Pochat and Olivera, the directors of the film, stated:

> Peculiar and unimaginable circumstances led us to get hold of the book around 2007, and we could not let it go. This story had remained hidden for years in our family, and once we got over the impression that it made on us, we decided, basically, out of necessity, to make it known through all means available to us. That was how, as filmmakers, we took on the difficult but exciting task of making a documentary film about Hipólito and Mika's amazing story. After four years of hard work, we are glad to say that we have achieved our goal.[17]

3 A Film-worthy Life

Road movies are considered a typical subgenre of North American culture and modern life, associated with notions of mobility and freedom. The form has been adopted by documentary filmmakers who focus on stories in which mobility and movement play a central role. Additionally, these documentaries

15 Gabay, "Identidad, género": 37.

16 In addition to the academic studies cited, it is also worth mentioning *Mika* by novelist Elsa Osorio, published in 2012, which caught the attention of Argentine and Spanish print media, websites, and online newspapers.

17 "Mika *Mi Guerra de España* Un documental de Fito Pochat y Javier Olivera," *Rayo Verde, Blog* (2014). In another interview, they explain: "Not only is this a powerful story, but it is also connected to us: we are Mika and Hipólito's grandnephews-in-law. But we did not know about them. The Etchébèhere family was very conservative. Hipólito was the *red* black sheep, wrongly influenced by Mika, who was of Jewish origin." Fátima Margu, "Mi guerra de España, documental sobre una capitana," *Fronterad Revista Digital 8* (2014), https://www.fronterad.com/mika-mi-guerra-de-espana-documental-sobre-una-capitana.

are enriched by visual techniques used in fictional movies.[18] These devices express various forms of mobility: handheld shots moving through memorable places; long traveling shots with cameras mounted on cars, buses, trains, and even airplanes; long shots of photo albums, and so on.[19] This new subgenre is a consequence of hybridization processes observed in audiovisual production. However, since their inception, road movies have drawn from the documentary aesthetic to show the typical environments and events of the time.[20]

Throughout the second half of the 1990s, there was an increase in documentary film production during the rise of the New Argentine Cinema. A new group of young directors tackled various political and social issues, driven by the possibilities offered by new technologies. The economic crisis that arose in December 2001 led to a significant increase in the production of documentaries about events occurring throughout the country. Some of these films recorded the recovery of factories by their workers, as in *FASINPAT*, by Incalcaterra (Buenos Aires, 2004); *Grissinópoli*, by Doria (Buenos Aires, 2004); and *Ocupar, resistir, producir*, Colectivo (Rosario, Santa Fe, 2005). At the same time, films that set aside classical narration in favor of subjective documentary were also produced: thus, documentary filmmakers addressed both individual and historical memory from a personal perspective. This genre expresses an internal point of view by which the filmmaker represents the historical world. This subjectivity can mix discursive elements that at first sight appear contradictory, such as general versus particular and political versus political; in other words, even though the expressive dimension focuses on individual stories, there is also a social dimension that serves as a subjective response. Social and political issues are addressed from a private and individual perspective.[21] Some of these documentaries focus on the search model, and thus portray mobility with several variables. Leading examples of these first-person films are *Papá Ivan* (María Inés Roque, 2000) and *Un pasaporte húngaro* (Sandra Kogut, 2001), among others. Another common denominator of these documentaries is the centrality of the family as a place for exploring the identity of the filmmakers; *La televisión y yo* (Andrés Di Tella, 2002) and *Los rubios* (Albertina Carri, 2003),

---

18 Verónica Garibotto and Jorge Pérez, "Introduction—Reconfiguring Precarious Landscapes: The Road Movie in Latin America," in Verónica Garibotto and Jorge Pérez eds., *The Latin American Road Movie* (New York: Palgrave Macmillan, 2016), 1–28.

19 Nadie Lie y Pablo Piedras, "Identidad y movilidad en el cine documental latinoamericano contemporáneo: Familia tipo (2009) e Hija (2011)," *Confluencia* 30, no. 1 (2014): 72–86.

20 Santiago García Ochoa, "Redefiniendo fronteras: La poética de la road movie en American Greyhounds (Héctor Muniente y Mario Aranguren, 2010)," *Revista Anual de Historia del Arte* (2018): 95–10.

21 Valenzuela, "Giro subjetivo": 1.

are examples that address family relationships and household-related themes. However, other documentaries examine family relationships and personal stories without linking them directly to the traumatic events of the recent past. Instead, they address stories of individuals who are not particularly extraordinary but are shaped partly by issues rooted in a broader sociocultural context[22]—hence, my analysis of the film *Mika, mi guerra de España*, which adopts characteristics of the road movie and rediscovers and reconstructs segments of Micaela Feldman-Etchebéhère's life, highlighting issues related to historical, political, and social memory. The film leads the viewer, almost in circular fashion, through the various stages of Mika's life and transforms them into sites of memory. These *lieux de mémoire* are like workshops for the reconstitution of the past, understood as social products with specific cultural and political content, where memory actualizes and results in specificity.[23] However, in contrast to Pierre Nora, the sites of memory, or stops, in this road movie are constructed and materialized not from the institutional but from the personal and the subjective, representing a particular world view dictated by the commitment and ideology of the subject of analysis.

> Between July 17 and 18, 1936, a sector of the Spanish Army rose up against the Second Republic of Spain. This act is considered the onset of the Spanish Civil War.

This opening statement from *Mika, mi guerra de España* fades into a close-up of the protagonist, Micaela Feldman Etchébèhere, who is preparing for an interview with TVE (Televisión Española) that was scheduled to be aired in 1984.[24] Alternating between time-worn images in black and white and others

---

22 Lie y Piedras, "Identidad y movilidad": 73.

23 Pierre Nora, "Between Memory and History: Les Lieux de Memoire," *Representations 26* (1998): 7–24.

24 According to the filmmakers, they were so surprised when they found that interview and another one done in 1973 that they felt the need to reconsider the entire film. The interview, in French, was carried out by an old partisan who had traveled to Paris as a result of the publication of Mika's autobiography. The journalist's name was Paolo Gobetti, and the material was found in the Archivio Nazionale Cinematografico Della Resistenza Onlus (ANCR) in Torino, Italy. "The interview was part of an investigation into Trotsky's murder. Mika was a close friend of the married couple Marguerite and Alfred Rosmer; Rosmer was Trotsky's secretary and was identified as the one responsible for allowing Ramón Mercader to get close in order to murder him. Mika granted the interview to TVE thinking that she would be discussing her participation in the Civil War, but when she became aware of the interviewer's intent of taking her down that other path she decided to end it abruptly. Therefore, the documentary includes only a few moments of that interview. It

in color, the documentary presents interviews that convey, through Mika's own words and gestures, her political conviction, ideological commitment, activism, and especially her determination to live in accordance with her ideas. The film also includes historical records; visits to places where the events took place; thoughts that Mika published in her autobiography, voiced by Argentine actress Cristina Banegas; and the account of her nephew Arnold Etchebéhère.[25] Arnold, guiding the filmmakers through various sites of memory, intervenes in the collective historical experience through his own experience, acting as a link between collective and personal memory.[26] Through his narrative, he expresses his personal point of view, as well as his feelings, thus providing a subjective and polemical version of history.[27]

Mobility facilitates Arnold's journey, structures the film space, and partly determines the visual aesthetic of this "revolutionary road movie,"[28] which, similar to the construction of memory itself, includes contradictions, and especially silences, conflicts, and instances of gathering and integration.[29]

Benegas' voiceover, reading parts of the book, serves as a narrative device that not only brings the main character and the viewer together, but also adds subjectivity to the account. The film begins on the banks of the River Seine in Paris, the first stop in this journey and the city where Mika spent most of her life. Arnold Etchébèhere recalls that, after she passed away, he considered scattering her ashes in Atienza, the Spanish locality where Hipólito lost his life. However, he explains that ultimately he decided to cast them into the mythical waters of the Seine. In the second sequence, or stop, the viewer arrives in Spain, where the story begins, without regard for the strict chronology of events. During the

was never aired, and we accessed it through TVE's archive." Correspondence between Fito Pochat, co-director of the film, Jonas Aharoni, and Margu, "Mi guerra de España."

25 Arnold maintained a close relationship with Mika, whom he met as an adult. An important detail is that Arnold lived in Spain, where he went into exile in 1976. Etchebéhère, who died in February 2017, was a pediatrician and, as stated in a feature article published in *Getafé*, "political commitment was his life, both in his home country and his adoptive country. He traveled to Sandinist Nicaragua and once he got there he never stopped prioritizing his commitment over his personal interests." Jesús Bejar, "Arnold, uno de los nuestros," *Getafe Capital*, April 3, 2017, https://www.getafecapital.com/2017/04/arnold-uno-los/.

26 Maurice Halbawchs, *On Collective Memory*, trans. Lewis Coser (Chicago: University of Chicago Press, 1992).

27 Gustavo Aprea, *Documental, testimonios y memorias: Miradas sobre el pasado militante* (Buenos Aires: Manantial, 2015).

28 Verena Berger, "Travelling Cinema—La road movie latinoamericana en el contexto global," in Robin Lefere and Nadia Lie eds., *Nuevas perspectivas sobre la transnacionalidad del cine hispánico* (Leiden, Brill Rodopi, 2016).

29 Elizabeth Jelin, *Los trabajos de la memoria* (Madrid: Siglo XXI, 2002).

first few minutes, Mika, through Benegas' voiceover, describes the mobilized people of Madrid, a city in commotion. The montage moves between personal photographs of the protagonist, newsreel images, and Etchebéhère's account, accompanied by an extradiegetic soundtrack that conveys the urgency and high degree of mobilization of the Madrilenian public, which, in Mika's words, "seized power but not the government ... in a moment in which the only lawfulness possible was revolutionary lawfulness."[30] From that revolution, which had always been the ultimate destination of her journey, the filmmakers take the viewer through a historical time warp to rediscover other settings, the puzzle pieces of Feldman-Etchebéhère's life.

The city of Buenos Aires is the third stop in this journey: various long shots of early twentieth-century buildings, its streets, a subway entrance, and a traditional café where the viewer once again meets Arnold, their guide on this trip. It is here that the origins of the revolutionaries are discussed. Although known for their membership of the Insurrexit group during the University Reform,[31] these young people, as noted, were born in the province of Santa Fe. "I have yet to meet another individual as dazzling as him," wrote Mika in her autobiography about her partner. She also highlighted her fierce political commitment:

> I began as an activist at a very young age, when I was almost a child. I made my debut with my anarchist comrades, and then became a Marxist by choice. I was expelled from the Communist Party after one year. I had always had a libertarian spirit.[32]

Hipólito, for his part, expressed his activism and political commitment through his job in the factories, where he spread revolutionary propaganda. His poor health, owing to the tuberculosis from which he suffered, in addition to his desire to make money for the trip he had planned to Europe, led to the couple spending some years in Patagonia, the fourth stop in this journey.[33] Extreme long shots of lakes, mountains, the car that Arnold drives through these

30 According to the interview carried out in Paris by Paolo Gobetti in 1973, Etchébéhère also highlights the solidarity between different factions and political groups. "Socialists, communists, anarchists ... all of us sang together the *Joven Guardia*, *Bandera Rosa*, anarchist songs, *La Internacional*" (06:39).

31 Together with personalities such as José Ingenieros and Alfonsina Storni, a friend of Mika's. *Mika, mi guerra de España* (12:19).

32 Ibid. (12:37).

33 There, Mika worked as a dentist, and Hipólito made dental prostheses. With their mobile clinic, they went around Patagonian villages and made their home in the city of Esquel (16:40).

desert-like southern highways, portray the iconography of the road as a metaphor of life, a characteristic of road movies. In Mika's words, Patagonia was "very tempting," but the strong convictions she had had since her youth eventually led her to leave the seemingly peaceful south of Argentina for the turbulence of pre-Nazi Germany, the fifth stop in the journey of their lives.

Long shots of buildings, trains going around the big city, factories, and constant movement set the stage for the viewer and convey the maelstrom of the metropolis in the present; the film then goes back to the year 1932, a time when the workers' movement was on the rise. The stop in Germany combines archive footage of mass worker mobilization with towering buildings in the present, as well as documentary scenes reflecting the power and grandeur of Nazism even in its early years. In her text, Mika expresses her disagreement with the policies of the Communist Party. "We saw how sixty years of worker struggles were about to crumble in Germany … Everything was going to disappear." The fifth stop ends with Arnold's visit to Rosa Luxemburg and Karl Liebknecht's grave, in the Socialist cemetery of Berlin. The stone reads as follows: "The dead do not allow us to forget." Thus, the film draws a parallel between Rosa Luxemburg and Mika by emphasizing their commitment and ideological convictions.[34]

Revolutionary Spain is the sixth stop in the film. After a narrative maneuver by the filmmakers, who open a sort of parenthesis in which they provide important details about the Feldman-Etchebéhère couple, the film returns to the beginning of the story, to Mika's interview in 1984. There follows archive footage showing the mobilization of Republican troops and the hope of the revolutionary people. Mika's account, through the voice of Benegas, discusses Hipólito once again, highlighting his superiority over other men, and his charm, which made him "more dazzling every day." The 150 men of the POUM join the Fifth Regiment that is headed for Atienza, a little town in the province of Guadalajara, Hipólito's final destination and the beginning of the legendary Captain Mika. In the present, Arnold guides the viewer to the small settlement. The pastoral images of its green fields and the castle of Atienza towering over the hill are interrupted by a long sequence that portrays the tension of the battle in which Hipólito lost his life; the film then fades to black and gives way to Mika's words: "… a bullet ripped him apart like lightning ripping through a tree."[35] The film then travels to the town of Sigüenza, in the Serranía de Guadalajara Comarca province of Guadalajara, Castile-La Mancha, to the seventh stop in this journey, and its magnificent cathedral, which provided shelter

34 It also highlights the Jewishness of both women, their critical stances, and their confidence in the revolution of the masses or the people.

35 *Mika, mi Guerra de España* (36:46).

and a refuge for both civilians and Republican soldiers. The camera pauses at the majestic arcades, the towers, and a Romanesque rosette, interspersed with archive footage of the same cathedral amidst the combat. In many ways, the Cathedral of Sigüenza represents what Mika and her POUM comrades thought about the war, and especially about the Republican leadership. In Mika's words:

> Let us not forget that the Siege of the Alcázar had taken place at that time,[36] so we wanted to take part in a Red Alcázar to show that we Reds could also hold our position in a cathedral and fight, fight until the end.[37]

The same camera focuses on the wall through which Mika and about ten soldiers escaped to the forest and made their way through a river to reach the long road leading back to Madrid.[38] Hipólito's fall in battle was a defining moment in Feldman-Etchébère's personal life and is also a turning point in the film. Mika returned to Paris to recover after her loss. The train takes Arnold to this eighth stop. He visits the small apartment where Hipólito and Mika lived during their time in Paris. Mika's memories about those years of political readings, debates, and cultural enrichment are intermingled with thoughts on war, merged with activist commitment and the revolutionary conviction that there is a need to change the world. After a short stay in the city, Mika returned to Madrid, determined to lead her column of the POUM, and took part in the battles of Moncloa, Pineda de Húmera, and Cerro de Aguila. Micaela and Hipólito had joined the POUM because they identified politically with the opposition to Stalinism; in Argentina, they were expelled from the Communist Party; in Germany, they were critical of the official political line; and in France, they had ties with leftist opposition groups. On Mika's return to battle, she became acquainted with each militia member under her command and upheld their morale; moreover, many women came from other columns to join hers.[39]

36 A battle of great symbolic value for the Franquist government. At the beginning of the Civil War, militarized civilians rebeled against the Republican government and stayed with their families in the building known as the Alcázar of Toledo. The siege lasted two months and Francisco Franco became the savior of the Alcazar and the undisputed leader of the uprising. Daniela Aronica, "El sitio del Alcazar de Toledo en la historiografía (pro) franquista: un mito construido entre crónica y leyenda," *Spagna Contemporanea* 25, no. 52 (2017): 91–105.

37 Stated in the interview with Gobetti in 1973 (39:03).

38 On one hand, the cathedral becomes a symbol of the resistance, but on the other—as expressed in the film—it evidences the command's failure to respond.

39 Violeta Bruck, "Un viaje en busca de la revolución," *Ideas de Izquierda. Revista de Política y Cultura* 4, October 28, 2013, http://www.laizquierdadiario.com/ideasdeizquierda/wp-content/uploads/2013/10/47.pdf.

When Mika left the front in mid-1938, she joined the anarchist organization Mujeres Libres (free women), teaching many militia members how to read and write.[40] This period coincided with the consolidation of Stalinism and the beginning of persecution of the opposition, which included militants of POUM. Mika recalls that she was imprisoned because a comrade of hers had yelled, “Long live Trotsky!” three times in a café. She stresses that while in prison even the guards called her “Captain.” Later, Arnold visits the city of Barcelona, the ninth stop in the film. The camera traverses the narrow passages of the Gothic Quarter of the city and stops for a close-up of the plaque commemorating the disappearance of POUM leader Andreu Nin.[41] “The minority Communist Party was growing with the support of the Soviet Union. Supporters of POUM were accused of being counterrevolutionaries,” explains Arnold in his account. Thus, the film discusses the internal struggles between Trotskyists and Communists that weakened the Republicans.[42] After spending five months locked up in the Lycée Français of Madrid, Mika returned to Paris in 1939. In December of the same year, she left for Buenos Aires to escape the Nazi invasion of France. Once in Buenos Aires, she contributed to *Argentina Libre*, a weekly anti-Fascist publication, with articles on the war in Europe and the situation of Argentina, as well as to *Sur*, a literary magazine founded by Argentine writer Victoria Ocampo in 1931, which published the works of both Argentine and foreign writers for over four decades.[43] In 1946, she returned to Paris and found it devastated by the horrors of the war. She reunited with old friends and began working as a translator.

4 Conclusion

Mika, mi guerra de España is a journey in search of a continuous revolution. The closing titles discuss Mika’s participation in the May 68 protests in France, as well as in one of the first marches organized in that country against the

40 *Mika, mi guerra de España* (01:05:03).

41 His disappearance sparked an extensive debate that continues to this day, with opposing versions regarding responsibility for his murder. See Pelai Pagès i Blanch, “El asesinato de Andreu Nin, más datos para la polémica,” *Ebre 38*, no. 4 (2010): 57–76.

42 For more information on the matter, see Bernard Bolotan, *The Spanish Civil War: Revolution and Counterrevolution* (Chapel Hill/London: University of North Carolina Press, 1991).

43 Horacio Tarcus, “Hipólito Etchebéhère y Mika Feldman, de la reforma universitaria a la guerra civil española Historia de una pasión revolucionaria,” *El Rodaballo. Revista de política y cultura* 11/12 (2000).

Argentine dictatorship in 1976. The poster advertising the film also highlights the political commitment of the protagonist: it shows her name in the top section, written in red Cyrillic font. Below, in red, there is an illustrated version of an iconic photograph of Mika as captain, clad in a trench coat and seemingly determined to defend her principles and ideology.

The film ends just like it began: Arnold, the guide on this journey, reexamines the photographs of these two exceptional figures who played a central role in universal historical events.

Both the documentary and the book reject the idea of the analytical category of ethnicity; thus, the Jewish origins of Feldman-Etchebéhère are not a parameter in the diegetic text. However, the film illuminates the role of Jewish-Argentines in the Spanish Civil War and their political and social commitment in one of the first major international conflicts of the twentieth century.

Bibliography

Aguado, A. (2011) "Memorias de la Guerra Civil e identidades femeninas antifranquistas," *Amnis 2*.

Alpert, M. (1999) "Una trompeta lejana: Las brigadas en la guerra de España: Una reconsideración sesenta años después," *Espacio, Tiempo y Forma, Serie V, H.ª Contemporánea, t. 12*: 225–38.

Aprea, G. (2015) *Documental, testimonios y memorias. Miradas sobre el pasado militante*, Buenos Aires: Manantial.

Aronica, D. (2017) "El sitio del Alcazar de Toledo en la historiografía (pro) franquista: un mito construido entre crónica y leyenda," *Spagna Contemporanea 25*, no. 52: 91–105.

Ballesteros Garcia, R.M. (2008) "El efecto de Cronos. Brigadistas olvidadas por la historia," *Aposta. Revista de Ciencias Sociales 38*: 1–40, www.apostadigital.com.

Bejar, J. (2017) "Arnold, uno de los nuestros," *Getafe Capital*, April 3, https://www.getafecapital.com/2017/04/arnold-uno-los/.

Berger, V. (2016) "Travelling Cinema—La road movie latinoamericana en el contexto global," in Robin Lefere and Nadia Lie, eds., *Nuevas perspectivas sobre la transnacionalidad del cine hispánico*, Leiden: Brill Rodopi.

Bolotan, B. (2000) "Los partidos de izquierda en la Guerra Civil Española," in Raanan Rein ed., *They Shall Not Pass: The Spanish Civil War, 1936–1939*, Tel Aviv: Zamora-Bitan.

Bruck, V. (2013) "Un viaje en busca de la revolución," *Ideas de Izquierda—Revista de Política y Cultura 4* October 28, http://www.laizquierdadiario.com/ideasdeizquierda/wp-content/uploads/2013/10/47.pdf.

Feierstein, R. (2006) *Historia de los judíos argentinos*, Buenos Aires: Galerna.

Gabay, C. (2016) "Identidad, género y prácticas anarquistas en las memorias de Micaela Feldman y Etchebéhère," *Revista Forma 14*: 35–56.

García Ochoa, S. (2018) "Redefiniendo fronteras: La poética de la road movie en American Greyhounds (Héctor Muniente y Mario Aranguren, 2010)," *Revista Anual de Historia del Arte*: 95–104.

Garibotto, V., and J. Pérez (2016) "Introduction—Reconfiguring Precarious Landscapes: The Road Movie in Latin America," in Verónica Garibotto and Jorge Pérez eds., *The Latin American Road Movie*, New York: Palgrave Macmillan.

Gould, C.C. (2007) "Transnational Solidarities," *Journal of Social Philosophy 38*, no. 1: 148–64.

Graham, H. (2003) "Mujeres y cambio social en los años 30," *Historia del presente 1*, no. 2: 9–24.

Halbawchs, M. (1992) *On Collective Memory*, trans. Lewis Coser, Chicago: University of Chicago Press.

Jelin, E. (2002) *Los trabajos de la memoria*, Spain: Siglo XXI.

Lie, N., and P. Piedras (2014) "Identidad y movilidad en el cine documental latinoamericano contemporáneo: Familia tipo (2009) e Hija (2011)," *Confluencia 30*, no. 1: 72–86.

Lines, L. (2009) "Female Combatants in the Spanish Civil War: *Milicianas* on the Front Lines and in the Rearguard," *Journal of International Women's Studies 10*, no. 4: 168–87.

Lugschitz, R. (2012) *Luchadoras en España: Mujeres extranjeras en la Guerra Civil Española*, Vienna: Lit Verlag.

Margu, F. (2014) "Mi guerra de España, documental sobre una capitana," *Fronterad Revista Digital*, August 8, https://www.fronterad.com/mika-mi-guerra-de-espana-documental-sobre-una-capitana.

Nora, P. (1998) "Between Memory and History: Les Lieux de Memoire," *Representations 26*: 7–24.

Ofer, I. (2009) *Señoritas in Blue—The Making of a Female Political Elite in Franco's Spain: The National Leadership of the Sección Femenina de la Falange (1936–1977)*, Eastbourne/Portland: Sussex University Press.

Rein, R. (2014) "Metas nacionales y étnicas en la lucha transnacional: argentinos-judíos en solidaridad con el bando republicano durante la Guerra Civil Española," *Investigaciones y Ensayos, Buenos Aires: Academia Nacional de la Historia 61*.

Rein, R. (2014) "A Trans-National Struggle with National and Ethnic Goals: Jewish-Argentines and Solidarity with the Republicans during the Spanish Civil War," *Journal of Iberian and Latin American Research 20*, no. 2: 171–182.

Smith, J., and H. Johnston eds. (2002) *Globalization and Resistance: Transnational Dimensions of Social Movements*, Lanham, MD: Rowman & Littlefield.

Tarcus, H. (2000) "Hipólito Etchebéhère y Mika Feldman, de la reforma universitaria a la guerra civil española Historia de una pasión revolucionaria," *El Rodaballo. Revista de política y cultura* 11/12.

Valenzuela, V. (2011) "Giro subjetivo en el documental latinoamericano: De la cámara-puño al sujeto-cámara," *La Fuga* 12: 1–7.

Zaagsma, G. (2017) *Jewish Volunteers, the International Brigades and the Spanish Civil War*, London: Bloomsbury.

CHAPTER 6

The Spirit of 1776 in 1948: American Machalnikim and Israel's War of Independence

Amy Weiss

In early 1948, Colonel David "Mickey" Marcus, an American Jewish West Point graduate and decorated army officer, arrived in Palestine as a foreign military expert. David Ben-Gurion, chair of the Jewish Agency Executive, had eagerly sought expertise from abroad in order to transform the Yishuv's disparate militias, including the Haganah, into a single, organized national army. Marcus, who was initially tasked with recruiting an American general and serving on his team of officers, ultimately accepted the position when no other qualified candidate stepped forward.[1] His decision to take the job stemmed from the belief that the Zionist enterprise served as an extension of the postwar system that sought to end ethnic strife. Akin to Revisionist rhetoric that likened Zionists in Palestine to revolutionaries in the American colonies, Marcus viewed those helping to establish a Jewish state as trailblazers. He believed that the Jewish community in Palestine could benefit from military assistance from abroad, just as American revolutionaries received aid from foreign army officers.[2] Writing to his wife Emma, who remained in New York that winter, Marcus further compared Palestine with the United States, viewing both locations as crucial Jewish immigrant refuges. "Palestine to them is what *America* was to us and our parents," declared Marcus, who considered the former a "land of security" and the latter a "land of opportunity."[3] These observations underscored Marcus' belief that American military assistance could export the opportunities American Jews experienced in the United States to the land of Israel. They further revealed how service in a foreign army reified American Jews' own narratives of belonging and loyalty to the United States.

The changing definition of who counted as a foreign volunteer during Israel's War of Independence underscored the issue of belonging for American Jews.

1 Uri Milstein, *History of Israel's War of Independence* Vol. 1 (Lanham, MD: University Press of America, 1996), 322–24.

2 Derek Penslar, *Jews and the Military: A History* (Princeton, NJ: Princeton University Press, 2013), 232–33.

3 Letter from Mickey Marcus to Emma Marcus, American Jewish Historical Society (AJHS), Center for Jewish History (CJH), I-501, box 14, folder 9, February 11, 1948 [emphasis in original].

 | DOI:10.1163/9789004462540_007

The Israel Defense Forces (IDF), the successor to the Haganah, established an administrative unit in 1948 to assist foreign fighters who came temporarily to aid the Jewish state. It was called Machal, the Hebrew acronym for the term "overseas volunteers," and these foreigners often referred to themselves as machalnikim. Although the Gahal unit (named for the Hebrew acronym for "overseas recruits") consisted mainly of recent immigrants who intended to reside in Palestine, and later Israel, long-term, the Yishuv administration did not always distinguish neatly between volunteer and immigrant, confounding these classifications.[4] Israel's Ministry of Defense conducted its first official accounting of Machal members in June 1949, determining that their ranks included approximately 2,400 individuals. By then, however, a ceasefire had already been declared and many of the machalnikim had since returned home, whether to the United States or to the several other countries from where they originated.[5] Historians have since estimated the numbers ranged somewhere between 3,500 and 4,500 volunteers, while the World Machal veterans' organization has suggested that the total number likely approached 4,500.[6] The dates of a machalnik's wartime service also contributed to the difference in overall numbers. Most of these volunteers came to Israel after May 1948. Individuals who arrived prior to Israel's establishment on May 14, including those that aided the Haganah with illegal immigration or with supplying war materiel, often fell outside the official Machal designation.[7] Broadening the definition of Machal to include all temporary overseas soldiers, military specialists, medical volunteers, and other support staff who assisted with the establishment of Israel, allows for a more nuanced understanding of the foreign assistance rendered to the Jewish state.

A variety of motivations, including Zionist ideals, compelled Americans to volunteer in the 1948 war, yet legal ramifications served as a potential deterrent.

---

4 Penslar, *Jews and the Military*: 228–29. On the collective remembering and forgetting of IDF soldiers in the 1948 Arab-Israeli War, see: Emmanuel Sivan, "To Remember Is to Forget: Israel's 1948 War," *Journal of Contemporary History* 28 (1993): 341–59.

5 Nir Arielli, "When Are Foreign Volunteers Useful? Israel's Transnational Soldiers in the War of 1948 Re-examined," *Journal of Military History* 78 (April 2014): 704.

6 Derek Penslar, "Rebels without a Patron State: How Israel Financed the 1948 War," in Rebecca Kobrin and Adam Teller eds., *Purchasing Power: The Economics of Modern Jewish History* (Philadelphia, PA: University of Pennsylvania Press, 2015), 188; Moshe Naor, "Israel's War of Independence as a Total War," *Journal of Contemporary History* 43, no. 2 (2008): 241–57; and World Machal Website, http://www.machal.org.il/index.php?option=com_content&view=article&id=302&Itemid=357&lang=en (accessed November 5, 2020).

7 See, for instance, letter from Akiva Skidell to Michel Finegood, AJHS, CJH, box 16, folder 10, February 27, 1987.

An estimated 1,100 Americans, mainly Jews, assisted the IDF.[8] While this number represented a sizeable percentage of the overall number of individuals within the Machal unit, it accounted for an infinitesimally small proportion of the American Jewish population of the 1940s.[9] Nationality and neutrality acts passed during the interwar and World War II periods sought to limit Americans' involvement in foreign military campaigns. The Nationality Act of 1940 determined that Americans could lose their citizenship if they joined an overseas military without the approval of the US government.[10] These acts thus highlighted the precarious citizenship status of any Americans who took up arms for the Jewish state, risking possible criminal charges and legal setbacks. While it may never be known how many individuals chose not to volunteer because of these concerns, those that did often sought to balance the need to conceal their overseas military involvement with their desire to publicly share stories of their service.

Examining Machal rhetoric on American identity and national belonging offers a new assessment of the experience of foreign volunteers in Israel's military. Scholars have recently given considerable attention to machalnikim, but their research has often addressed these recruits' overall wartime experiences without analysis of American religious or cultural influences.[11] Instead, their philanthropic donations, assistance with illegal immigration to Palestine, and acquisition of war materiel have been discussed.[12] Describing the mechanics of war is a necessary piece in understanding the totality of American volunteer experiences, yet such narratives often overlook the ways in which these individuals' motivations for serving were deeply connected to their identities

---

8 Yaakov Markovitzky, *Machal: Overseas Volunteers in Israel's War of Independence* (Jerusalem: Israel Information Center, 2003), 15.

9 Jonathan Sarna, *American Judaism: A History* (New Haven, CT: Yale University Press, 2004), 375.

10 Nir Arielli, *From Byron to Bin Laden: A History of Foreign War Volunteers* (Cambridge, MA: Harvard University Press, 2018), 134.

11 Nir Arielli, "Recognition, Immigration, and Divergent Expectations: The Reception of Foreign Volunteers in Israel during and after the Wars of 1948 and 1967," *Journal of Modern European History* 14, no. 3 (2016): 374–89; Arielli, "When Are Foreign Volunteers Useful?": 703–24; and David Bercuson, *The Secret Army* (New York, NY: Stein and Day, 1983).

12 Derek Penslar, *Jews and the* Military: 225–53; Penslar, "Rebels without a Patron State": 171–91; and Aviva Halamish, "American Volunteers in Illegal Immigration to Palestine, 1946–1948," *Jewish History* 9, no. 1 (1995): 91–106. For research that addresses American machalnikim during and after the 1948 war, and upon which this chapter is built, see: Amy Weiss, "1948's Forgotten Soldiers? The Shifting Reception of American Volunteers in Israel's War of Independence," *Israel Studies* 25, no. 1 (Spring 2020): 149–73.

as Americans. The study of American machalnikim's concerns regarding issues of citizenship and belonging, both before and after the 1948 war, shows that these volunteers mediated their anxieties about the legal ramifications of their service with their Zionist and humanitarian interests in aiding Jewish communities abroad.

Just as American volunteers shaped their own understandings of their 1948 wartime and postwar experiences, so too did various agents of the United States influence Machal encounters.

American consuls, Zionist representatives, legal advocates, and intelligence agents all informed the ways in which volunteers understood their relationship to the United States. These officials recognized the legal limbo faced by individuals serving in a foreign war, at times assisting volunteers with the proper paperwork needed to re-enter the United States and at others interrogating them about their overseas military service. Israeli state and military officers likewise influenced machalnikims' connections not only to the Jewish state, but ironically also to the United States. They hoped these volunteers would decide to remain in Israel long term in order to assist with the social and economic development of the country. Even if they chose to return home, however, Israeli leaders viewed Machal members as playing a prominent role in the country's public relations outreach to the United States.

The overall lack of prosecution of American machalnikim revealed a reality that at times differed from the volunteers' perceptions of their place within American society. The absence of a legal response explains in part why many volunteers, including Mickey Marcus, invoked the spirit of 1776 in their discourse about the 1948 Arab-Israeli War. Despite anxieties about their extra-legal activities in Israel, many machalnikim aligned themselves with the revolutionary figures from the eighteenth century who also demanded independence from Britain. This positioning, however, put them at odds with Palestinians, who referred to the 1948 war as the Nakba, or catastrophe.[13] These disparate narratives demonstrate the complex influences associated with American volunteers' assessment—and reassessment—of their experiences during Israel's War of Independence.

---

13 For information on competing narratives of the 1948 war, see Avraham Sela and Alon Kadish, "Israeli and Palestinian Memories and Historical Narratives of the 1948 War—An Overview," *Israel Studies 21*, no. 1 (2016): 1–26.

1 Preparing for Independence

The Haganah, the forerunner of the IDF, looked to expand its military personnel and weapons cache prior to the start of the 1948 Arab-Israeli War. It anticipated a clash with Palestinian Arabs, especially given the Zionist movement's resolve to establish a Jewish state and the British government's intent to withdraw its mandate. Britain's regulation of Haganah access to weapons compelled the Jewish leadership in Palestine to look to the Soviet Union and the United States for its arms. Haganah operatives collaborated in particular with prominent American Jews to create an arms-trafficking operation. The first meeting of this clandestine enterprise took place on July 1, 1945, in the New York apartment of Rudolf Sonneborn, an American Jewish oil industry tycoon.[14] Participants included David Ben-Gurion and Henry Montor, director of the United Jewish Appeal, along with seventeen other Jewish businessmen and Zionists. Together, they formed the "Sonneborn Institute," an illegal arms-smuggling operation masquerading as a legal charity organization. When the likelihood of war with neighboring Arab armies became more certain in 1947, the Haganah began investing not only in weapons but also in its armed forces. Al Schwimmer, an American navigator previously employed by TWA, began securing pilots and planes for the Yishuv.[15] The Haganah also recruited approximately 250 American Jews, some experienced sailors, to participate in Aliyah Bet, the transfer of Jewish immigrants from Europe to Palestine in violation of the British naval blockade.[16]

David "Mickey" Marcus served as the most well-known and highly-ranked American volunteer recruited by the Haganah. Born on the Lower East Side of Manhattan in 1902, Marcus received an early childhood religious education before attending public high school and then enrolling at West Point. He practiced law and served as a district judge in New York before training Army Rangers for their deployment in the Pacific during World War II. He later held a senior staff officer position in the US Army's Civil Affairs Division. The Haganah initially asked to consult with Marcus regarding their search for a military expert to transform the Yishuv's many militias into a single national army. Yet, when none of the qualified candidates agreed to the position, Marcus volunteered. He arrived in Palestine in February 1948 under the assumed name

14 Letter from Rudolf G. Sonneborn to Rabbi Herbert Friedman, American Jewish Archives, MS-763, box 4, folder 4, May 11, 1949.

15 Ricky-Dale Calhoun, "Arming David: The Haganah's Illegal Arms Procurement Network in the United States, 1945–1949," *Journal of Palestine Studies* 36, no. 4 (2007): 22–32; Penslar, "Rebels without a Patron State": 181, 184.

16 Halamish, "American Volunteers": 91–92.

Mickey Stone.[17] Although top US military brass, including Major General John Hilldring, knew of Marcus' overseas appointment, the use of an alias highlighted his concerns about the legal murkiness of taking on such a prominent role in a foreign military.[18]

Prior to Marcus' arrival, a civil war had broken out in Palestine. On November 29, 1947, the United Nations Special Committee on Palestine voted in favor of partitioning the region into Jewish and Arab states and establishing an international zone around the city of Jerusalem. The uncertainty caused by this decision prompted Jews and Arabs to prepare for war. One of Marcus' first priorities was the production of handbooks outlining military organization and logistics for the soon-to-be created Israel Defense Forces. American Jews closely followed the news coming out of Mandatory Palestine. Although most Jews considered themselves firmly rooted in the United States, without plans to permanently move elsewhere, they nonetheless cared deeply about the future of a new Jewish state. They especially feared that the fate of European Jewry during World War II might befall the Jews of Palestine.[19]

Marcus' military priorities shifted when the 1948 Arab-Israeli War began. The Yishuv leadership declared Israel's independence on May 14, 1948, and a full-scale regional war erupted the following day when armies from Egypt, Iraq, Jordan, and Syria invaded. Marcus then focused his efforts on developing an armored force in the Negev before being tasked with commanding the Jerusalem front. The Arab Legion's advances near Jerusalem threatened the Jewish community there with severance from any military or medical assistance. Marcus, now bearing the rank of major general, helped create the "Burma Road," a secondary access route through the Judean hills to provide the besieged city with necessary supplies. On the evening of June 10, a Palmach guard stationed in Abu Ghosh mistakenly killed Marcus when he called out to him in Hebrew demanding that he identify himself; Marcus, however, did not understand the language and failed to respond.[20]

American Zionists reimagined Marcus' accidental death as the ultimate sacrifice of a decorated US and Israeli military commander. The Americans for Haganah, an organization formed in 1947 to support the Jewish army in Palestine, published a biweekly newspaper called *Haganah Speaks!* (originally,

---

17 Penslar, *Jews and the Military*: 230–32; see also Ted Berkman, *Cast a Giant Shadow: The Story of Mickey Marcus, a Soldier for All Humanity* (Philadelphia, PA: Jewish Publication Society of America, 1967).

18 Milstein, *History of Israel's War of Independence*: 323.

19 Hasia Diner, *The Jews of the United States, 1654 to 2000* (Berkeley, CA: University of California Press, 2004), 222, 281.

20 Penslar, *Jews and the Military*: 251–52.

Americans for Haganah). Its front-page coverage of Marcus' death in its June 16 issue reported that he "died in action." The paper provided no further details on his passing, other than lamenting that his death occurred shortly before the war's first ceasefire went into effect on June 11. It also quoted David Ben-Gurion as declaring: "American Jewry will be proud of its great and gallant son who has given his life for the liberation of Israel."[21] Together, this information implied that Marcus lost his life in action, fighting in defense of Israel, when his death was actually caused by friendly fire outside of battle. Americans for Haganah embellished the nature of Marcus' death to underscore the significance of his Zionist commitment. Despite the legal ambiguity of volunteering in a foreign military, Americans for Haganah openly mourned the loss of David Marcus, using his given name instead of his assumed identity. The newspaper's monthly circulation reached between 30,000 and 100,000 individuals, including Zionist supporters, corporate donors, and American Jewish organizational leaders.[22] Although the readers of *Haganah Speaks!* knew of Marcus' service to the Jewish state, the circulation numbers suggest that American volunteerism in the newly established Israel Defense Forces was still not a well-known phenomenon among most Americans.

American Zionist and Haganah officials nonetheless sought to publicly commemorate Marcus' death. On June 16, 1948, a nine-member committee met to make arrangements for Marcus' funeral. The committee included Teddy Kollek, then a Haganah representative in Washington, DC, and later a long-serving mayor of Jerusalem; Emanuel Neumann, president of the Zionist Organization of America; Gottlieb Hammer, a leader of the Jewish Agency for Israel's American division; and Colonel Charles Stevenson, a US Army representative. They determined that an honor guard would be present when Marcus' body arrived at LaGuardia Field (later renamed LaGuardia Airport) and that the military caisson would be draped with an American flag. The committee's indecision as to whether an Israeli flag should also be displayed highlighted their sensitivity to the United States' ambiguous attitude toward overseas military volunteers. The funeral service took place at Union Temple in Brooklyn, immediately followed by interment at West Point, located roughly forty miles north of New York City. Marcus' headstone described him as "a soldier for all humanity," a statement that suggested his service exceeded the boundaries of nationality. It also absolved any concerns about formally acknowledging his

21 "American Hero's Genius Shaped New Israel Army," *Haganah Speaks!* 2, no. 10 (June 16, 1948): 1, 7.

22 Penslar, *Jews and the Military*: 252.

Israeli military experience.[23] The committee tentatively planned a public ceremony in Central Park that July at which they sought high-ranking officials, such as the New York City mayor, the New York governor, and General Dwight Eisenhower to give eulogies. They also considered radio and television coverage of Marcus' illustrious career, noting that a documentary should be made about his life.[24] The writer Ted Berkman later wrote *Cast a Giant Shadow*, a 1962 biography of David Marcus, and the screenplay for a 1966 film of the same name. The public nature of these actual and proposed celebrations of Marcus' life revealed that the manner of his death, while sensationalized, received less attention than the numerous achievements of his lifetime. Although friends and family did not hide Marcus' involvement in Israel's War of Independence, they nonetheless took care as to how they framed his overseas service.

2 Wartime Encounters

While the Haganah, and later the IDF, recruited Americans like David Marcus for their military expertise, a variety of motivations compelled other Machal volunteers to assist the nascent Jewish state. Career soldiers, individuals who made a living participating in one war after the next, sought adventure and a steady salary.[25] A December 1948 report sent to David Ben-Gurion described many Machal volunteers as "non-Zionists" who were "idealists, adventurers, misfits that have never been able to settle down after the last war, and mercenaries."[26] Just as many volunteers, however, were guided rather by humanitarian concerns, a sense of Jewish solidarity, or Zionist inclinations for which they received a small stipend.[27] Ira Feinberg, born in New Jersey in 1930 and a member of the Young Judea Zionist youth group before he traveled to Palestine in February 1948, recalled the desire to "help rescue the persecuted

23 In 1985, a plaque inscribed with the words "In recognition of his sacrifice for the establishment of the state of Israel" and the seals of the United States Army and the Israel Defense Forces were added to Marcus' headstone.

24 Memorandum from Lester B. Stone to Arthur Schwartz, Gottlieb Hammer, et al. "Resume of Preliminary Plans for Ceremonies Relating to the Return of Mickey Marcus' Body and Creation of Some Form of Marcus Memorial Foundation," Central Zionist Archives (CZA) A123/403, June 17, 1948.

25 In 1949, rumors spread throughout Israeli army camps that Indonesian army recruiters were looking for American volunteers to fulfill military contracts. "Report: Recruiting Soldiers for Indonesia," AJHS, CJH, I-501, box 8, folder 2, January 28, 1949 [Hebrew].

26 Lou Harris and Lee Harris, "Report on Machal," CZA, L62/87, December 29, 1948.

27 "Machal Specialists Conditions of Service," CZA, L62/67, n.d.

Jews of the world and bring them to freedom."[28] When asked decades later what motivated Isadore Cohen to volunteer, he explained: "I believed that the Jews were going to be annihilated by the Arabs."[29] The IDF's education of its soldiers during the 1948 Arab-Israeli War further instilled lessons about victimization and militarism that shaped the way in which volunteers and recruits viewed the role of Palestinian Arabs.[30] Yet, the rescue of Jewish displaced persons after World War II, along with the fear of another genocide of Jews in the wake of the Holocaust, compelled Americans also to join Machal ranks.

Proximity to the 1948 war motivated others to serve. When fighting broke out, many volunteers were already living in the region as students at the Hebrew University of Jerusalem. They were either benefitting from a stipulation of the GI Bill that paid for American veterans' tuition and living expenses at foreign universities in order to ease the influx of servicemen enrolling in universities in the United States, or participating in study abroad programs. Freida (Yocheved) Zeira, née Drillick, a New York-born Orthodox Jew active in the Zionist youth group Hapoel HaMizrachi, was one such student. "There might very well be a draft of the girls this week if there is no cease fire," she wrote in a letter on June 7, 1948, but noted, "anyway, I want to join up."[31] Zeira served as a nurse in a military hospital in Jerusalem from 1948 to 1949, ultimately deciding to remain in Israel after the war.[32] Zipporah Porath, also of New York, likewise left her studies at the Hebrew University of Jerusalem to serve as a medic and later worked in the Israel Air Force's intelligence unit.[33]

A desire to make history served yet as another reason for joining Machal. Ralph Lowenstein, a freshman at Columbia University during the 1947–48 academic year later reflected on his decision to volunteer with the IDF in the summer of 1948, despite a lack of prior military experience. As an American Jew originally from Virginia, he declared, "it was 1776 in this country," linking Israel's War of Independence with American colonists' late eighteenth-century decision to breakaway from Great Britain. "No other Jew, even though Jews

---

28 Ira Feinberg, "Survey of Americans and Canadians Who Volunteered to Serve in the Defense of Israel, 1947–1949," AJHS, CJH, I-501, box 10, folder 22, June 8, 1992.

29 Isadore Cohen, "Survey of Americans and Canadians Who Volunteered to Serve in the Defense of Israel, 1947–1949," AJHS, CJH, I-501, box 9, folder 38, March 29, 1992.

30 Shay Hazkani, "Political Indoctrination of Soldiers in the IDF, 1948–1949," *Israel Studies Review* 30, no. 1 (Summer 2015): 20–41.

31 Letter from Frieda Drillick to [?] and Freddie [?], AJHS, CJH, I-501, box 10, folder 12, June 7, 1948.

32 Freida (Drillick) Zeira, "Questionnaire for Americans and Canadians Who Volunteered to Serve in the Defense of Israel, 1947–1949," AJHS, CJH, box 10, folder 12, June 27, 1995.

33 Zipporah Porath, *Letters from Jerusalem, 1947–1948* (Jerusalem: Association of Americans and Canadians in Israel, 1987).

fought in subsequent wars, like the Six Day War, no one else could ever have fought in a war of independence," he explained.[34] Lowenstein considered his participation in Machal an exceptional experience that would forever link him to Israel's foundational history. Like David Marcus, he, too, viewed Israel's founding through the lens of the American Revolution. This rhetoric, especially as Lowenstein explained it decades after the 1948 Arab-Israeli War, ascribed an ahistorical connection between these wars of independence fought centuries apart, yet related the significance of American revolutionary activity to American participation in the 1948 war.

The IDF officially formed the Machal unit as more and more overseas volunteers made their way to Israel after its establishment in May 1948. The military differentiated between men and women in this department from other recruits or volunteers because they anticipated that their stay in the nascent country would be temporary. Believing it would have less influence over machalnikim than soldiers who would be remaining in Israel after the war, the IDF concluded that its rules could not be guided by compulsory edicts, but rather should be a "union of love" between the volunteers and their commanding officers.[35] Lee Harris, the civilian advisor for Machal, expressed this attitude in November 1948 when he requested the discharge of a Machal volunteer awaiting a court martial for breaking army regulations. "This is no reason for application for discharge not to be processed," he noted, while conceding that any decision of the Machal personnel office would be enacted only after the trial. Harris also wanted to streamline the bureaucracy of the IDF by establishing a central location for the Machal office in Tel Aviv that would house representatives from the United States, Canada, and South Africa under one roof. He also requested that branches of the Machal personnel and welfare office be set up in Haifa and Jerusalem to better accommodate the American volunteers scattered throughout the country.[36] The IDF chief of staff acknowledged the needs of machalnikim when he requisitioned the San Remo Hotel in Tel Aviv to provide a recreational center for volunteers on leave.[37]

While the IDF debated the influence it had on American volunteers, machalnikim still believed the United States held significant authority over their citizenship status. These volunteers experienced a legal limbo in which they had joined a foreign army at a time when loyalty to the United States served as a

34 Efrem Sigel, "They Needed to Be There," *The Jerusalem Post Magazine*, May 8, 2008, 26, found in AJHS, CJH, I-501, box 16, folder 18.

35 Report on Machal by Lou Harris and Lee Harris, CZA, L62/87, December 29, 1948.

36 Letter from Lee Harris to Mr. Hammer, CZA, L62/82, January 31, 1949.

37 Letter from the Chief of Staff to Lee Harris, CZA, L62/69, October 29, 1948.

paramount Cold War interest. In addition to outlining nationality and citizenship eligibility, the Nationality Act of 1940 expressly declared that Americans could lose their citizenship if they participated in a foreign country's military without permission of the US government.[38] Given the isolationist stance of the United States in the aftermath of World War I and the new world order brought about by the end of World War II, the American government sought to ensure its citizens remained loyal to its foreign policy objectives. Although the United States recognized Israel's establishment within minutes of the pronouncement, no clear legal guidelines existed regarding American participation in the IDF or any other overseas armed forces at the time.

Volunteers therefore amended the conditions associated with their service to avoid actively breaking the legal code. When Al Schwimmer clandestinely recruited planes and pilots for Israel, he enlisted Hal Livingston, a former sergeant in the American air force during World War II, to assist in the process.[39] Livingston himself soon became a recruiter, recalling: "I would telephone the selectee, [and] vaguely query him concerning his interest in a flying job ... if he expressed interest I invited him to visit the Service Airways office."[40] Of course, the "flying job" to which he referred involved either transporting aircraft to Israel or participating in the Israeli Air Force and the "Service Airways office" was also a cover. By never revealing the true nature of the proposed work, Livingston was attempting to protect himself and others from potential arrest and to guard against the entire operation from being shut down. In other instances, some volunteers used forged Displaced Persons passports as they traveled from European ports to Israel, thus dodging citizenship questions by using fake documents. Other machalnikim reported their insistence on not taking an oath in support of the IDF in order to avoid being accused of violating a stipulation in one of the United States neutrality acts that prevented Americans from swearing allegiance to a foreign government or military.[41]

The case of Larry Hoffman, an American volunteer from New York, demonstrated the potential legal difficulties machalnikim could encounter if they were caught using falsified identity papers. State Department officials flagged Hoffman's 1948 application for a passport to travel to New York from France when they learned that he had changed his destination to Israel without

38 Arielli, *From Byron to Bin Laden*: 134.

39 See Harold Livingston, "Survey of Americans and Canadians Who Volunteered to Serve in the Defense of Israel, 1947–1949," AJHS, CJH, I-501, box 14, folder 1, April 6, 1992.

40 Quoted in David Malet, *Foreign Fighters: Transnational Identity in Civil Conflicts*. (New York, NY: Oxford University Press, 2013), 140.

41 Report on Machal by Lee Harris and Lou Harris; Nathaniel Kaplan to Canadian-American Club, July 28, 1948.

updating his application. In a signed affidavit from June of that year, Hoffman explained that he had decided instead to travel to Israel for business reasons and did not realize he needed to update this information. The agent assigned to his case pointedly asked Hoffman if he planned on joining the Israeli army, to which he expressly declared that he had no intention of doing so.[42] Although detained, Hoffman received the passport necessary for him to travel abroad and by November 1948, regardless of whether his statement had been truthful or not, he had enlisted in the IDF.[43] American officials seemed not to closely follow Hoffman's whereabouts until he attempted unsuccessfully to re-enter the United States through Canada on a forged British passport. Perhaps concerned that using his own passport would further arouse suspicion about his activities in Israel, he sought to hide his connections to the IDF by using false identification. The American consulate in Saint John, in the province of New Brunswick, questioned whether Hoffman had "expatriated himself under Section 401(b) of the Nationality Act of 1940" and requested that American immigration officials ascertain his citizenship status.[44] Hoffman, after much interrogating, returned to his home in New York, but had his passport privileges revoked. He would need to petition the State Department to travel abroad should he wish to do so in the future.[45] Decades later a bitter feud broke out among a handful of Machal members over Hoffman's service in Machal during the 1948 Arab-Israeli War, demonstrating not only the high legal stakes for American volunteers, but also revealing the ways in which volunteers considered themselves arbiters of Machal history and sought to determine who could participate in the narrative.[46]

The IDF also recognized the potential legal difficulties Americans could encounter as volunteers in a foreign military. Israeli representatives in charge of the Machal unit at times requested that customs officials in Israel not stamp American service members' (valid) passports upon their entry into the country in order to expedite their eventual return to the United States. In a December 1948 report sent to David Ben-Gurion, Machal advisors Lou Harris

42 Signed affidavit by Elmer Lawrence Hoffman, AJHS, CJH, I-501, box 12, folder 10, June 16, 1948.

43 Larry Hoffman, "American Veterans of Israel Questionnaire," AJHS, CJH, I-501, box 12, folder 10, n.d.

44 Memorandum "Transmitting Application for Registration as an American citizen of Elmer Lawrence Hoffman," AJHS, CJH, I-501, box 12, folder 10, March 4, 1949.

45 Airgram from Ralph Miller, AJHS, CJH, I-501, box 12, folder 10, March 18, 1949; letter from B.B. Shipley to Elmer L. Hoffman, AJHS, CJH, I-501, box 12, folder 10, August 24, 1949.

46 Letter from Larry Hoffman to Chaverim, AJHS, CJH, I-501, box 12, folder 10, October 14, 1986.

and Lee Harris noted that "instances have occurred of stamping of passports with [an] Israeli visa, which so incriminates the man concerned, that he may loss [*sic*] his livelihood." The report concluded that this practice "must stop" in order to protect Americans from possible legal or criminal charges.[47] The US consulate likewise understood the dilemma machalnikim faced, informing the IDF that it would assist any American in need of a passport or an extension of the passport's expiration date. Although the United States declared it illegal for American citizens to serve in a foreign army, the US consulate nonetheless expressed its willingness to assist American volunteers so that they could return home with the proper legal documentation. This tension between following the legal code and circumventing it in order to assist Americans overseas showed how US representatives recognized the tenuousness of citizenship and sought to ease those difficulties. It also demonstrated the United States' budding diplomatic and cultural relationship with Israel.[48]

The IDF, as well as American Machal members, benefitted from US consular assistance. The Israeli military regarded machalnikim as important public relations agents upon their return to the United States. They therefore worked within the malleable legal framework established by consulate officials to assist volunteers with safe passage back home, further helping to ensure that their service to Israel would be seen in a positive light among the American public. During a late 1948 press conference on machalnikim, an IDF spokesperson explained to journalists that American volunteers "will have much to do with moulding public opinion in those Jewish communities on whose continued financial and political support Israel will of necessity have to lean for some time." IDF officials thus suggested that all American volunteers either be allowed to participate in organized tours of the country or be granted two weeks paid leave prior to their discharge so that they could experience Israel beyond military life. The IDF spokesperson also requested Israelis' assistance in making sure that Machal members felt welcome while they were in Israel so that they could later speak positively about their reception by Israeli society.[49] Their temporary volunteer status, however, necessarily marked them as outsiders. A 1949 survey of American Machal members reported that of the 387 respondents, 55 percent of them held unfavorable opinions of both Israel and its inhabitants. These machalnikim considered that Israelis placed too

47 Report on Machal by Lou Harris and Lee Harris.

48 Michelle Mart, *Eye on Israel: How America Came to View the Jewish State As an Ally* (Albany, NY: State University of New York Press, 2006), 23–52.

49 Memorandum from Akiva Skidell on meeting with the American Consul, CZA, L62/67, March 10, 1949 [Hebrew].

much significance on a person's country of origin, including foreign volunteers from the United States. They also believed that residents did not truly comprehend the sacrifices made by machalnikim who risked their lives to serve in the IDF.[50] Although this survey represented the opinions of a fraction of volunteers and was conducted after many of them had already returned home, it nonetheless revealed the social difficulties associated with integrating into a new society when temporarily participating in a foreign military.

Whereas Americans prepared to return home at various points throughout the 1948 war, Israel hoped many of these volunteers would remain in the country indefinitely. The Machal Settlement Department estimated that approximately 200 American volunteers might stay if they received additional financial assistance beyond what the government had made available to them. Akiva Skidell, head of the Machal unit, and Shimon Gafni, head of the Machal Settlement Department, urged the Jewish Agency to inform its leaders abroad of this urgent fundraising campaign. They explained that a lack of financial aid for Machal veterans would "damage the campaigns, if they return [to the United States] disgruntled because they were unable to get the assistance necessary to enable them to remain here."[51] To further entice volunteers to remain in Israel, the IDF offered housing assistance and job placement in "new industries which are badly in need of top-notch specialists (especially those being established by individuals from the Mahal countries)."[52] The Israeli government, along with the military, sought potential immigrants from countries, including the United States, who they perceived were in a better position to assist with the economic development of the Jewish state. The handful of volunteers who did remain in Israel or who returned to the US only to ultimately settle in Israel, often grew up in Zionist households, in which a desire to the live in the land of Israel had been a predisposition even prior to their military service.

3 Returning Home

Most American Machal members returned to the United States after their temporary service to Israel ended, some after ceasefire agreements had been

50 "Machal Comes to Israel," Israel Democracy Institute, 1949, https://www.idi.org.il/articles/4622 (accessed November 21, 2020).

51 Memorandum from Machal Settlement Department to Mr. Nahum Goldman, "Settlement of American Mahal Veterans in Israel," CZA, Z6/109, n.d.

52 Facts on Mahal Veteran Settlers, CZA, Z6/109, May 8, 1949.

reached and others at staggered times throughout 1948 and 1949, based on either prearranged terms of service or personal or family pressures to come home. Henry Lowenstein, the father of 18-year-old Machal member Ralph Lowenstein, wrote to Lee Harris on November 13, 1948, requesting that his son be discharged from the Israeli military promptly so that he could continue his studies at Columbia University for the start of the 1949 spring semester "before he has to go to the army here." The elder Lowenstein continued: "Ralph applied for his discharge some weeks ago … and was promised that he could get out when he wished, as he did not sign up for any specified time."[53] His father recognized that his son would soon have military obligations to the United States, given the situation prevailing in Korea during the Cold War. Indeed, Ralph served as a corporal in the 101st Airborne Division and the 261st Signal Company during the Korean War.[54] Yet, before he could return to his studies at Columbia or participate in the American military, Ralph Lowenstein needed the Israeli military to approve his request for discharge. In such cases, Lee Harris would insist that the IDF fulfill all such Machal applications. "I think it places the Board [for Discharges] in a ridiculous position, and only affronts Army discipline for the Board to refuse permission of a U.S. citizen to return home if he so insists," explained Harris.[55] The Machal administrator recognized the delicate balance between handling American personnel requests and IDF military needs during the war, since it wished to avoid any escalation of such matters that might involve the American consulate—or volunteers' family members.

Given the covert nature of their volunteerism, American machalnikim sought tangible evidence of their IDF service. They typically did not experience elaborate homecomings, in contrast to the public commemorations held after David Marcus' death.[56] Instead, in the immediate months after the 1948 Arab-Israeli War, these volunteers often sought more private ways to mark their experience in Israel. In the fall of 1948, the IDF chief of staff agreed to provide each Machal member with a letter of thanks upon their discharge from the military.[57] After his return stateside, Ralph Lowenstein requested formal discharge papers, stating: "I would treasure it, and it would mean a lot

53 Letter from Henry Lowenstein to Mr. Lee Harris, CZA, L62/83, November 13, 1948.

54 Ralph Lowenstein, "Questionnaire for Americans and Canadians Who Volunteered to Serve in the Defense of Israel, 1947–1949," AJHS, CJH, I-501, box 14, folder 2, December 1, 1996.

55 Memorandum from Lee Harris to Chairman, Board for Discharges, CZA, L62/57, September 16, 1948.

56 Email correspondence with Dr. Samuel Z. Klausner, November 8, 2020.

57 Letter from Lee Harris to Chief of Staff, CZA, L62/57, October 14, 1948.

to me."[58] Akiva Skidell quickly agreed, signing a brief statement certifying "that *Lowenstein Ralph* served in the Israeli Defence Army and was honourably discharged therefrom on December 24, 1948 [*sic*]."[59] Although Lowenstein entered Israel with false identification, he wanted to ensure that the IDF attached his real name to his service record so there would be evidence that he participated in Israel's War of Independence. This document reflected his desire to have his volunteerism recorded for posterity, rather than serve as a public record in the immediate postwar years.

American citizenship status and criminal punishments served as key concerns for other volunteers discharged from the IDF. Charlie Winters, a volunteer who smuggled aircraft to Palestine, served eighteen months in prison for his participation in this clandestine activity.[60] Although convicted of conspiracy and losing some of his civil rights, Al Schwimmer did not do jail time.[61] Julius Cubernek from Chicago served as a navigator in the US army during World War II and for the IDF during the 1948 war. He experienced no difficulties returning to the United States in April 1949, but remembered FBI agents in Chicago questioning him in 1953 about his IDF service. Cubernek indicated that he had lost his "passport privilege" until 1958 when his "right was restored." In 1965, he left for Israel, this time as an immigrant.[62] Returning to the United States thus held real consequences. While most American volunteers did not experience legal difficulties or criminal accusations, the few that did served as examples to others that the potential for punishment existed.

In order to avoid legal problems, many volunteers did not emphasize their participation in the 1948 war when they returned. Dr. Samuel Z. Klausner, who had traveled to Palestine in January 1947 to start graduate school at the Hebrew University of Jerusalem, joined the Haganah when fighting broke out at the end of that year. Klausner recalled: "My return to the United States was as clandestine as the departure had been." He explained that volunteers knew they were violating a "regulation prohibiting American citizens from serving

58 Letter from Lee Harris to Kiev Skidell, CZA, L62/80, August 14, 1949.

59 Honorable Discharge Statement for Ralph Lowenstein, AJHS, CJH, I-501, box 14, folder 2, August 18, 1949.

60 "Honoring Charlie Winters," Congressional Record, vol. 155, part 10 (2009), 13424, https://www.govinfo.gov/content/pkg/CRECB-2009-pt10/html/CRECB-2009-pt10-Pg13424.htm (accessed November 29, 2020).

61 President Bill Clinton pardoned Al Schwimmer in 2001. This act restored his civil rights. See Don Van Natta Jr. and Marc Lacey, "Access Proved Vital in Last-Minute Race for Clinton Pardons," *The New York Times*, February 25, 2001. The pardons of both Schwimmer and Winters revealed the shift in discourse in the United States about American machalnikim.

62 Julius Cuburnek, "Questionnaire for Americans and Canadians Who Volunteered to Serve in the Defense of Israel, 1947–1949," AJHS, CJH, I-501, box, 10, folder 5, July 24, 1995.

in a foreign army," but the "more alert [machalnikim] knew to ask the Israeli passport control officer to stamp their entry visa on a separate slip of paper."[63] This action enabled the volunteers to comply with the necessary entrance paperwork, but it also created plausible deniability should their whereabouts be questioned upon their re-entry into the United States. Sidney Berger of Silver Spring, Maryland said that he did not advertise his service and instead "juggled" his resume to account for the gap in his employment during the time he spent in Israel.[64] Jules Doneson of Farmington Hills, Michigan recalled leaving Israel under the assumed name "Yochan Danon" and resumed use of his given name when he changed planes en route to the United States.[65] The use of a false identity enabled Doneson and other volunteers to circumvent the stipulations of the Nationality Act, in the hope of drawing attention away from their legal identities and avoiding their incrimination as American citizens violating US law.

Questions about national belonging became further reinforced due to the war's casualty count. The IDF experienced close to 6,000 deaths during the war, a number that included 119 machalnikim, 28 of whom were from the United States.[66] The loss of volunteers was especially apparent in the embryonic Israeli Air Force. Machal members made up about 70 percent of that service branch and accounted for nineteen of its thirty-three combat deaths.[67] In comparison, about 15,000 Palestinians died during the war, along with approximately 1,400 members of the Egyptian army, several hundred from the Iraqi, Jordanian, and Syrian armies, and a few dozen from the Lebanese army. The number of war dead among Israelis and Palestinians accounted for nearly one percent of each community.[68] These figures embedded the 1948 Arab-Israeli War in the collective memories of all those on the battlefront and home front, including American volunteers. Their service—and survival—became even more significant in the face of the many machalnikim who died in defense of the Jewish state.

---

63 Email correspondence with Dr. Samuel Z. Klausner, November 8, 2020.

64 Sidney Berger, "Survey of Americans and Canadians Who Volunteered to Serve in the Defense of Israel, 1947–1949," AJHS, CJH, I-501, box 9, folder 11, April 4, 1992.

65 Jules Doneson, "Questionnaire for Americans and Canadians Who Volunteered to Serve in the Defense of Israel, 1947–1949," AJHS, CJH, I-501, box 10, folder 11, July 8, 1995.

66 "Machal War Dead," AJHS, CJH, I-501, box 4, folder 1, October 27, 1993.

67 Moshe Naor, *Social Mobilization in the Arab/Israeli War of 1948: On the Israeli Home Front* (New York, NY: Routledge, 2013): 19.

68 Anita Shapira, *Israel: A History* (Waltham, MA: Brandeis University Press, 2012), 158, 174; Benny Morris, *1948: A History of the First Arab-Israeli War* (New Haven, CT: Yale University Press, 2008), 406–7.

4 Conclusion

Decades later, participation in Israel's War of Independence remained at the forefront of many American volunteers' recollections from their past. Ralph Lowenstein, a half-track driver for the IDF in the second half of 1948, went on to have a career as a professor of communications at the University of Florida. Between 1992 and 2002, he wrote to surviving American Machal members or the relatives of deceased volunteers asking them to complete a fifty-two-item-questionnaire. The four-page document asked about the volunteer's family background (including his or her place of birth and religion, along with the same information for his or her parents), military experience, Zionist affiliation, educational background, and employment history. It then asked respondents to reflect on their role during the 1948 war, including answering questions such as: "Did you experience any problems with re-entering the US (or Canada)?" and "Has your citizenship status ever been questioned as a result of your service with Israel?" These questions underscored the ongoing discourse surrounding American volunteers' identity and sense of belonging, close to fifty years after their service in the IDF.

The responses, along with the desire to have volunteers even submit such questionnaires so long after their overseas service, demonstrated their interest in providing a more complete narrative of American experiences during the 1948 war. "The role of Americans and Canadians in Israel's War of Independence is little known either in the U.S. or Israel," Lowenstein acknowledged, despite considering their participation being considered "an essential part of the mosaic of victory." Given his belief of "incomplete" records on the American Machal unit maintained by the IDF, he recognized that "this research project may be the last serious chance to fill in the blanks for future generations."[69] The questionnaire, completed by approximately 500 volunteers, offers crucial information not only about the machalnikims' wartime experiences, but also about their lives before and after the 1948 war. This project, later preserved in the archives of the American Jewish Historical Society, provides an extensive assessment of volunteers' involvement in the IDF, one that could not be fully ascertained in the immediate postwar period given the legal limbo associated

69 Letter from Ralph Lowenstein to "Chaver," AJHS, CJH, I-501, box 14, folder 2, March 13, 1992; letter from Ralph Lowenstein to Dr. Michel Finegood, AJHS, CJH, I-501, I-501, box 14, folder 2, March 16, 1992.

with overseas military service.[70] In infusing an American perspective into narratives about Israel's War of Independence, machalnikim sought not only to record their own history, one they feared had not been fully captured at the time, but also to ensure that future generations of Americans and Israelis remembered them.

Bibliography

Arielli, N. (2018) *From Byron to Bin Laden: A History of Foreign War Volunteers*, Cambridge, MA: Harvard University Press, 2018.

Arielli, N. (2016) "Recognition, Immigration, and Divergent Expectations: The Reception of Foreign Volunteers in Israel during and after the Wars of 1948 and 1967," *Journal of Modern European History 14*, no. 3.

Arielli, N. (2014) "When Are Foreign Volunteers Useful? Israel's Transnational Soldiers in the War of 1948 Re-examined," *Journal of Military History 78* (April).

Bercuson, D. (1983) *The Secret Army*, New York, NY: Stein and Day.

Berkman, T. (1967) *Cast a Giant Shadow: The Story of Mickey Marcus, a Soldier for All Humanity*, Philadelphia, PA: Jewish Publication Society of America.

Calhoun, R.-D. (2007) "Arming David: The Haganah's Illegal Arms Procurement Network in the United States, 1945–1949," *Journal of Palestine Studies 36*, no. 4.

Diner, H. (2004) *The Jews of the United States, 1654 to 2000*, Berkeley, CA: University of California Press.

Halamish, A. (1995) "American Volunteers in Illegal Immigration to Palestine, 1946–1948," *Jewish History 9*, no. 1.

Hazkani, Sh. (2015) "Political Indoctrination of Soldiers in the IDF, 1948–1949," *Israel Studies Review 30*, no. 1 (summer).

Malet, D. (2013) *Foreign Fighters: Transnational Identity in Civil Conflicts*, New York, NY: Oxford University Press.

Markovitzky, Y. (2003) *Machal: Overseas Volunteers in Israel's War of Independence*, Jerusalem: Israel Information Center.

Mart, M. (2006) *Eye on Israel: How America Came to View the Jewish State As an Ally*, Albany, NY: State University of New York Press.

Milstein, U. (1996) *History of Israel's War of Independence*, vol. 1, Lanham, MD: University Press of America.

70 "Transfer of Machal and Aliyah Bet Archives to the American Jewish Historical Society," *American Veterans of Israel Legacy Corporation Newsletter* (Winter 2010), 1, http://www.israelvets.com/newsletters/AVILC-2010-Winter-newsletter.pdf (accessed November 29, 2020).

Morris, B. (2008) *1948: A History of the First Arab-Israeli War*, New Haven, CT: Yale University Press.

Naor, M. (2008) "Israel's War of Independence as a Total War," *Journal of Contemporary History 43*, no. 2.

Naor, M. (2013) *Social Mobilization in the Arab/Israeli War of 1948: On the Israeli Home Front*, New York, NY: Routledge.

Penslar, D. (2013) *Jews and the Military: A History*, Princeton, NJ: Princeton University Press.

Penslar, D. (2015) "Rebels without a Patron State: How Israel Financed the 1948 War," in Rebecca Kobrin and Adam Teller eds., *Purchasing Power: The Economics of Modern Jewish History*, Philadelphia, PA: University of Pennsylvania Press.

Porath, Z. (1987) *Letters from Jerusalem, 1947–1948*, Jerusalem: Association of Americans and Canadians in Israel.

Sarna, J. (2004) *American Judaism: A History*, New Haven, CT: Yale University Press.

Sela, A., and A. Kadish (2016) "Israeli and Palestinian Memories and Historical Narratives of the 1948 War—An Overview," *Israel Studies 21*, no. 1.

Shapira, A. (2012) *Israel: A History*, Waltham, MA: Brandeis University Press.

Sivan, E. (1993) "To Remember Is to Forget: Israel's 1948 War," *Journal of Contemporary History 28*.

Weiss, A. (2020) "1948's Forgotten Soldiers? The Shifting Reception of American Volunteers in Israel's War of Independence," *Israel Studies 25*, no. 1 (Spring).

CHAPTER 7

"To Get a Conviction a Morality Officer Has to Practically Crawl into Bed with a Hooker": Phil Givens, the Police, and the Battle for Toronto's Yonge Street

David M.K. Sheinin

In the spring and early summer of 1977, Toronto was on tenterhooks. Violent crime shook the Yonge Street "strip," a few city blocks in the downtown core that featured pornographic bookstores, strip clubs, illicit drug sales, and sex work. The strip was also a fading centre of Toronto's "Gay village." In July, two men raped and murdered twelve-year-old Emanuel Jaques, the son of Portuguese immigrants, who worked regularly as a shoeshine boy on the strip. In short order, the two men accused of the killing were released from prison on their own recognizance, as was a man accused of raping three teenage girls at knifepoint, and a fourth charged with raping a twenty-one-year-old in her apartment. Of the alleged Jaques murderers, the *Toronto Sun* newspaper editorialized that it was "astonishing that the pair would be turned loose on the community." Those charged with "crimes of perversion" were "the vilest of all. Society should not be jeopardized by unleashing people who may have weird, uncontrollable appetites that are slaked or satisfied by violating children."[1]

The Jaques killing coincided with the appointment of Phil Givens as chair of the Metropolitan Toronto Police Board, which oversaw the city police but had long taken a hands-off approach to day-to-day police operations. Givens fell instantly into a political maelstrom. A newly vociferous and politicized Gay liberation movement in Toronto had begun confronting police violence against LGBT people,[2] a struggle now complicated by the dangerous conclusion reached by many Torontonians that the Jaques killing was somehow the fault of LGBT people. There was growing public sentiment against the violent

1 "Vile crimes," *Toronto Sun*, September 6, 1977, 10; Gilberto Fernandes, "Beyond the 'Politics of Toil': Collective Mobilization and Individual Activism in Toronto's Portuguese Community, 1950s–1990s," *Urban History Review/Revue d'histoire urbaine* 39, no. 1 (2010): 59–72.

2 Catherine J. Nash, "Toronto's Gay Village (1969–1982): Plotting the Politics of Gay Identity," *The Canadian Geographer* 50, no. 1 (2006): 1–16; Tom Warner, *Never Going Back: A History of Queer Activism in Canada* (Toronto: University of Toronto Press, 2002), 94–95.

 | DOI:10.1163/9789004462540_008

treatment of Black Torontonians by the police. In August 1979, police officers shot Albert Johnson to death in his home. Two police constables were charged with manslaughter in the case. "I never realized just how close to the surface one would find racial tensions," Toronto mayor John Sewell told an audience at the Toronto Liberal Forum in September 1979. Dismayed, he quoted from a letter he had received from an elderly couple unhappy with the mayor's criticism of police violence and in reference to members of the Black community: "If they don't like our laws ... why do they come here, we don't need them."[3] Two years into Givens' term as chair, tensions between the police and the Black community reached a peak when on October 14, 1979, 1200 people rallied against racism and police violence at City Hall. A small number of demonstrators began fighting with members of the neo-Nazi Nationalist Party of Canada who had arrived to oppose the demonstration with placards reading "White Unity" and "Boat them Back." One counter-demonstrator appeared in a gorilla suit carrying a sign stating, "I demand equal rights too." In the aftermath of the melee, police charged eight of the anti-racism demonstrators with assaulting police officers and breach of the peace.[4]

Meanwhile, in late 1977, a crowd of twenty bystanders watched and chanted, "Kill the cop" as five attackers kicked and punched an officer trying to separate a man and woman fighting in the street. David Keefer was only one of 988 officers assaulted by members of the public that year. In another case, an officer answering a domestic quarrel call found two women fighting. When he tried to separate them, they threw him down a flight of stairs. Other residents who joined the fray knocked him out.[5] Through it all, Givens worked tirelessly to explain police conduct to a divided city. In March 1978, in an echo of police positions, Givens told a reporter that before the Jaques murder, the police had done everything possible to crack down on sex work on the Yonge Street strip, as though sex workers were partially to blame for the murder of a child. At the same time, Givens invoked a longstanding police grievance that community-police conflict was the result of the federal government tying the hands of officers with legislation that prevented them from doing their jobs. The police fight against sex work, he argued, was hampered by the removal of "vagrancy" from the Criminal Code. Now, irate and still commenting on the

3 John Sewell, "Notes for Speech to the Toronto Liberal Forum," September 19, 1979, Collection of John Sewell, Toronto.

4 Michael McAteer, "Eight Arrested as 1,200 Hold Rally against 'Racism and Police Violence,'" *Toronto Star*, October 15, 1979, A3.

5 Pat McEnly, "Hostile Crowds Make Life Grim for Policemen," *Toronto Star*, January 30, 1978, A3.

Jaques murder, Givens told the media, "To get a conviction a morality officer has to practically crawl into bed with a hooker."[6]

This chapter focuses on Phil Givens' term as Police Commission chair from 1977 to 1985. Givens led Canada's largest urban police force during a period of enormous change in Toronto during which the police came under continuous criticism from the media, local politicians, immigrant community leaders, and many others for violence, racial discrimination, the mistreatment of LGBT Torontonians, and rogue behaviour. When Givens was appointed commission chair, John Sewell saw him as an unknown quantity on policing. However, immediately and throughout his tenure, Givens became a staunch, loud, socially conservative defender of the police on the most sensitive of issues. What changed between 1963 and 1966, when Givens served as a liberal mayor of Toronto, and 1977 when he took over as commission chair? How did Givens, a well-known Toronto progressive and a civil rights standard bearer in the 1960s, become the most eloquent defender of the indefensible—police brutality and discrimination against minority communities?[7]

Only half-joking, John Sewell suggested that Givens wasn't the only Torontonian whose politics shifted right on ascending to the commission chair; people seemed to assume police positions once they joined the commission. But Givens' transformation represented more than a shift from a popular liberal (and Liberal Party member) as Toronto mayor in the mid-1960s, to a more conservative tough-on-crime Police Commission chair in the late 1970s.[8] His views did not change suddenly or radically. He continued to self-identify as a Liberal. But there is more to it. Givens' experience offers insights into the shifting character of what it meant to be progressive in Toronto, the changing face of the Liberal Party of Canada, the nature of municipal politics in Toronto, and how Givens' longstanding understanding of civil rights and support for Israel may have influenced his thinking on policing.[9]

1 Givens and Toronto Politics

The second Jewish mayor of Toronto (1963–1966), Phil Givens was born in 1922 and grew up in the city's iconic Jewish neighbourhood along and off Spadina

6 Pat McEnly, "Givens Says Clean-Up Began before Jaques," *Toronto Star*, March 21, 1978, A7.

7 John Sewell, *How We Changed Toronto: The Inside Story of Twelve Creative, Tumultuous Years in Civic Life, 1969–1980* (Toronto: Lorimer, 2015), 195.

8 Author's interview with Gillian Cosgrove, Toronto, Canada, May 11, 2020.

9 Ibid.

Avenue, south of Bloor Street. The son of a garment factory worker, he graduated from law school in 1949. Throughout his adult life, Givens self-identified as a "liberal" in the Canadian context, emerging in the 1960s as a staunch defender of civil rights and of a modern, pro-immigration, open society. During the early Cold War, the Liberal Party could rely on strong support from Jewish voters in Toronto and other Canadian cities, as it could from Italian-Canadians and other first- and second-generation immigrant families who identified the Liberals as more welcoming of newcomers to Canada, and less likely to discriminate against them than the Progressive Conservative Party. A gregarious, witty politician, Givens became mayor at a time when anti-Semitic strictures on the opportunities for Jewish Canadians were quickly fading. He cut a popular figure in a rapidly modernizing Toronto.[10]

Despite strong Liberal Party support, Jewish Canadians, like everyone else, voted across the political spectrum. They were crucial to electing the Communist J.B. Salsberg to the Ontario legislature from 1943 to 1955 for the St. Andrews electoral district. In a city of ethnic politics and neighborhoods, this was the so-called "Jewish ward" and included Givens' childhood Spadina neighbourhood.[11] But in 1955, Allan Grossman won the riding from Salsberg for the Progressive Conservative Party and held the seat for twenty years. The starkness of the Salsberg-Grossman shift in St. Andrews was atypical. But it points to a slow rightward transition in Jewish-Canadian electoral support. When Phil Givens began his term as mayor, the Liberal Party could count on the "Jewish vote" in much of Toronto. But by the early 1980s, that was no longer the case. The journalist Jeffrey Simpson attributed the shift to the evolving Israel policy of the two major parties—more hesitancy from the Liberals, and more consistent support from the Progressive Conservatives.[12] But this is an incomplete explanation. In Toronto, the rightward shift had much more to do with the changing nature of the city, the growing success of Jewish integration into Canadian life (and the attendant decline among Jews in their adherence

10 "Givens Becomes Toronto's Acting Mayor," *Toronto Star*, November 20, 1963, 1, 9; Gerald Tulchinsky, "Ben Lappin's Reflections on May Day Celebrations in Toronto's Jewish Quarter," *Labour/Le Travail 49* (2002): 211–15; Thirstan Falconer, "Andrew Thompson, Liberal Party Reforms and the Engagement of Ethnocultural Communities, 1957 to 1961," *Canadian Ethnic Studies 50*, no. 3 (2018): 111–29; Christo Aivalis, *The Constant Liberal: Pierre Trudeau, Organized Labour, and the Canadian Social Democratic Left* (Vancouver: UBC Press, 2018), 53–68; Jatiner Mann, "The Introduction of Multiculturalism in Canada and Australia, 1960s–1970s," *Nations and Nationalism 18*, no. 3 (2012): 483–503.

11 Author's interview with David Crombie, Toronto, June 15, 2020.

12 Jeffrey Simpson, "How the Political Shift among Jewish voters plays in Canada," *Globe and Mail*, September 28, 2011, https://www.theglobeandmail.com/opinion/how-the-political-shift-among-jewish-voters-plays-in-canada/article4199522/.

to the Liberal Party as their "defenders"), and to social mobility and the identification of more Canadian Jews with Canadian middle- and upper-middle class Progressive Conservative politics. That meant somewhat greater fiscal and social conservatism than their Liberal counterparts in a polity where the difference between center-left and center-right was far less significant than in 2021.[13]

In the 1960s, unlike many American cities that were in the early stages of industrial decline, "white flight," and growing social violence, Toronto was booming. Givens' term as mayor coincided with the rapid evolution of the city from a quiet town with a prohibition on stores opening on Sundays to a city heavily influenced by growing immigrant communities and by the 1960s global counter-culture.[14] Givens positioned himself as forward looking, determined to bring progress and the hallmarks of modernity to the city. The contrast between Givens and his predecessor in the mayor's office, Donald Summerville, and what it meant for the changing character of Toronto, was sharp.

> While Summerville, as a boy in the east end could listen to his politician father's dinner table tales of how the Protestant Orange Tory Establishment ran Toronto, Givens was walking around the corner, past the mattress factory on Euclid Ave. where the noise of the machines and trucks filled the air all night long and the lint from the felting machines got in your nostrils, to find the shop where you could buy day-old bagels for a nickel.[15]

As an adult, Givens spoke frequently about anti-Semitic violence he had experienced as a child on Toronto's streets. He also remembered his father taking him to hear Communist Party members J.B. Salsberg and Sam Carr speak publicly. He earned money for university working summers at the Steel Company of Canada and at Canada Packers in hard manual jobs, though Givens never

---

13 Ester Reiter, "Secular *Yiddishkait*: Left Politics, Culture, and Community," *Labour/Le Travail* 49 (2002): 12–46; Gerald Tulchinsky, "Family Quarrel: Joe Salsberg, the 'Jewish' Question, and Canadian Communism," *Labour/Le Travail* 56 (2005): 149–73.

14 See Stuart Henderson, "Toronto's Hippie Disease: End Days in the Yorkville Scene, August 1968," *Journal of Canadian Historical Association* 17, no. 1 (2006): 205–34; David S. Churchill, "American Expatriates and the Building of Alternative Social Space in Toronto, 1965–1977," *Urban History Review/Revue d'histoire urbaine* 39, no. 1 (2010): 31–44.

15 Ron Haggart, "Phil Givens Wasn't Sure He Wanted to be Mayor." *Toronto Daily Star*, November 21, 1963, 7; Christian Champion, "Courting 'Our Ethnic Friends': Canadianism, Britishness, and New Canadians, 1950–1970," *Canadian Ethnic Studies* 38, no. 1 (2006): 23–46; William J. Smyth, *Toronto, the Belfast of Canada: The Orange Order and the Shaping of Municipal Culture* (Toronto: University of Toronto Press, 2015), 244–71.

espoused socialist causes. He won an election to Toronto City Council in 1953 against a Communist opponent and in 1957 ran unsuccessfully as a Liberal for a seat in the federal parliament. Appointed mayor in 1963 by the city council after the sudden death of his predecessor, Givens won the mayoralty in a December 1964 election for a two-year term.[16]

Givens' 1960s vision of modernity in downtown Toronto focused first on development, by which he and his allies meant slum clearance in favor of grand, modern building projects. At the inaugural meeting of the Toronto City Council after his election as mayor, Givens stressed the need for downtown development as a means to relieve residential taxes in other parts of the city. He noted the $200,000 in building permits issued in 1964 and predicted more for the city in 1965.[17] It was not until immediately after Givens' mayoralty ended that Toronto progressives began to advocate for working-class home owners and tenants in the downtown core, and to question assumptions of the sort that Givens' "crawl-into-bed" remarks suggested regarding connections between poverty, purported morality flaws, and criminality. Givens never repudiated those links.[18]

The hallmarks of Givens' term of mayor were those of an urban 1960s Liberal. His vision for rapid growth and development depended on private investment. He was also far keener than Donald Summerville had been on higher property taxes in the service of a larger municipal government that would in turn provide services and infrastructure upgrades to the fast-growing city. Givens had strong support among apartment dwellers (with less concern over property taxes than home owners). His political Achilles heel was the sprawling east end of Toronto. While Givens could count initially on the backing of Italian and other immigrant communities west of Yonge Street, a British-heritage working class in the east end of the city had shifted their support between Progressive Conservatives and the social democrat New Democratic Party. In more prosperous neighborhoods, five to fifteen kilometres along Yonge Street to the

---

16 Henry Srebrnik, "Chasing an Illusion: The Canadian Jewish Communist Movement," in Matthew B. Hoffman and Henry F. Srebrnik eds., *A Vanished Ideology: Essays on the Jewish Communist Movement in the English-Speaking World in the Twentieth Century* (Albany: SUNY Press, 2016), 101–34.

17 "Givens Stresses Need for Redevelopment," *Globe and Mail*, January 5, 1965, 25.

18 John Sewell, *The Shape of the City: Toronto Struggles with Modern Planning* (Toronto: University of Toronto Press, 1993), 140–71; Richard White, *Planning Toronto: The Planners, the Plans, their Legacies, 1940–80* (Vancouver: UBC Press, 2015), 184–219.

north of the strip, middle- and upper-middle class Torontonians of British origin backed Givens soundly.[19]

As mayor, Givens was acutely aware of the identity politics of the day, and willing to adapt accordingly. In late 1964, Defense Minister Paul Hellyer announced that the Toronto-based Irish Regiment of Canada, a unit in the Canadian Armed Forces, would be deactivated. One of several Irish military units formed in the nineteenth century as rifle divisions with Irish subjects as members, by the twentieth century the Irish Regiment had become an ethnic mix of Canadians. Even so, Irish Canadians saw the unit as theirs and Givens opposed Hellyer on the basis of negative reactions to the deactivation order from Irish Torontonians.[20]

In 1966, Givens lost the mayoralty in a three-way race. "In no Toronto election of recent years," the journalist Ron Haggart wrote, "have voters responded so clearly to the dictates of their ethnic, religious and economic identities ... with the choice of [William] Archer (the executive man), Givens (the modern man), and [William] Dennison (the working man)."[21] Givens won strong majorities among middle- and upper-middle class downtown residents, and enormous majorities among Jewish Canadian middle-class and upper class voters. But it was not enough. Dennison drew unexpected strength from Italian-Canadian, Ukrainian-Canadian, and other ethnic communities.[22]

2 Left to Ottawa, Then Right to Toronto

Givens' brief tenure as a Liberal member of the federal parliament for the York West riding (1968–71) saw him adopt a new political imperative: housing rights for those without means. Gone was the boosterism of a fast development agenda. He now wanted something done about what he called "the gut issue of housing" and urban poverty. There were 16,000 families in Toronto on waiting lists for public housing, he told colleagues in Ottawa. Reflecting the growing interest of the federal Liberal Party in urban housing and an increased willingness in Ottawa to fund housing reform, he announced that Toronto's paradox was "a great deal of wealth on one side and the abject and grinding poverty of

19 Ron Haggart, "A Formula for Victory in December's Election," *Toronto Daily Star*, September 28, 1966, 7.

20 "Givens, Lamport Back Regiment: Hellyer Gets Toronto's Irish Up," *Globe and Mail*, November 11, 1964, 5.

21 Ron Haggart, "The Rich Liked Archer, Elected Dennison," *Toronto Daily Star*, December 6, 1966, 7.

22 "The Go-Go Mayor Went—Gracefully," *Toronto Daily Star*, December 6, 1966, 21.

FIGURE 7.1 Police and shoeshine boy after the Emanuel Jaques killing, Dave Morris, 1977

COURTESY OF TORONTO STAR PHOTO ARCHIVE, TSPA_0014579F, TORONTO PUBLIC LIBRARY

the rat-infested slums on the other." Veering still further left he pressed for a capital gains tax to fund urban housing renewal.[23]

As a member of the federal parliament, Givens had moved north to a new riding he now represented, along with many Jewish-Torontonians who had also moved north and into middle- and upper-middle class neighborhoods. His

23 John Zaritsky, "Canada Must Solve Urban Blight or 'Rue Day,' Givens Says," *Toronto Daily Star*, June 3, 1969, 24; Christopher Stoney and Katherine A.H. Graham, "Federal-Municipal Relations in Canada: The Changing Organizational Landscape," *Canadian Public Administration/Administration Publique du Canada* 52, no. 3 (2009): 371–94.

FIGURE 7.2 Police outside Albert Johnson's home after his killing, John Mahler, 1979
COURTESY OF TORONTO STAR PHOTO ARCHIVE, TSPA_0121538F, TORONTO PUBLIC LIBRARY

district was 30 percent Jewish. As a member of the provincial parliament from 1971 to 1977, Givens represented in succession two Toronto ridings that again reflected the ongoing northern movement of Jewish Torontonians, many of whom comprised his provincial constituencies. In 1971, Givens reverted to earlier positions on urban development and abandoned his publicized concern for urban housing projects. Now, as a backbench member of the provincial parliament, his thinking was increasingly out of step with urban progressives who had begun to advocate forcefully against big development projects at the expense of downtown neighborhoods. His politics were now more in keeping

FIGURE 7.3 Phil Givens: "We're taking the brunt of the criticism because cops are supposed to be angels and paragons of virtue," Boris Spremo, 1979.
COURTESY OF TORONTO STAR PHOTO ARCHIVE, TSPA_0050227F, TORONTO PUBLIC LIBRARY

with conservative sectors of the Liberal Party and reflected the class politics of the more prosperous neighborhoods he had come to represent.[24]

3 Commission Chair

On Givens' appointment to the Police Commission in late August 1977, perhaps because of his shifting politics as a parliamentarian, there was no consensus on what sort of a chair he would be. The president of the Metro Toronto Police

24 "Givens Stands Alone in York-Forest Hill as the Candidate in Favor of Expressways," *Toronto Daily Star*, October 5, 1971, 37.

Association, Al Evelyn, called him a "political hack." The *Toronto Sun* was positive, if circumspect. "Givens has been around (and around and around) and is unlikely to bear grudges or be petty … It is a good appointment." Givens received dozens of congratulatory letters. One came handwritten from Frances Dance of Concord, Ontario, just north of Toronto. "Mr. Davis," she wrote, in reference to provincial premier William Davis who had appointed Givens to the commission, "is apparently at last realizing that the people of Ontario want strong action taken against crime and inflation." Dance went on to write that 90 percent of Ontarians were in favor of a return to capital punishment[25] and that the police were "completely frustrated by the courts, lawyers, and citizens-rights nuts."[26]

By contrast, J.S. Midanik, president of the Canadian Civil Liberties Association (CCLA), was optimistic that Givens would be a defender of civil rights. Midanik recalled that as a young lawyer in the late 1940s he had attended a commission meeting when Robert Saunders was both mayor of Toronto and commission chair.

> I found out that the meetings were closed and … I should consider myself lucky to be in the room … I made some remark about open meetings to Saunders, who gave me a tongue lashing for about fifteen minutes … I complained to the reporters about the meeting being closed and they all looked at me like I was off my rocker.[27]

Midanik told Givens that he hoped the commission would change. With time, he went on only half in jest he hoped that even Alan Borovoy, the CCLA general counsel and most combative civil rights lawyer of his day, would be welcome at commission meetings.[28] "As for Alan Borovoy," Givens joked back in reply, "I mentioned his name the other day in my first Commission meeting and the alarm system did not go off in the building."[29] What Givens did not mention to Midanik was his conviction that by and large, the commission had to function

25 In 1976, the House of Commons passed Bill C-84 striking the death penalty from the Criminal Code.

26 "The New Bick," *Toronto Sun*, August 26, 1977, 10; Dance to Givens, September 16, 1977, 5–1, 3, Box 3, Philip Givens Fonds, 51, Ontario Jewish Archives, Toronto (OJA); Dance to Editor, *Globe and Mail*, September 16, 1977, 5–1, 3, Box 3, Philip Givens Fonds, 51, OJA.

27 Midanik to Givens, September 6, 1977, 5–1, 4, Box 3, Philip Givens Fonds, 51, OJA.

28 Midanik to Givens, September 6, 1977, 5–1, 4, Box 3, Philip Givens Fonds, 51, OJA; Dominique Clément, "An Exercise in Futility? Regionalism, State Funding, and Ideology as Obstacles to the Formation of a National Social Movement Organization in Canada," *BC Studies*, no. 146 (2005): 5–33.

29 Givens to Midanik, September 9, 1977, 5–1, 4, Box 3, Philip Givens Fonds, 51, OJA.

in secret. "A mandatory requirement to conduct *all* matters at commission meetings in the open would simply be an open invitation to disaster," he told commissioners privately on September 9, 1977, the same day he joked with Midanik about Borovoy.[30]

In October 1977, Givens gave a speech at the King Edward Hotel in Toronto. He was cautious, but outlined his support for conservative positions on crime and police conduct. In the face of criticism from John Sewell and others that as a body responsible for police policy and budgets, the commission should be overseen by the city council and should hold open-door meetings, Givens was adamant in refuting the need for change. This was the "invitation to disaster" that he had referenced to commission members a month earlier, a view he shared with senior police officers. Earlier, when interviewed on a radio program, following a flurry of calls for the reintroduction of the death penalty in Canada after the Jaques murder, Givens told his King Edward Hotel audience that he had answered callers that it was a federal matter over which he had no authority as commission chair. But when pressed, he had stated that he favored capital punishment. In his speech, Givens complained that he was then criticized as a "hanging judge," as though capital punishment were his top priority. Givens pointed out that capital punishment could be considered cruel and unusual if other penalties were adequate. He added that sometimes "the wrong man" is executed and that there was no evidence whether capital punishment functioned as a deterrent. Even so he added cryptically, "some people forget criminals are not civilized."[31]

Givens worried about the near future. He expected an increase in crime, and a diminishment in trust in the police. He argued that Toronto could not be insulated for much longer from the global uptick in terrorism, hijackings, kidnappings, and bombings. He warned of political violence reaching Toronto, and of the rise of the fascist National Front in Britain which fomented lawlessness and disorder. "There are groups," Givens added, "actively seeking to undermine confidence in the police." They call police "pigs." He noted a column in "a weekly Communist newspaper" that had said Givens "used to be a good guy" when he was young but that now, as commission chair, he was a bad guy. The police were the enemy. "They break people's heads in strikes—take

30 Givens to Board of Commissioners of Police, September 9, 1977, 5–1, 5, Box 3, Philip Givens Fonds, 51, OJA.

31 Givens, Notes for a Speech, Quarter Century Service Association, King Edward Hotel, Toronto, October 28, 1977, 5–1, 5, Box 3, Philip Givens Fonds, 51, OJA.

away their freedoms—terrorize the public.... Givens as Chairman of the Police [is] *ipso facto*, evil."[32]

Shortly after Givens was appointed chair, Walter Pitman, president of Ryerson Polytechnic Institute, issued a report for the city on racism and violence. There were seventeen recommendations, including an end to height and weight requirements for new police recruits (seen as a limitation on immigrant community recruits); training on cultural differences among immigrant groups; and the creation of a police complaints office. Givens answered Pitman privately. He took strong exception to Pitman's statement that "many police irritated the South Asian minority whenever the opportunity arose."[33]

On Pitman's conclusion that "some policemen have attended to their duties with an unfriendly manner that is cold and obviously uncaring," Givens responded that an officer's training stressed the need for a calm, cool, and collected response. What Pitman saw as uncaring, Givens regarded as avoidance of favoritism. Where Pitman accused the police of being seen "as less hesitant to use force against black civilians," Givens saw lies invented by writers in the newspapers *Contrast* and *Islander* (each dedicated to informing the Black community) and noted, erroneously, that Toronto had not "suffered acute racial tensions for several years." Pitman wrote of low morale in the police force. Givens countered that "repetitive bail, lack of sentencing, [the] quick release of criminals, all these tend to frustrate the police." For Pitman, there was "no active co-operation between the public and the police." Givens saw the opposite. "Citizens have proved beyond any shadow of doubt that they support law and order in the community and are very helpful to the police." Givens cited police crime prevention programs to Pitman which, since 1959, had resulted in 2,056 civilian citations from the commission for civic responsibility. Pitman urged the promotion of officers to encourage diversity in the force. Givens responded that merit, not "color or race" would determine promotions.[34]

As did Givens, police chief Harold Adamson both anticipated and rejected Pitman's criticism. In the draft of a speech to the Toronto Downtown Businessmen's Council on January 13, 1976, Adamson identified what officers believed was the source of their growing conflicts with the community.

> One of our biggest problems in recent years has been that after we've apprehended the offender the law has far too often permitted him to go

---

32 Ibid.

33 Givens to Pitman, November 1, 1977, 5–1, 5, Box 3, Philip Givens Fonds, 51, OJA.

34 Ibid.

> free awaiting trial, on the age-old assumption that a man is innocent until proven guilty.[35]

For Adamson, this had led to rising crime and what he called "street difficulties." The change had been prompted by the federal Bail Reform Act (1972) which had been "influenced too much by the academic viewpoint, and not enough by those with first-hand practical knowledge of criminals," that is, the police. Results included increased pressure on the police when thousands failed to appear for trial—8,727 in 1972 increasing to 11,258 in 1975. In addition, when a parolee was arrested and charged with a crime, release was contingent on cash bail. Now, parolees arrested for a criminal offense could no longer be held in custody. Adamson echoed the concerns of police officers that since the Bail Reform Act, people released after having been arrested "go and commit further crimes. In other words, despite the good intention of its architects, the Bail Reform Act has, in many cases resulted in the professional criminal being allowed to carry on his profession."[36] Adamson and Givens spoke with one voice.

As Givens had pointed out in the aftermath of the Jaques murder, in 1972, the vagrancy sections of the criminal code were repealed. While police departments "warned" politicians of the consequences, "theoreticians, who could not believe that people would still want to beg in a country where welfare was so easily available, refused to listen." Body rub parlors and what Adamson called their offshoots—adult movie theatres, nude dancing, nude encounters, and more—began appearing in downtown Toronto that year and "mushroomed" quickly. He argued that many were fronts for the sex trade. A recent provincial legislative change allowed that for the purpose of licensing, a "body-rub parlour" would include any business where a body rub was performed, save licensed medical establishments. These, and a host of other licensing requirements, opened the door for the police to claim legitimacy in raiding body rub parlors in the downtown Gay community, to conflate sex trade-related crimes with the absence of illicit activity in most such establishments, and to harass patrons violently on a range of false charges, while often exposing to the public and their families men who were in the closet.[37]

---

35 Harold Adamson, "Policing in the Downtown Area," January 13, 1976, 5–1, 5, Box 3, Philip Givens Fonds, 51, OJA.

36 Ibid.

37 Ibid.; Catherine J. Nash, "Consuming Sexual Liberation: Gay Business, Politics, and Toronto's Barracks Bathhouse Raids," *Journal of Canadian Studies/Revue d'Études Canadiennes* 48, no. 1 (2014): 82–105.

In November 1977, on the advice of the Pitman report, the Ontario solicitor general introduced a bill in the provincial legislature to establish an independent citizens' review board to handle complaints against the police. Givens told members of the Ontario Electric Association in January 1978 that, "civilians will be like Monday morning quarterbacks second guessing what the police should have done on Saturday night. The policeman's attitude might be 'what the hell' and he might decide not to get involved and turn away and treat it [crime] like any other citizen." He tore into the report, arguing that it made police officers seem bigots. The proof, he argued, was scant. "I wouldn't hang a dog on evidence like that. I'm responsible for the morale of 6,000 policemen and policewomen and I've got to look after their interests too."[38]

In July 1978, Givens wrote a list of what he called the "ten greatest problems" facing Canadian law enforcement agencies. As he wrote, the list ballooned to sixteen and included the deterioration of any consensus among Canadians on moral norms; public apathy over "disturbances, maladjustments, or injustices" as long as "their ox is not being gored"; the offense taken by many when caught committing a minor legal misdemeanor; and demands for rights while neglecting the obligations of citizenship. He also lamented the weakening of laws designed to benefit the weak and innocent but which benefited criminals, and the extreme leniency of the courts. He underlined the threat of the "kind of senseless 'red brigade' terrorism" in Europe and more grimly still, the rise of economic discontent and political divisiveness resulting in a loss of confidence in authority, and its likely attendant effect on policing and community tranquility.[39]

In private comments for a police force audience, Givens replied snidely to Pitman.

> The Pitman Report was presented and reported as if racism was being discovered in Toronto for the first time like Columbus discovered America ... I remember personally being chased through the streets of Toronto with racial taunts, and being assaulted for racial reasons. This did not stop me from subsequently becoming a lawyer [or] being elected Mayor of Toronto ... I remember outright race riots being led by swastika

38 "Civilian Board Could Hamper Police—Givens," *Toronto Star*, January 12, 1978, A3; "Pitman Report Makes Police 'Appear Bigots,'" *Toronto Star*, January 12, 1978, A3.

39 "Phil Givens' List of the Ten Greatest Problems Facing Law Enforcement Agencies in Canada," July 1978, 5–1, 6, Box 3, Philip Givens Fonds, 51, OJA; Jeremy Brown, Pagurian Press, to Givens, July 5, 1978, 5–1, 6, Box 3, Philip Givens Fonds, 51, OJA.

> clad hoodlums in Kew Beach and involving violence between hundreds if not thousands of people.[40]

This was the crux of Givens on civil rights. He had experienced profound anti-Semitism and other forms of discrimination in Toronto, which he viewed as both recent and newly eliminated. In *Noble and Wolf v. Alley* [1951] S.C.R. 64, the Supreme Court had struck down deed covenants restricting land sales to Jews or Blacks. In *Re Drummond Wren* [1945] O.R. 778, the Ontario High Court had struck down a restrictive covenant in a Toronto lot that prohibited the sale of land to "Jews or persons of objectionable nationality." These and other judicial decisions were reinforced by provincial legislation that protected all citizens' rights to buy property and compete fairly for jobs.[41]

For Givens, Toronto's civil rights triumph had been fought and won. His own trajectory was a testament to social and cultural change in Canada. In the late 1970s, he could not grasp the struggles of the Toronto Gay and Black communities in that context.[42] Even so, while Givens saw the civil rights battles of Canadian Jews as instructive for the larger positive state of civil rights in Canada, he saw two threats to Jewish Canadians that he did not see for other groups. Givens' repeated references to the threat of Red Brigades-like terrorists in Canada (although such groups did not operate there) came as a consequence of international developments since his time as mayor. These included the Six Day War, the Munich Olympics massacre, and the Yom Kippur War as

40 Givens, "Remarks on Presentation of the Pitman Report," 1977, 5–1, 5, Box 3, Philip Givens Fonds, 51, OJA.

41 Givens, "Remarks on Presentation of the Pitman Report," 1977, 5–1, 5, Box 3, Philip Givens Fonds, 51, OJA; Monda Halpern, "The 'Malestrom' at Christie Pits: Jewish Masculinity and the Toronto Riot of 1933," *Canadian Jewish Studies/Études juives canadiennes 28* (2019): 12–33; Richard Harris, "From 'Black-Balling' to 'Marking': The Suburban Origin of Redlining in Canada, 1930s–1950s," *The Canadian Geographer/Le Géographe canadien 47*, no. 3 (2003): 338–50.

42 There is an irony here for Givens, other 1960s Jewish progressives, and indeed many Canadians during the Cold War. Many were acutely aware of and in many cases engaged in struggles for racial equality in the United States, a country whose history of discrimination they found entirely at odds with Canada's putatively non-racialized past. In 1948, for example, the Communist United Jewish Peoples Order-affiliated Toronto Jewish Folk Choir invited US civil rights leader Paul Robeson to sing with them at Toronto's Massey Hall. In 1962, Toronto's Holy Blossom Temple invited Martin Luther King Jr. to speak. Martin Bauml Duberman, *Paul Robeson: A Biography* (New York: Knopf, 1989), 399; Laurel Sefton MacDowell, "Paul Robeson in Canada: A Border Story," *Labour/Le Travail 51* (2003): 177–221; Rosanne Waters, "African Canadian Anti-Discrimination Activism and the Transnational Civil Rights Movement, 1945–1965," *Journal of the Canadian Historical Association 24*, no. 2 (2013): 386–424.

existential threats to Israel, as well as the shocking 1970 murder of Québec deputy premier Pierre Laporte and the kidnapping of British High Commissioner James Cross by Front de la Libération du Québec (FLQ) terrorists. In addition to terrorism, Givens began to worry about a resurgence of anti-Semitism in Toronto and elsewhere in Canada. At the 1969 annual breakfast for the Toronto Men's Regional Council of B'nai B'rith, Givens told the audience: "People tell us that we Jews are paranoid. We come by it honestly. We live in times reminiscent of the 1930s." On the subject of Québec nationalism and the threat to national unity, Givens argued that Jews had "learned by experience that when anything happened to disunify a country, we were the minority group that suffered first."[43]

Celebrating Canada's progress on civil rights in 1977, Givens cited concrete change that might have been unimaginable a generation earlier while reminding people of the ongoing danger to Jewish Canadians. Because of changes to federal law, "people of all races" staffed private and public offices in Toronto at all levels. Such offices had "the appearance of miniature U.N. buildings." At the same time, Givens recalled John Beattie, leader of the Canadian Nazi Party, who had led an infamous rally at Toronto's Allen Gardens in May 1965. Cementing his close ties to Harold Adamson, Givens argued that had then-staff inspector Adamson not been present that day,

> there probably would have been bloodshed, because there were many people in the park who were being provoked by the racists, who bore concentration camp tattoo numbers on their forearms, and who were terrified by what was going on. This same Harold Adamson is now our Chief [of Police], and he is the same person who, at the [anti-police] demonstrations held ... a few weeks ago, had his name carried on signs calling for his removal.[44]

Givens recalled that as recently as the mid-1960s there had been "very serious and extensive swastika dobbings on synagogues ... which terrified people with concentration camp and European backgrounds. The police acted splendidly as they [staked] out these properties and ultimately apprehended the perpetrators."[45]

---

43 "Urges Concern for Nation: Givens Warns Jewish Audience They Would Be First to Suffer in Canada Unity Crisis," *Globe and Mail*, February 10, 1969, 5; Jean-Paul Brodeur, "La Crise d'octobre et les commissions d'enquête," *Criminologie* 44, no. 1 (2011); 45–66.

44 Givens, "Remarks on Presentation of the Pitman Report," 1977, 5–1, 5, Box 3, Philip Givens Fonds, 51, OJA.

45 Ibid.

In keeping with the largely male establishment political culture of 1970s' Toronto, and in a manner that erased the distinction between his Jewish identity and the non-Jewish identities of his powerful colleagues in city governance, Phil Givens' idea of humor incorporated discriminatory tropes. Givens' quips reassured some as a reaffirmation of what they saw as a kind of sensible traditionalism. To progressives, Givens seemed increasingly tone deaf, out-dated, and worse. In 1979, the journalist Patricia Hluchy wrote a short piece in keeping with a new cultural touchstone of the feminist moment, querying which important figures in the city *still* asked their secretaries to bring them coffee. The results were divided along political lines, with the more progressive, younger, hip political figures on board with feminist politics. Prime Minister Joe Clark and Toronto Mayor John Sewell fetched their own coffee. Metro Chairman Paul Godfrey, a commission ally of Givens noted that no one on his staff had "ever complained" about bringing him coffee or his lunch. While he noted that he did not drink coffee, Givens admitted that he did "occasionally ask his secretary to bring him a glass of milk, which, he said, she does ungrudgingly."[46]

In contrast to his defense of the Irish Regiment of Canada as mayor, Givens was dismissive of identity politics as commission chair. In a 1982 letter congratulating Bruce McCaffrey on his appointment as Minister of Citizenship and Culture, Givens shared his disdain for some of the communities that had been his antagonists over the preceding five years.

> I guess your new job means your [*sic*] going to have to make nice to all the ethnic, cultural and minority groups. Now don't forget, the Gays are now a legitimate minority group with a distinct culture. Its [*sic*] very important for you to know that. Susan Fish will keep reminding you. Have a Gay time.[47]

The jab at Susan Fish, a progressive member of Toronto City Council in the late 1970s, a critic of Givens as commission chair, and now, like McCaffrey, a member of the provincial parliament, was an allusion to her gender. It was still uncommon for women to serve in important political positions.[48]

In 1984 Givens interrupted a commission discussion to tell Commissioner Jane Pepino, "You turn me on." Laughter from the three other male commissioners came quickly to an end when Pepino, the only woman on the commission

46 Patricia Hluchy, "Coffee Politics Brewing Up a New Storm," *Toronto Star*, October 17, 1979, A24.

47 Givens to McCaffrey, February 15, 1982, 5–1, 10, Box 3, Philip Givens Fonds, 51, OJA.

48 Ibid.

board and a former Ontario Human Rights Commission representative, called the joke unacceptable and offensive. It got worse for Givens. Pepino bested him in the media, stating that while "it's never too late to teach an old dog new tricks," she would not ask for an apology because "Givens comes from a generation when such remarks were acceptable."[49] While they were not always on opposite sides of police-related debates, Pepino and Givens had clashed a year earlier in a case where Givens stepped forward to defend police practice and the position of the police force hierarchy in regard to sexual assaults. Pepino led a city task force on violence against women. Her report stressed what was quickly becoming the norm in North American cities: teams of detectives specially trained to address sex crimes not only as a criminal manner, but in a way that offered counselling and psychological assistance to victims. In part to facilitate victims coming forward, she also urged the formation of sexual assault treatment centers with doctors trained to address rape trauma. Givens, on the other hand, supported the police, and was unsympathetic to Pepino's recommendations. Givens backed the police argument that victims of sexual violence would be more forthcoming with officers from their own neighborhoods than with members of a specially trained squad.[50]

By early 1980, Givens was under even greater scrutiny than in the past. The journalist Gillian Cosgrove wrote that Givens had "survived almost 30 years in politics as mayor of Toronto, federal MP and provincial MPP ... He once was known as a champion of civil rights, [but] is now being called 'the police commissioner of some of the people all of the time'." A Working Group on Police-Minority Relations was formed in March 1979, representing more than twenty civil rights groups. The group demanded that Givens not be reappointed chair at the end of his three-year term in August 1980. Mayor John Sewell and the prominent Catholic archdeacon Arthur Brown also called for Givens' ouster, and Alderman Allan Sparrow described the idea of his reappointment as "the greatest possible insult" to Toronto minorities.[51]

But Givens was reappointed in July 1980 and became even more combative. "We've taken the brunt of the criticism," he intoned, "because cops are supposed to be angels and paragons of virtue. You can't expect everyone to be satisfied." It was not simply that Givens seemed tone deaf on the issues of racism, bias, and violence in the police force. Critics argued that without Givens'

49 "Turn-On Comment Oroves a Turnoff," *Globe and Mail*, September 21, 1984, M2; Joanne Philpott, "A Commissioner Puts a Word in for the Police," *Globe and Mail*, December 6, 1984, L10; "Givens Made Sexist Remark, Pepino Says," *Toronto Star*, September 21, 1984, A2.

50 Alden Baker, "Task Force at Odds over Rape Teams," *Globe and Mail*, July 22, 1983, 5.

51 Gillian Cosgrove, "Minorities Put Police Leader under the Gun," *Toronto Star*, April 5, 1980, B6.

leadership on reform, officers were simply doubling down on longstanding abusive behaviors. Ellert Frerichs, a United Church of Canada minister and an executive member of the Liaison Group on Law Enforcement and Race Relations accused Givens of blindly defensive behavior with regard to the police and having developed a fortress mentality that eliminated the chance for strong community-police dialogue.[52] On Givens' appointment to a second term, Charles Roach, an African-Canadian civil rights lawyer announced that Givens' presidency of the Canadian Zionist Federation made him unpopular among Arab Torontonians. Alfred Hamilton, publisher of the newspaper *Contrast*, serving the Black community in the city, also expressed his disappointment at the news. Dudley Laws, executive director of the Universal African Improvement Association and a long-time critic of Givens, called it a "deliberate attempt by the province to undermine the concerns of the Black community."[53]

In the early 1980s, tensions continued to rise between the police and the Gay community. In 1981, two city aldermen, Patrick Sheppard and David White, reported to City Council that officers had destroyed property and physically abused many of the 309 men charged during raids on four bathhouses on February 5. Police entered the bathhouses carrying crowbars and sledgehammers. They photographed naked men, and called them "queers, faggots, and fairies." Officers wrote in ink on some men's arms to identify them and performed illicit rectal searches. The fifty-eight-page report contradicted the police report of events that described the officers as having acted professionally.[54]

At an annual Catholic "communion breakfast" the following month, Givens made his views clear. With 250 police officers present, Givens and Emmett Cardinal Carter, archbishop of Toronto, spoke at the gathering that preceded a service at St. Michael's Cathedral. "A bathhouse," he stated to loud applause "is a residence in a pretty town in England of the same name, Bath." To more applause and laughter, Givens noted that, "an Irishman's definition of a queer is someone who prefers women to booze." Givens cited the recent chants of "Givens Go, Givens Go" of those demonstrating against the bathhouse raids. "I've been thinking where I could go." When Givens mentioned San Francisco

52 Ibid.

53 "Givens Is Back on the Job—and Critics Aren't Happy," *Toronto Star*, July 1, 1980, A2.

54 Ian Mulgrew, "Police Abused Men in Bathhouse Raids, Two Aldermen Say," *Globe and Mail*, February 27, 1981, 5; Gerald Hannon, "Taking It to the Streets," *The Body Politic*, 71 (March 1981): 9, 12, 16; George Jonas, "Operation Bathhouse," *Toronto Life*, May 1981, 35; Jamie Bradburn, "Historicist: Racism and Homophobia in the Pages of a Police Magazine," *Torontoist*, June 4, 2016, https://torontoist.com/2016/06/historicist-all-the-news-views-that-arent-fit-to-print/.

as a possible destination, a city known for its large Gay community, there was more laughter in the audience.[55] There were raids on two more bathhouses in June 1981 raising the number of those arrested as bawdy house "found-ins" to 341, and another raid in April 1983 leading to twenty new arrests. Of the first group of 304 that went to trial, only 37 were found guilty. One received a suspended sentence and the rest were given absolute or conditional discharges. The message from the judiciary was clear. Givens and the police were now out of step with a judicial system increasingly unsympathetic to police attacks on the Gay community.[56]

4 Conclusion

Givens mixed his Zionism with his political positions easily while in municipal, provincial, and federal office and in a way that would be impossible today. He came to Zionism as a child and, as did many of his generation, he equated the founding of Israel with the same principles that guided his term as mayor of Toronto—modernity and progress. Givens chaired the building committee for the new Toronto Zionist Centre, whose cornerstone was laid in 1962. In 1968, he described Theodor Herzl much as he might have described himself. "He was what they called an assimilationist," wrote Givens, unhappy with those who confused what he regarded as modern Judaism with the pejorative use of "assimilation." "A successful writer, a fashion plate, a bon vivant, whose social mobility was free and unimpeded," Givens continued on Herzl.[57]

Givens was a prominent Canadian voice for Israel throughout the 1960s and 1970s. A year after having been voted out of the mayor's office, Givens hosted a packed house at the O'Keefe Centre theatre in Toronto for the "Chanukah Festival for Israel." The borscht belt comedian Henny Youngman was there along with his iconic violin and according to journalist Joshua Cohen, Givens was the only person on stage to give Youngman "a run for his money." When the audience applauded Givens loudly at one point he asked them, in

---

55 Diane Ellson, "Bathhouse References Get Laughs from Police," *Globe and Mail*, March 2, 1981, 5.

56 "The Bathhouse Files," *Globe and Mail*, April 22, 1983, 6; David Townsend, "The 'Blood Libel' and the Spectator's Eye in Norwich and Toronto," *Journal of the Canadian Lesbian and Gay Studies Association/Revue de la Socieété canadienne des etudes lesbiennes et gaies*, 1 (1999): 1–24; Arnold Bruner, "Out of the Closet: Study of Relations Between the Homosexual Community and the Police" (unpublished report to Toronto City Council, September 24, 1981).

57 Philip Givens, "A Light in Our Lives," *Globe and Mail*, April 27, 1968, A9.

reference to his 1966 electoral loss for the mayoralty, "Where were you at the last election?"[58]

In October 1975, 1500 Jewish Torontonians rallied in protest against the United Nations labelling Zionism a form of racism. They heard federal members of parliament John Fraser and John Roberts, and provincial parliamentarian Phil Givens condemn the resolution (one the Canadian government had opposed). The Canadian Zionist Federation, of which Givens was president at the time, organized the protest. "They are not just insulting us," he told the crowd "they are trying to expel us."[59] Four months later, in February 1976, the Canadian Zionist Federation took out a full-page ad in the *Jerusalem Post* announcing its Fourth National Convention, to be held in Israel. "Israel is Zionism—Zionism is Israel: From two jumbo jetloads of Canadian Zionists!" read Phil Givens' opening statement in the ad. In 1978, there was growing concern in the Montreal Jewish community over French Canadian nationalism in Québec and the vociferous anti-Zionism of some members of the separatist Parti Québécois provincial government. In November 1978, as president of the Canadian Zionist Federation (and chair of the Police Commission), Givens weighed in. Montreal Jews, he announced, "were petrified with fear." Givens tied anti-Zionist rhetoric in 1970s' Québec to the rabid anti-Semitism of mainstream political leaders in the province in the 1930s. It "rings a bell," Givens argued.[60]

Givens frequently mixed his roles as commission chair and his presidency of the Canadian Zionist Federation (1973–85). In September 1982, he wrote to Arie Ivtsan, inspector general of the Israel Police, on behalf of Murray B. Koffler, chairman of Super Pharm, Israel, and its parent company, the Canada-based Shoppers Drug Mart. It was a letter of introduction for Koffler framed as a discussion of policing and crime prevention. However, the point of the letter was to request a meeting between Ivstan and a Super Pharm security officer, Gadi Stern, to coordinate "loss prevention" at Super Pharm. "The world Jewish Community continues to support all your efforts," Givens wrote in conclusion.[61]

---

58 Joshua Cohen, "Phil Givens Runner-Up with Comic Youngman," *Globe and Mail*, December 18, 1967, 16.

59 Bob Graham, "Metro Jews Told UN's stand on Zionism Is Anti-Semitic," *Toronto Star*, October 29, 1975, B1.

60 Ya'Acov Reuel, "Quebec's Jewish Problem," *Jerusalem Post*, November 17, 1978, SM6; Robert Fulford, "Blaming the Jews: Is Anti-Semitism Marching Again in Canada? Everywhere There Are Disturbing Signs," *Saturday Night*, November 1980, 5–8; Ben G. Kayfetz, Board of Directors, Canadian Jewish Congress, "Only Yesterday," speech to the Toronto Jewish Historical Society, May 1972, 5–1, 9, Box 3, Philip Givens Fonds, 51, OJA.

61 Givens to Ivtsan, September 15, 1982, 5–1, 10, Box 3, Philip Givens Fonds, 51, OJA.

Givens understood civil rights through a prism of judicial and legislative change that had ended anti-Semitic restrictions on Jewish Torontonians, but at the same time in a context of new threats to Toronto Jews, to Israel, and to civic order in Canada and elsewhere. He viewed that progress as seminal and applicable to all Canadians so that when members of minority groups complained of discrimination in the 1970s, Givens balked. Their complaints made no sense to him in light of his view that progress on civil rights for Jews was universally applicable in Canada. Givens also viewed police forces in Canada, Israel. and elsewhere as a bulwark against the sort of terrorism, criminality, and disorder that he viewed as historically threatening to a stable modern society, and even more particularly to Jews in such societies.

In his defense of the police, Liberal Party politics also shaped Givens' thoughts. In 1969, as minister of justice, Pierre Trudeau introduced a bill in the House of Commons that would be passed as the Criminal Law Amendment Act (1968–69), S.C. 1968–69, c. 38. It decriminalized homosexuality.[62] "There's no place for the state," he stated, "in the bedrooms of the nation." But Trudeau was ahead of his times in regard to how Canadians viewed homosexuality. Most police chiefs in Canada had opposed the change because, as they stated at the time, they believed homosexuality intersected with prostitution, drug use, and other crimes. For most Canadian police officers, to be Gay was to be linked to a range of criminal behaviors. Trudeau's shift left on homosexuality, on the legalization of abortion, on the decriminalization of contraceptive sales, and on other social issues as justice minister, then as prime minister left many Liberals with a wider range of political options within a party that now incorporated a broader political span. Many like Givens rejected Trudeau's social advances even as they supported his tough-on-terrorist actions against the FLQ in 1970. Givens' shift left on housing policy as a federal parliamentarian, then right again once he returned to Toronto in the provincial parliament was likely the result of his adapting to the new, more complex norm in Liberal politics. But more important, while Givens continued to identify as a Liberal, like many of his generation he remained out of step with the seismic shifts on social policy that Trudeau led. As did police positions on criminality, sex work, police-community relations, and homosexuality, Givens represented the views of at least half of Toronto's population, including many middle-class urban Liberals who were uneasy with rapid social change in Canadian cities.[63]

62 Robert Wright, *Trudeaumania: The Rise to Power of Pierre Elliott Trudeau* (Toronto: HarperCollins, 2016), 121–26.

63 Tom Hooper, "Queering '69: The Recriminalization of Homosexuality in Canada," *Canadian Historical Review* 100, no. 2 (2019): 257–73; Tom Hooper, "Was Homosexuality

US president Ronald Reagan famously quipped that he had not left the Democratic Party in 1962; the Democratic Party had left him. This was similarly true for Givens and for many Torontonians. In the 1960s, and culminating in Pierre Trudeau's 1968 development of a Canadian Charter of Rights, the Liberal Party emphasized civil rights and Canada's strength in immigrant communities as the underpinnings of a just, democratic society, focused on middle-class ideals of progress and social change.[64] By the early 1970s, in addition to the shift on homosexuality and other social issues, Liberals had begun to adopt the cause of discrimination against Black Canadians, immigrant Canadians from Southeast Asia, and the Gay community. As mayor Givens had famously dismissed the notion that Canada had a "race" problem. In the 1970s and early 1980s, as Givens' generation and that of his children, moved from the old Jewish ward north, to more middle-class communities, the shift left in Liberal Party politics on social issues did not resonate with many, even as the Progressive Conservative Party became more explicitly welcoming of immigrant Canadians and stronger in its backing of Israel.

In 2016, a decade after Givens died, a small group gathered in the rain at Toronto City Hall to celebrate what some remembered as Givens' most lasting achievement. It was the fiftieth anniversary of the installation of The Archer, a Henry Moore sculpture. For Phil Givens, the piece had been a triumphant capstone to Finnish architect Viljo Revell's newly opened New City Hall. Revell had included public art in the design as integral to the composition of the public square on which the new buildings were erected. Givens enthusiastically joined Revell in choosing the Moore piece and promised $120,000 in public funds for the purchase. Givens considered himself a connoisseur of modern art. He was a follower of Harold Town, Jack Bush, and other members of the Toronto "Painters Eleven" group of abstract artists, and a regular visitor at the Greenwich Gallery and other cutting edge art hot spots in Toronto. The first mayor who occupied the New City Hall, Givens boxed up and sent into storage dozens of works by Owen Staples and other early twentieth-century Toronto landscape artists, paintings he regarded as out of date in his modern Toronto. When his advisers began telling him that Torontonians were unhappy with the price tag for Moore's metal sculpture that didn't "look" much like an archer, Givens responded, "I don't care what the public thinks."[65] He eventually bowed to pressure and abandoned the idea of funding the project publicly. Even

Really Decriminalized in 1969?" *Now*, November 21, 2019, https://nowtoronto.com/news/homosexuality-canada-pride-1969/.

64 Wright, *Trudeaumania*: 153–55.

65 Author's interview with David Crombie, Toronto, June 15, 2020.

though he raised the purchase price from private donors, voters punished him at the polls in 1966. According to many, Givens lost the mayoralty because voters remained livid over wasteful spending on the sculpture.[66]

What people missed about Phil Givens, having seen him as an early 1960s' progressive on urban issues, was that in 1965 "progressive" for Givens and many others meant modernizing the city by slum clearance, big development projects, modern architecture, abstract art, and educating residents on what was good for them aesthetically—not progressive 1970s' ideas on consultative politics, urban poverty, fighting expressways, and the links between policing and violence. In addition, what many in the 1970s could not see about Givens on civil rights was framed by a dramatic shift in Toronto from the 1960s to the 1970s. Givens was a 1960s' man before the 1960s had *become* the 1960s. Despite the sharp cultural differences that Roy Haggart saw between Givens and his predecessor as mayor, Donald Summerville, they may have had far more in common than Givens and his 1970s' critics. Givens' understanding of problems was shaped in the first instance by dramatic changes for the Jewish community in his own lifetime and through his election as mayor. He never wavered in believing that those changes were a triumphal story universally applicable to all Canadians. This dovetailed with the emergence of Canadian multicultural policy precursors in the 1960s at the same time that the Canadian government altered immigration rules to eliminate discrimination against people from the underdeveloping world, but *before* the arrival of large numbers of immigrants from two of the communities most at odds with policing in the 1970s, Southeast Asian and Black Torontonians. Stated simply, multiculturalism as federal Liberal government policy was never imagined or implemented to include LGBT people or people of color, even as Givens, like many Canadians saw multiculturalism as a confirmation of the universality of the twentieth-century narrative of Jewish Canadian advancement. In the end, John Sewell may also have been right about people who joined the commission and changed their politics: Givens *became* a member of the police. As he told journalist Gillian Cosgrove in 1980 when she pressed him a little too hard on civilian oversight of the police: "We are cops. You can't expect us to sit down and have a goddam democratic meeting during a shoot-out or a burglary."[67]

---

66 Julian Mirabelli, "Henry Moore's 'The Archer' Celebrates 50th Anniversary," *Urban Toronto*, October 27, 2016, https://urbantoronto.ca/news/2016/10/henry-moores-archer-celebrates-50th-anniversary.

67 Gillian Cosgrove, "Minorities Put Police Leader under the Gun," *Toronto Star*, April 5, 1980, B6; Thomas M. Prymak, "The Royal Commission and Rudnyckyj's Mission: The Forging of Official Multiculturalism in Canada, 1963–71," *University of Toronto Quarterly 88*, no. 1

Bibliography

Aivalis, C. (2018) *The Constant Liberal: Pierre Trudeau, Organized Labour, and the Canadian Social Democratic Left*, Vancouver: UBC Press.

Anwar, A. (2014) "Canadian Immigration Policy: Micro and Macro Issues with the Points Based Assessment System," *Canadian Ethnic Studies 46*, no. 1: 169–79.

Brodeur, J.-P. (2011) "La Crise d'octobre et les commissions d'enquête," *Criminologie* 44, no. 1: 45–66.

Champion, C. (2006) "Courting 'Our Ethnic Friends': Canadianism, Britishness, and New Canadians, 1950–1970," *Canadian Ethnic Studies 38*, no. 1 (2006): 23–46.

Churchill, D.S. (2010) "American Expatriates and the Building of Alternative Social Space in Toronto, 1965–1977," *Urban History Review/Revue d'histoire urbaine 39*, no. 1: 31–44.

Clément, D. (2005) "An Exercise in Futility? Regionalism, State Funding, and Ideology as Obstacles to the Formation of a National Social Movement Organization in Canada," *BC Studies 146*: 5–33.

Duberman, M.B. (1989) *Paul Robeson: A Biography*. New York: Knopf.

Falconer, T. (2018) "Andrew Thompson, Liberal Party Reforms and the Engagement of Ethnocultural Communities, 1957 to 1961," *Canadian Ethnic Studies 50*, no. 3: 111–29.

Fernandes, G. (2010) "Beyond the 'Politics of Toil': Collective Mobilization and Individual Activism in Toronto's Portuguese Community, 1950s–1990s," *Urban History Review/Revue d'histoire urbaine 39*, no. 1: 59–72.

Halpern, M. (2019) "The 'Malestrom' at Christie Pits: Jewish Masculinity and the Toronto Riot of 1933," *Canadian Jewish Studies/Études juives canadiennes 28*: 12–33.

Harris, R. (2003) "From 'Black-Balling' to 'Marking': The Suburban Origin of Redlining in Canada, 1930s–1950s," *The Canadian Geographer/Le Géographe canadien 47*, no. 3: 338–50.

Henderson, S. (2006) "Toronto's Hippie Disease: End Days in the Yorkville Scene, August 1968," *Journal of Canadian Historical Association 17*, no. 1: 205–34.

Hooper, T. (2019) "Queering '69: The Recriminalization of Homosexuality in Canada," *Canadian Historical Review 100*, no. 2: 257–73.

MacDowell, L.S. (2003) "Paul Robeson in Canada: A Border Story," *Labour/Le Travail 51*: 177–221.

Mann, J. (2012) "The Introduction of Multiculturalism in Canada and Australia, 1960s–1970s," *Nations and Nationalism 18*, no. 3: 483–503.

(2019): 43–63; Arif Anwar, "Canadian Immigration Policy: Micro and Macro Issues with the Points Based Assessment System," *Canadian Ethnic Studies 46*, no. 1 (2014): 169–79.

Nash, C.J. (2014) "Consuming Sexual Liberation: Gay Business, Politics, and Toronto's Barracks Bathhouse Raids," *Journal of Canadian Studies/Revue d'Études Canadiennes 48*, no. 1: 82–105.

Nash, C.J. (2006) "Toronto's Gay Village (1969–1982): Plotting the Politics of Gay Identity," *The Canadian Geographer* 50, no. 1: 1–16.

Prymak, T.M. (2019) "The Royal Commission and Rudnyckyj's Mission: The Forging of Official Multiculturalism in Canada, 1963–71," *University of Toronto Quarterly 88*, no. 1: 43–63.

Reiter, E. (2002) "Secular *Yiddishkait*: Left Politics, Culture, and Community," *Labour/Le Travail* 49: 12–46.

Sewell, J. (2015) *How We Changed Toronto: The Inside Story of Twelve Creative, Tumultuous Years in Civic Life, 1969–1980*, Toronto: Lorimer, 2015.

Smyth, W.J. (2015) *Toronto, the Belfast of Canada: The Orange Order and the Shaping of Municipal Culture*, Toronto: University of Toronto Press, 2015.

Srebrnik, H. (2016) "Chasing an Illusion: The Canadian Jewish Communist Movement," in Matthew B. Hoffman and Henry F. Srebrnik eds., *A Vanished Ideology: Essays on the Jewish Communist Movement in the English-Speaking World in the Twentieth Century*, Albany: SUNY Press, 101–34.

Stoney, C., and K.A.H. Graham (2009) "Federal-Municipal Relations in Canada: The Changing Organizational Landscape," *Canadian Public Administration/Administration Publique du Canada* 52, no. 3 (2009): 371–94.

Townsend, D. (1999) "The 'Blood Libel' and the Spectator's Eye in Norwich and Toronto," *Journal of the Canadian Lesbian and Gay Studies Association/Revue de la Socieété canadienne des etudes lesbiennes et gaies 1*: 1–24.

Tulchinsky, G. (2002) "Ben Lappin's Reflections on May Day Celebrations in Toronto's Jewish Quarter," *Labour/Le Travail* 49: 211–15.

Tulchinsky, G. (2005) "Family Quarrel: Joe Salsberg, the 'Jewish' Question, and Canadian Communism," *Labour/Le Travail* 56: 149–73.

Warner, T. (2002) *Never Going Back: A History of Queer Activism in Canada*, Toronto: University of Toronto Press.

Waters, R. (2013) "African Canadian Anti-Discrimination Activism and the Transnational Civil Rights Movement, 1945–1965," *Journal of the Canadian Historical Association 24*, no. 2: 386–424.

White, R. (2015) *Planning Toronto: The Planners, the Plans, their Legacies, 1940–80*, Vancouver: UBC Press.

Wright, R. (2016) *Trudeaumania: The Rise to Power of Pierre Elliott Trudeau*, Toronto: HarperCollins.

CHAPTER 8

"We Had Our Own Problems and So We Had Our Own Bitachon": Jewish Self-Defense in Uruguay, 1960–1987

Raanan Rein

The kidnapping of Nazi war criminal Adolf Eichmann in Buenos Aires in May 1960 by Israeli Mossad agents ignited a wave of violent anti-Semitism in Argentina which, in turn, prompted the consolidation of a Jewish self-defense framework, known as "the Irgún" (the Hebrew word for "organization").[1] As has often occurred, these events had wide repercussions in neighboring Uruguay. Jewish-Uruguayans also faced a series of anti-Semitic attacks and the dilemma of how to react to this violent aggression. Many Jewish-Uruguayan youngsters decided, as their Jewish-Argentines counterparts did, to react in kind. They would defend community institutions and petition national and municipal authorities to punish the anti-Semitic hooligans, but they would also teach the thugs a lesson and show them that Jews would not be passive victims anymore. They set up their own self-defense organization, known as "the Bitachon/ Bitajón," or "Misgeret" (the Hebrew words for "security" and "framework," respectively)—or simply, "La Orga."[2]

Due to the small size of the country (2.5 million people in the 1960s) and of the Jewish community (estimated at fewer than 40,000 in the 1960s),[3] the local self-defense did not involve as many youngsters as in Argentina, but it was well organized and probably tied more strongly to the Israeli embassy than its Argentine counterpart. If in Argentina, Mauricio (Tata) Furmanski

1 Raanan Rein, "Defying Traditional Shtadlanut: Jewish Self-Defense in Argentina," in Katalin F. Rac and Lenny U. Valerio eds., *Jewish Experiences in the Americas: Local Histories through Global Lenses* (Gainesville: University Press of Florida, forthcoming, 2021).

2 The first Misgeret was established by the Mossad in Morocco in 1956, after the country obtained its independence and Israel launched the Jewish-Moroccan immigration project. See Ian Black and Benny Morris, *Israel's Secret Wars: A History of Israel's Intelligence Services* (New York: Grove Press, 1991), 176.

3 U.O. Schmelz and Sergio DellaPergola, "The Demography of Latin American Jewry," *American Jewish Yearbook 85* (1985): 51–102, 62; Rafael Porzecanski, *El Uruguay judío: Demografía e identidad* (Montevideo: Trilce, 2006); Maya Shorer-Kaplan, "Ethnic Migration in Comparative Perspective: A Case Study of Jewish Migration from Uruguay to Israel and Other Countries, 1948–2010," *Hagira—Israel Journal of Migration 6* (2016): 25–56.

 | DOI:10.1163/9789004462540_009

was considered the key figure in Jewish-Argentine self-defense, in Uruguay members of the Ring family, especially Julio Ring, played a central role in the organization's activities during the 1960s and early 1970s.

Analyzing briefly three key moments in the development and activities of Jewish self-defense in Uruguay, this chapter looks at the initial organization of these semi-underground groups in the wake of Eichmann's abduction in Argentina in the early 1960s; the impact on their activities of the elimination of the Latvian war criminal Herberts Cukurs in Uruguay in 1965 by Mossad agents; and the addition of Arab targets to their operations, especially in the years following the 1967 Six Day War.

Current historiography has steered clear of the subject, presumably fearing possible damage to Israeli-Latin American relations, or to the image and status of Jews on the continent.[4] Consequently, relevant documents are not accessible in Israeli archives. A search for "Jewish self-defense" in the catalog of Jerusalem's Central Zionist Archives, for example, results in items referring to early twentieth century Russia and Eastern Europe and none at all to Latin America. An oral history project was therefore necessary for the purpose of this research.

This chapter is an attempt to fill the vacuum as far as the history of Jewish self-defense is concerned, highlighting the unique characteristics of the organization in Montevideo, as well as the fact that: a) although the organization was politically, ideologically, and perhaps also socially, very heterogeneous, as in Argentina, the influence of revisionist Zionism on self-defense activities was more pronounced than in its neighboring country; and b) Israel's role and involvement in these quasi-military groups in Montevideo was more substantial than in the continent's other countries.

1 Panic and the Establishment of Self-defense Groups

On June 13, 1960, three homemade bombs were planted in synagogues around Montevideo; two of them were deactivated before detonation.[5] Subsequently, two Uruguayans were arrested as a result of an explosion in a house where three youths were handling explosives. The third, César Giorgetti, a 17-year-old

4 Notable exceptions are a few pages in Fernando Amado, *Mandato de Sangre* (Montevideo: Sudamericana, 2012), and the recent book of a former Jewish self-defense activist in Venezuela, Isaac Cherem, *Judíos sin kipa* (n.p., 2019).

5 "Atentados antisemitas," *La Mañana*, June 14, 1960; "Murió un joven terrorista y dos están gravemente heridos," *El Plata*, June 14, 1960; "Terrorismo en la ciudad," *Acción*, June 14, 1960.

Argentine affiliated with the Alianza Nacionalista Argentina, was killed by the explosion. One of the arrested youths had a swastika tattooed on his hand.[6] The group had transnational connections, mainly in Argentina, and had planned a series of anti-Semitic attacks, including one against the Israeli embassy in Montevideo. The local press reported that one of their leaders in Argentina was a German nicknamed "Willie." Extreme right-wing organizations in Argentina, especially the Movimiento Nacionalista Tacuara, headed by Alberto Ezcurra Uriburu, and the Guardia Restauradora Nacionalista, had close ties with Uruguayan nationalists.[7]

Later in June 1960, a second accidental explosion caused the death of the person who was preparing the explosives. There were also several cases of swastikas painted on the walls of Jewish institutions and shops owned by Jews.[8] A publication of the small right-wing nationalist Liga Oriental Antisemita (LOA) began to circulate in May 1960. This was one of several extremist organizations in the heterogeneous mosaic of the Uruguay Right. As a result of all these incidents, largely a spillover from Argentina, the Montevideo police decided to tighten security around Jewish institutions.[9]

Israel's ambassador to Uruguay, Yitzhak Harkavi, arrived in South America in the El Al plane that would bring Eichmann to Israel. His testimony, albeit inaccurate in places, is very illustrative:

> I arrived [in Montevideo] in May 1960, with my family; it was just after the capture of Eichmann in Argentina. The night before the presentation of my credentials we went ... to sleep ... around midnight, I received a call from Oscar Sapolinsky, one of my attaches, who [told] me: "Harkavi, a bomb exploded in the Sephardic community, I think there are no victims, but there is physical damage to the building ..." An hour later Sapolinsky, who was already in the city center, call[ed] again and [told] me that unexploded bombs were also found in the Hungarian Jewish community

6 Magdalena Broquetas, "A Propósito de Las Repercusiones Del 'Caso Eichmann': Antisemitismo y Anticomunismo en Uruguay (1960–1962)," *Encuentros Uruguayos 3* (2010): 47–63, 49. A year-and-a-half after the event, the owner of the house who had been severely injured in the June 1960 explosion, expressed regrets over his participation in anti-Semitic activities. See "A Repentant Anti-Semite [Hebrew]," *Davar*, December 1, 1961, 5.

7 Ernesto Bohoslavsky and Magdalena Broquetas, "Vínculos locales y conexiones transnacionales del anticomunismo en Argentina y Uruguay en las décadas de 1950 y 1960," *Nuevo Mundo Mundos Nuevos*, May 2017, http://journals.openedition.org/nuevomundo/70510; DOI: 10.4000/nuevomundo.70510 (accessed October 1, 2019); Juan Manuel Padrón, *Ni yanquis, ni marxistas! Nacionalistas* (La Plata: UNLP, 2017).

8 "Otro atentado de los nazis," *El Popular*, June 20, 1960.

9 "Safeguarding Jewish Institutions in Uruguay [Hebrew]," *Ma'ariv*, June 17, 1960, 1.

> [center] and in the German-speaking community [center]. At two in the morning another call from Sapolinsky [informed] me ... that an explosion in a shed had killed a youngster who was handling the bombs. It was like this throughout the night.[10]

According to Graciela and Victor Ben Dror (Blit), in their history of Socialist-Zionism in Uruguay, already in 1961 "some of the Zionist youth were trained in self-defense and had even purchased weapons and practiced with them," following the anti-Semitic violence that plagued Argentina and Uruguay in the wake of Eichmann's capture.[11] There is similar testimony from right-wing Zionist sources too. In the mid-1970s, Julio Ring of the Revisionist Betar movement wrote in his curriculum vitae submitted to the Israeli Mossad:

> In late 1961, along with several other members of the [Betar] movement, I was recruited to the command of the local framework [self-defense organization] due to the intensification of anti-Semitism and racism. Devotedly, I erased swastikas and anti-Semitic graffiti [from the walls of Jewish institutions], and guarded synagogues and [Jewish] public assemblies. Most often in the central district. In 1962 [his first year of medical school] I moved to the intelligence arm of the framework and engaged in surveillance and intelligence gathering on [suspected] targets.[12]

Following the execution of Eichmann in Israel in June 1962, there was another wave of anti-Semitic incidents in both Argentina and Uruguay. A small group of Uruguayans gathered in Plaza Cagancha, in the center of Montevideo, for a ceremony in honor of Eichmann. They left a bouquet of flowers and a note saying: "The Juventudes Nacionalistas Orientales in memory of A. Eichmann, soldier of the cause of the peoples murdered by international Judaism."[13] A few days later, four youths burned the Israeli flag and left a pamphlet, stating: "In memoriam of Adolf Eichmann ASSASSINATED by international Jewry,

---

10 Itzjak Harkavi, "En Uruguay durante la época de la captura de Eichmann," in *Historia viva: Memorias del Uruguay y de Israel* (Jerusalem: Hebrew University, 1989), 130–31; author's zoom conversation with A. S., April 2020.

11 Graciela and Victor Ben Dror (Blit), *Radical Socialist Zionism on the Banks of Rio de la Plata: Mordechai Anielevich Movement in Uruguay, 1954–1976* [Hebrew] (Rehovot: Havatzelet Group, 1986), 61. Still, Aharon Melika, envoy of the Shomer Hatzir movement in Montevideo, 1965–69, did not recall any active participation of Socialist-Zionist youth in self-defense activities. Author's electronic correspondence with Graciela Ben Dror, April 2020.

12 Julio Ring's curriculum vitae in handwriting submitted to the Mossad in 1974.

13 "A. Eichmann: Hicieron un insólito intento de homenaje," *La Mañana*, June 3, 1962.

Uruguayan youth pays tribute to him." Nobody was arrested.[14] Soon after, a Molotov cocktail was thrown at a branch of the Communist Party. The local press accused the LOA and the Frente Estudiantil de Acción Nacional (FEDAN) of responsibility for the attacks, highlighting the fact that their ties with the Argentine Tacuara attested to "the execution of a plan of joint attacks on both banks of the River Plate."[15]

The Israeli embassy received numerous telephone threats, such as: "This is Eichmann's son speaking, señora de Harkavi. If your husband goes out on the street tomorrow, we will kill him. Take care. Don't go outside."[16] Soledad Barrett, a 17-year-old Paraguayan exile and leftist activist, was kidnapped by a group of thugs who tried in vain to force her to shout "Viva Hitler" in the car. They tattooed a swastika on each of her thighs and threw her out of the vehicle.[17] This brutal act was similar to one less than three weeks earlier in Buenos Aires directed against the Jewish-Argentine student Graciela Sirota.[18]

But the incident that gave the final push to the establishment of a Jewish self-defense organization in Uruguay was the beating to death in July 1962 of Samuel Zalzberg, a Holocaust survivor, in his Montevideo office.[19] Although some people linked his murder to robbery, it triggered strong reactions among the public, and among Jewish-Uruguayans in particular.

As in neighboring Argentina, a growing number of Jews decided to relocate to Israel, and a self-defense organization was set up. Jewish parents also joined forces to establish a nationally accredited private Jewish day school where pupils would not be vulnerable to anti-Semitic attacks. The result was Montevideo's Jewish integral school, founded in February 1962. There, the children were instructed in both the required Uruguayan curriculum and Jewish religion and culture. One of the founding fathers of the school was the

---

14 "Fascists in Uruguay Mourning Eichmann [Hebrew]," *Kol Ha'am*, June 12, 1962, 1; "Discriminación racial," *El Debate*, June 11, 1962.

15 "Así empezó el fascismo," *Acción*, June 10, 1962. The first issue of the FEDAN, *Centinela*, negated the Jewish Holocaust: "Juicio o parodia," November 13, 1961.

16 Harkavi, "En Uruguay": 132.

17 "Thighs of a Uruguayan Girl Tattooed with a Swastika [Hebrew]," *Hatzofe*, July 10, 1962, 4; "Uruguay Takes Steps to Track Down Anti-Jewish Terrorists," *Jewish Telegraphic Agency*, July 25, 1962. According to one report, of the eight men and women kidnapped and branded with swastikas in Montevideo during those weeks, one was the Jewish physician Dr. Máximo Handel, and most of the others were employed by Jews, either in stores or as domestic servants. See *Jewish Telegraphic Agency*, July 30, 1962, 1.

18 Raanan Rein, *Argentine Jews or Jewish Argentines? Essays on Ethnicity, Identity and Diaspora* (Boston: Brill, 2010), chap. 8.

19 Jewish Telegraphic Agency, July 30, 1962, 1. See also "Anti-Semitic Violence Flares in South America," *Rhode Island Herald*, August 10, 1962; *New Jersey Jewish News*, August 3, 1962.

FIGURE 8.1 Founding members of the self-defense organization in a Montevideo café, early 1960s. On the left sits Mario Kronenberg (with the mustache) and to his right sits Julio Ring.
COURTESY OF JULIO RING'S FAMILY

well-respected dentist Mario Kronenbeg, who was also one of the founders of the Jewish self-defense organization.[20] Other leading figures in the early stages of self-defense were Julio Mario Blankleder, Mario Julio Yudka Ring, a cousin of the abovementioned Julio Ring, and Jorge Hacas, "the only one who came from the Zionist Left."[21]

Unlike Argentina, the establishment of a self-defense organization in Uruguay was no secret and the Israeli press gave it wide coverage, in both left-wing dailies identified with the government and right-wing dailies of the opposition. *Davar*, the organ of the Histadrut, the General Organization of Workers in Israel, announced on its front page that armed patrols of Jewish self-defense groups had been seen in the Uruguayan capital, guarding Jewish homes and synagogues. The newspaper quoted a spokesperson of the organized Jewish community, who referred to this initiative of guarding Jewish lives and properties.[22] *Herut*, the mouthpiece of the eponymous right-wing

20 Author's telephone interview with his son, Gustavo Kronenberg, April 2020.
21 Author's interview with C. G. April 2020.
22 "Self-Defense Set Up by Montevideo Jews Following the Murder of Elderly Jew [Hebrew]," *Davar*, July 26, 1962, 1.

party representing Revisionist Zionism, also highlighted the establishment of Jewish self-defense units in Montevideo.[23]

Elections were to be held in November 1962, and the political divide spawned more extremist groups. As of 1958 the country had been governed by the conservative and nationalist Blanco party. The more liberal and reformist Colorado party had ruled Uruguay for almost a century, until the Blancos' victory. But the traditional two-party system seemed incapable of dealing with the economic difficulties, increasing social demands, and political polarization. In 1958, and even more so in 1962, left-wing groups improved their performance in the polls.

Although not to the same extent as in Argentina, anti-Semitic incidents created an atmosphere of panic among Jewish-Uruguayans. A Jewish man shot a taxi driver in July 1962, when the latter slowed down in a relatively deserted street, fearing that the driver might try to kidnap or attack him, as had happened to several Jews in the previous days.[24] Twenty-two-year-old Pinchas Said, a leading figure in self-defense activities, died in such circumstances. Apparently, he entered into an argument with a bus driver and was stabbed to death. Pinchas' remains were brought by his parents for burial in Israel in March 1963 on board the *Flaminia*.[25]

Jewish-Uruguayans were less used to violent racism and xenophobia compared with their Argentine neighbors. Uruguay witnessed only a few Nazi-inspired anti-Semitic manifestations during the 1930s. Discrimination against Jews was reflected mainly in restrictions on their immigration to the country, as with most countries at the time.[26] The organized Jewish community was hesitant to publicly label these restrictions as anti-Semitic.[27] Jews developed a variety of institutions spanning a range of political orientations that facilitated their integration into the host society, while maintaining their ethnic identity. The main organizations were Zionist in orientation and celebrated the creation

23 "Montevideo Jews Set Up Self-Defense Units Following the Murder of Samuel Zalzberg [Hebrew]," *Herut*, July 26, 1962, 1.

24 "Panic among Uruguayan Jews Following the Murder of Elderly Man: Self-Defense Being Organized—Students Strike Tomorrow [Hebrew]," *Hatzofe*, July 26, 1962, 1.

25 "Parents from South America Respect Son's Will [Hebrew]," *Ma'ariv*, April 1, 1963, 16; "Remains of Youngster Murdered in Anti-Semitic Attack Brought on Olim Ship [Hebrew]," *Hatzofe*, April 1, 1963, 4.

26 Rosa Perla Raicher, *Uruguay, La Comunidad Israelita y el Pueblo Judio* (Montevideo: Universidad de la Republica-Universidad Hebrea de Jerusalem, 2003), 94. See also Clara Aldrighi, et al., *Antisemitismo en Uruguay: Raíces, discursos, imágenes (1870–1940)* (Montevideo: Trilce, 2000); Alfredo Alpini, *La derecha política en Uruguay en la era del fascismo, 1930–1940* (Montevideo: Universidad de la República, 2015).

27 Miguel Feldman, *Tiempos difíciles: Inmigrantes judíos en Uruguay, 1933–1945* (Montevideo: Universidad de la República, 2001).

of the State of Israel. Uruguayan ambassador Enrique Rodríguez Fabregat played a key role in the 1947 United Nations Special Committee on Palestine, and his report was an important prelude to the General Assembly decision on the partition of Palestine and the establishment of a sovereign Jewish state. It was no coincidence that in December 1948 Uruguay became the first country in Latin America in which Israel established a diplomatic legation.[28]

Jews were an integral part of Uruguayan society by the early 1960s, when the Eichmann affair provoked a two-year anti-Semitic campaign. The new kind of anti-Semitic attacks that it generated should be viewed against the background of national circumstances, characterized by increased political violence and social tensions due to labor movement strikes and police repression within the broader context of the Cold War and the Cuban Revolution.[29] The growing polarization of Uruguayan politics, as well as increased social unrest, was also linked to the economic downturn. Prosperity, which had characterized Uruguay's first decade of the postwar period, came to an end.[30] All economic indicators pointed to the country's stagnation and the deterioration of living standards of the working and middle classes. In this context, the action of anti-Communist armed bands was a salient characteristic, and the supposed inefficiency of the police in investigating and bringing the attackers to justice was suspicious. In Uruguay, as elsewhere, self-defense activities originated in a lack of trust in the national security forces to protect the Jews from anti-Semitic elements (hence the title of this chapter).

The 1960s also marked growing political involvement of young Latin Americans, from both the Left and the Right, Jewish and non-Jewish, all seeking to differentiate themselves from the older generation and willing to take risks and embark on "heroic" activities. Both the aggressive attacks on Jews and the organization of Jewish self-defense reflected this atmosphere. The use of violence gained increasing legitimacy among revolutionary youth of all shades.

University students in Uruguay, like their peers in Argentina, were highly politicized and not a few were involved in nationalist and/or anti-Semitic

---

28 The delegation included Yaacov Tsur, Yizhak Navon, and Judith Berman. See their memoirs: Yaacov Tsur, *Credential No. 4: First Diplomatic Mission in South America* [Hebrew] (Tel Aviv, 1981), 10–40; Issac Navon and Iehudit Berman Melamed, in *Historia viva*, 99–104.

29 Magdalena Broquetas, "'Los frentes del anticomunismo': Las derechas en el Uruguay de los tempranos sesenta," *Contemporánea. Historia y Problemas del Siglo XX* 3 (2012): 11–29.

30 Rosa Alonso y Carlos Demasi, *Uruguay 1958–1968: Crisis y estancamiento* (Montevideo: EBO, 1986); Benjamín Nahum, et al., *El fin del Uruguay liberal, 1959–1973* (Montevideo: EBO, 2007).

activities.[31] The students' role in the anti-Semitic campaign was clear to contemporary observers. Reporting for the *American Jewish Yearbook*, Nelson Pilosof referred to "neo-Nazi student organizations and publications" which were active from July 1961 to August 1962. In his report, he noted that anti-Semitic incidents included threats, tattooing people with swastikas, burning Israel flags, bombing Jewish organizations, and vandalism against street signs bearing the name "State of Israel." National authorities and the press publicly condemned the attacks and numerous organizations and individuals expressed solidarity with the Jews. Leading cultural and religious figures organized an assembly at the national library to repudiate anti-Semitism. High school and university students also went on strike to demonstrate solidarity with Jewish-Uruguayans.[32]

As young people were the main protagonists of these racist incidents, first-generation Jews born in Uruguay were deeply affected by these episodes. R. S. was one of them. Born in Montevideo in 1944, his parents were originally from Galitiza and he was raised in a traditional Jewish home in Montevideo. He attended the Yavne Jewish School and was active in the Zionist movements Dror and Hashomer Hatzair. His experiences during those agitated times created lifelong memories. He would carry a knife hidden in his bag so he would be ready in the event of an attack. As a high-school sophomore, he went with one of his friends, "Chito," to beat up a classmate who had painted "Juden Raus" on the school wall. The fact that the culprit was with his entire gang did not deter them. Proudly he told me: "We were two against twenty."[33] The following day he was about to be expelled from school, but the principal, Hugo Fernández Artucio, an anti-fascist intellectual and author of the books *Nazis en Uruguay* (Buenos Aires, 1940) and *The Nazi Underground in South America* (New York, 1942), defended him.

31 Magdalena Broquetas, "A Propósito de Las Repercusiones Del 'Caso Eichmann'"; Broquetas, "Los Frentes Del Anticomunismo."

32 Nelson Pilosof, "Uruguay," in *American Jewish Yearbook* (New York: American Jewish Committee, 1964), 288–92, 289–90.

33 Author's interview with R. S., Hertzliya, June 2019. On student mobilization, see Vania Markarian, *Uruguay, 1968: Student Activism from Global Counterculture to Molotov Cocktails* (Oakland, CA: University of California Press. 2017); María Eugenia Jung, "La Universidad de la Repúbica como enemigo interno: La reacción de las derechos uruguayas, 1958–1973," *Anuario IEHS 32*, no. 2 (2017): 149–70.

2 Betar Uruguay: Leading Jewish Self-Defense Initiatives

Established in Montevideo in 1932 by Jewish immigrants from Eastern Europe in the post-World War I period, the Uruguayan chapter of Betar gradually became an important Zionist youth movement and the leading force in self-defense initiatives. By the 1960s it was recognized as such even by Socialist Zionists. Many Jewish young people—of all political and ideological hues, in both the U.S. and South America—were much influenced by the Revisionist Zionist thinker Vladimir (Ze'ev) Jabotinsky who, during the first decades of the twentieth century, warned of the dangers facing Jews in the Diaspora and called for the establishment of a sovereign Jewish state in Palestine. He envisioned a new type of Jew: proud, generous, and fierce.[34] Born in Odessa in 1880, he was among the founders of the Jewish Self-Defense Organization there in 1903. As the prospects of additional pogroms increased, the Jewish Self-Defense Organization aimed to safeguard Jewish communities throughout Russia.[35]

With Joseph Trumpeldor, Jabotinsky co-founded the Jewish Legion of the British army in World War I to fight against the Ottomans who controlled Palestine at the time. After being discharged from the British army in September 1919, he began training Jews in warfare and the use of small arms in Palestine. Jabotinsky died of a heart attack on August 3, 1940, while visiting a Jewish self-defense camp in Hunter, New York, run by Betar, the Revisionist Zionist youth movement founded in 1923 in Riga, Latvia.[36]

Menachem Begin, Jabotinsky's closest protégé, continued the latter's trajectory through the Irgun Tzva'i Leumi (known by its Hebrew acronym Etzel), which struggled against British rule in Palestine in the mid-1940s. Begin's book, *The Revolt*, had wide influence. The book was translated into Spanish and published in Buenos Aires in 1951, together with a letter handwritten by Begin to Argentine readers.[37] Although Begin's political party, Herut, remained a minority in community institutions, the Israeli leader and his struggle against the British fascinated many Jewish youngsters in South America. Begin also visited

---

34 Colin Shindler, *The Triumph of Military Zionism: Nationalism and the Origins of the Israeli Right* (London: I.B. Tauris, 2009), 116–32.

35 Shlomo Lambroza, "Jewish Responses to Pogroms in Late Imperial Russia," in Jehuda Reinharz ed., *Living with Antisemitism: Modern Jewish Responses* (Hanover and London: Brandeis University Press, 1987), 253–74.

36 For a recent biography of Jabotinsky, see Hillel Halkin, *Jabotinsky: A Life* (New Haven, CT: Yale University Press, 2014). See also Eran Kaplan, *The Jewish Radical Right: Revisionist Zionism and Its Ideological Legacy* (Madison, WI: University of Wisconsin Press, 2005).

37 Menajem Begin, *La rebelión en Tierra Santa* (Buenos Aires: Santiago Rueda, 1951).

this region more than any other senior Israeli politician. In September 1962, after he returned once again, he called, from Rio de Janeiro, for Jewish youth in the Southern Cone to form self-defense units "as an effective response to anti-Semitic hooligans," using "all necessary means at their disposal."[38] He repeated this message at a meeting held in Montevideo, in ambassador's Harkavi's presence, with thirty-three community leaders, mostly from Betar and the Revisionist party.[39] It is no coincidence that by then Jewish defense groups in Argentina called themselves "the Irgún." While in Argentina, however, Socialist Zionists took the leading role in Jewish self-defense, in Montevideo, Revisionist Zionists played first fiddle.

Samuel Hendel's testimony highlights Betar's early activism. Born in Galitzia in 1923, he arrived with his family in Uruguay in 1928 and, at the age of 15, joined the local branch of Betar. In the following years he would become a leading figure in the local Zionist Revisionist party, as well as in the organized Jewish community, becoming president of Uruguay's Comité Central Israelita in 1974. In his testimony about the late 1930s and the 1940s, he recalled:

> We also did all kinds of paramilitary exercises, shooting practice for example. This was before the [Second World] War. We had a large empty swimming pool, and we practiced shooting there. We began to practice self-defense tactics and some of the Japanese martial arts.[40]

Interestingly, Betar was not only a pioneer in instilling the idea of Jewish self-defense and the need to confront anti-Semitism perpetrated by extreme right-wing nationalists but also in seeing Arab-Uruguayans as a potential threat to Jewish life in the country. Hendel relates:

> We also intervened in the year '47 against some outbursts of activity by the Lebanese. They organized some demonstrations, especially when Jews gathered for activities in the Ateneo. They appeared on more than one occasion [to hinder the activity]. They belonged to the Sociedad Libanesa, of Colonia Street. They carried out street actions, driving by in a car and shouting [nationalist slogans].[41]

---

38 "M. Begin Calls on Jewish Youth in South America to Organize Self-Defense [Hebrew]," *Herut*, September 10, 1962, 1.

39 Author's zoom interview with A. S., April 2020.

40 Samuel Hendel, "Actividades de un joven sionista revisionista montevideano," in *Historia viva*: 90–92.

41 Ibid.

Betar was never the leading Zionist youth movement in Uruguay in terms of membership but its ideological influence was widespread.

3 The Cukurs Affair

Not long after Eichmann's execution in Israel, another Nazi criminal was exposed in neighboring Brazil. His fate, however, would be sealed on the outskirts of Montevideo, on February 23, 1965. Herberts Cukurs had been a Latvian aviation ace during the 1930s, earning the nickname "the Latvian Charles Lindberg." During World War II he played a leading role in the massacre of the Jews of Riga, personally committing appalling atrocities.[42] Among the crimes that survivor testimonies attributed to him were the killing of three hundred Jews herded into a synagogue which was then set alight, the drowning of twelve hundred Jews in the Kuldiga Lake during winter, and, finally, in November 1941, leading thirty thousand Riga Jews into the forests, where they were undressed and shot by German soldiers. Thus, he also earned the less sympathetic moniker of "the butcher from Riga."

At the end of the war, he fled to Brazil, via France, using a Jewish woman, whom he had protected during the war, in order to create a cover story as a savior of Jews. Cukurs settled first in Rio de Janeiro and indeed managed to attract sympathy from local Jewish-Brazilians.[43] As rumors about his wartime atrocities began to spread, however, he decided to move to a more distant location. For a period of ten years he managed to exist quietly, making his living by operating a tourist boating venture and an old seaplane, until he was publicly exposed by Brazilian Jewish organizations which organized several demonstrations against him.[44]

Calls for his expulsion from the country were rejected due to "insufficient evidence." Israel also considered requesting his extradition.[45] Following the Eichmann kidnapping, Cukurs was well aware of the danger to his life and asked for protection from the Brazilian government and permission to carry a

---

42 Andrej Angrick and Peter Klein, *The Final Solution in Riga: Exploitation and Annihilation, 1941–1944* (Oxford: Berghahn Books, 2009).

43 Anton Künzle and Gad Shimron, *The Execution of the Hangman of Riga: The Only Execution of a Nazi War Criminal by the Mossad* (Hertfordshire: Mitchell Vallentine, 2004), 45.

44 Michael Bar Zohar and Nissim Mishal, *Mossad: The Greatest Missions of the Israeli Secret Service* (New York: Ecco, 2012), chap. 11.

45 "Justice Minister Reviews a Request by Knesset Member Nurock to Demand the Extradition of the Annihilator of Latvian Jews [Hebrew]," *Herut*, September 1, 1960.

gun.[46] He was assigned six policemen to guard him, but his bid for naturalization was rejected by Brazilian president Juscelino Kubitchek.[47]

During the first half of the 1960s, the West German government had to make a decision regarding the prosecution of Nazi perpetrators: either bring an end to war crimes investigations because of the Statute of Limitations for acts of manslaughter and murder, or find ways to allow prosecutors additional time to gather evidence and initiate proceedings against remaining suspects. The prescribed cut-off point for dealing with Nazi acts of murder, May 8, 1965, became a bone of contention. In 1964, Prime Minister Levi Eshkol and Meir Amit, head of the Israeli Mossad, believed that apprehending another major war criminal would put pressure on German legislators to vote in favor of continuing legal procedures against people suspected of committing war crimes, lest groups of volunteers take upon themselves their removal.[48] Amit commissioned Mossad agent Yaakov Meidad to eliminate Cukurs. Meidad travelled to Brazil, disguised as an Austrian businessman named Anton Kunzle, and managed to gain Cukurs' confidence with promises of involving him in major investments in tourist projects in South America.

There were at least two differences between the operation to kidnap Eichmann and the one to capture Cukurs. Eichmann had forged a fake identity under the name of Ricardo Klement and remained in the shadows; Cukurs, however, had never hidden his identity and was an easier target. At the same time, the pressure to expel Cukurs put him and the Brazilian security forces on the alert. Another complication for the Mossad was the fact that the death penalty was legal in Brazil, hence its agents would be at risk in the event of mission failure. An additional consideration was the possibility that the Jewish-Brazilian community would be exposed to neo-Nazi revenge. Thus, democratic Uruguay offered a more convenient theater of operations. The Mossad thought the local Jewish community would be less vulnerable to violence, and the death penalty did not exist there.

Lured by a business project, Cukurs traveled to Uruguay and was taken by Meidad to a house in a remote suburb of Montevideo. After a short and violent struggle with a group of Mossad agents, he was shot twice in the head and his

46 "After Eichmann Capture Latvian War Criminal in Brazil Wants Protection," *The Sentinel*, June 30, 1960, 3; "Urge Brazil to Deport Latvian," *Bnai Brith Messenger*, December 16, 1960, 5.

47 "Former Nazi," *Bnai Brith Messenger*, July 22, 1960, 23; "Brazil Says 'No' to Ex-Nazi's Bid," *The Sentinel*, October 20, 1960, 3.

48 At the end of WWII there were indeed initiatives of this kind. See Dina Porat, *The Fall of the Sparrow: The Life and Times of Abba Kovner* (Redwood City, CA: Stanford University Press, 2009).

body was stuffed into a trunk with a note in English explaining the reasons for his murder. Following information sent to a news agency in West Germany by a group identified as "Those Who Shall Never Forget," the Uruguayan police found Cukurs body.[49] The police referred to an "act of Jewish vengeance" and the press reported that, "an Israeli Government employee was involved in slaying Herberts Cukurs."[50] However, an Israeli foreign ministry spokesman declared this claim was "unfounded" and shortly afterward the Uruguayan government apologized for the accusations.[51] In order to create more confusion, rumors were spread in Brazil that Cukurs was killed by former Nazis who feared he would betray information about the location of other Nazis, especially Josef Mengele.[52] Since the Uruguayan police decided within a few days that it had no clue about the perpetrators, there was no damage to Uruguayan-Israeli relations.[53]

Cukurs was the only Nazi war criminal killed by the Mossad. The operation's goal was achieved and the Federal Republic of Germany decided to continue with the investigation of and judicial processes against suspected Nazis. As in the case of Eichmann's capture in Buenos Aires, the Mossad could not have eliminated Cukurs without the help of local Jewish-Uruguayans. Their role in what they called "our [Uruguayan] Eichmann affair" has always been hinted at but never explicitly discussed,[54] either in order to glorify the Mossad or not to provide further pretexts for anti-Semitic attacks.

Nevertheless, Cukurs' killing sparked another wave of anti-Semitic incidents in Montevideo similar to that following the Eichmann affair.[55] Once again, the

---

49 For the police report, see Policía de Montevideo, Dirección de Investigaciones, Departamento de Inteligencia y Enlaces, "Memoria annual de Investigaciones y Enlaces Año 1965," 156. I would like to thank Magdalena Broquetas for sharing this document with me.

50 "Police Charge Cukurs Murder Suspect Is Employee of Israeli Government," *The Sentinel*, March 18, 1965, 5; "Organization of Anti-Nazi Revengers in Uruguay Exterminated Cukurs, the Murderer of Latvian Jews [Hebrew]," *Ma'ariv*, March 7, 1965; "Israeli Commando Executes Nazi Criminal Cukurs [Hebrew]" *Lamerchav*, March 8, 1965.

51 "Formal Apology for the Allegations," *Lamerchav*, March 11, 1965.

52 "Nazi Murdered by Nazi 'Mafia'?" *Bnai Brith Messenger*, March 26, 1965, 28; "Israeli Diplomat Took Part in Cukurs Killing [Hebrew]," *Al Hamishmar*, March 10, 1965.

53 "Uruguayan Police Ceases Investigation of Cukurs Case," *Lamerchav*, March 11, 1965; "Uruguayan Police Closes Cukurs Case and Hand It Over to Interpol [Hebrew]," *Hatzofe*, March 11, 1965.

54 See, for example, Dan Raviv and Yossi Melman, *Spies against Armageddon* [Hebrew] (Tel Aviv: Yedioth Aharonot, 2012), 131–33. Gustavo Kronenberg recalls, as a seven-year-old child, secret meetings taking place with his father at his home on the eve of Cukurs' elimination.

55 "Anti-Semitic Glowing Embers in Montevideo [Hebrew]," *Herut*, March 10, 1965; "Uruguayan Jews Worrying about the Future [Hebrew]," *Al Ha-Mishmar*, March 21, 1965.

economic deterioration of that year, including a drop in real wages, a rise in unemployment, and the collapse of the banking system, contributed to social unrest. Shots were fired at one synagogue and a bomb exploded in another, causing minor damage. The Sociedad Hebraica Macabi on General Máximo Tajes Street was attacked, swastikas and graffiti were smeared on walls, and anonymous telephone calls threatened Jewish leaders with death or warnings that synagogues and community institutions would be bombed.[56] Suspicious parcels were sent to Jews. One Jewish store owner received a wooden box containing a bomb which fortunately did not explode, and another received a letter bomb.[57] The sender of the letter was identified as Everildo Héctor Díaz Santos, who had ties to the Tacuara movement.[58]

The Comité Central Israelita placed the responsibility for these events on what they characterized as a small but virulent underground Nazi group, claiming that it had connections with Nazi movements abroad and was aided by the Arab League.[59] It is clear, however, that the commotion and reactions that followed the assassination of Cukurs were lower in intensity than the response to the Eichmann affair. The annual report of the American Jewish Committee devoted to Latin American Jewish communities did not even mention the incidents that followed Cukurs' assassination.[60] An entire generation of Jewish-Uruguayans was impacted, however, explaining in part the efforts to keep an eye on the activities of neo-Nazi José Paladino two decades later.

While the Jews of Uruguay seemed to be in a safer position than they had been for years, the self-defense organization did not cease its activities. Its targets now included not only anti-Semitic extreme right groups but also cells from the small Arab-Uruguayan community—in Montevideo, Chuy, and Rocha—and groups supported by the Arab League.[61] By now, the ad-hoc

56 "Three Detention Orders Issued in Uruguay after Cukurs Elimination—Nobody Arrested [Hebrew]," *Ma'ariv*, March 8, 1965; "Bomb in Synagogue in Montevideo," *The Sentinel*, March 11, 1965, 5; "Nazis in Montevideo Shoot at Synagogue [Hebrew]," *Kol Ha-Am*, March 11, 1965.

57 "Bomb in Wooden Box Sent to Jew in Uruguay [Hebrew]," *Herut*, March 19, 1965.

58 Magdalena Broquetas, *La trama autoritaria: Derechas y violencia en Uruguay, 1958–1966* (Montevideo: Banda Oriental, 2014), 236.

59 "Uruguay," in *American Jewish Yearbook* (New York: American Jewish Committee, 1967), 306–8.

60 Abraham Monk and José Isaacson, *Comunidades Judías de Latino América: 1966* (Buenos Aires: Comité Judío Americano, 1966), 92.

61 Most Arab-Uruguayans were of Lebanese descent and were Christians; the rest were Muslims. Alberto Abdala, of Lebanese origin, was vice president of Uruguay, 1967–72. See Renzo Pi Hugarte, "Asimilación cultural de los siriolibaneses y sus descendientes en Uruguay," *Anuario de Antropología Social y Cultural en Uruguay* (2005): 53–58.

groupings of the early 1960s had become a well-structured, hierarchical organization with strong connections to the organized Jewish community, on the one hand, and to the Israeli embassy, on the other. At the beginning, the organization's main task was to provide protection for Jewish youth movements, community institutions, the Israeli embassy, and Israeli visitors. Soon, however, more systematic recruitment and well-prepared training programs were initiated, emulating in part the Jewish-Argentine experience.

Members of the organization were trained in the Israeli martial art *krav maga* (lit., "contact combat"). *Krav maga* trains people to wrestle with one or more attackers and includes a variety of techniques, using the body, as well as knives, axes, batons, stones, and bottles. An additional skill is the disarming of attackers carrying firearms. Over the years Jewish-Uruguayans practiced at the country site of the Sociedad Hebraica Macabi in Carrasco, in the woods on the outskirts of Montevideo, and in a camp in Colonia. Especially from the late 1960s on, they were sometimes sent to Macabilandia, the main training site for Argentine self-defense groups in the province of Cordoba. A smaller number received additional training in Israel.[62] At the same time, members of Jewish self-defense groups from Argentina, Venezuela, Paraguay, Brazil, Chile, and Peru came to Uruguay for training. "We had a Latin American network of Jewish self-defense."[63]

As in Argentina, Jewish self-defense in Uruguay was characterized by its hierarchical organization that included a small group of leaders, some of whom received a monthly salary. The volunteers were the backbone of the organization, and in general were young people aged between 17 and 24. The salaried leaders were generally older and had several years of training, some of it in Israel. Usually, they were the only members of the organization that carried guns. The recruitment process among Jewish youth in high schools and Zionist youth movements was mostly secret. At the first meeting two recruiters typically explained the commitments that each member took upon her/himself and the characteristics of their training. If the person agreed, they were contacted for initial training. The first two years of physical and theoretical lessons were the key to determining whether a person would become a permanent member of the organization. Men and women were trained under equal conditions in situations that simulated real violence and in strict secrecy; not even the family knew about a member's involvement. According to the testimony of a former member included in Fernando Amado's book, his was a "religious experience." Another person, a former community leader, told

62 Author's interview with C. G., April 2020.

63 Ibid.

Amado, a political scientist and politician, that on some occasions during their training a person was injured and the leader had to provide an explanation to the parents.[64]

4 Arab Activities in Montevideo as a New Target for the Bitachon

A major source of concern for the organized Jewish communities of Argentina and Uruguay throughout the 1960s was increased anti-Zionist and anti-Jewish propaganda disseminated by the embassy of the Arab League. The embassy was based in Buenos Aires and headed by a Tunisian, Hussein Triki, until the mid-1960s. The campaign was led then by Arab diplomats from various countries, especially after their defeat by the Israeli armed forces in the 1967 Six Day War. Arab representatives collaborated with extreme right-wing nationalist organizations in both countries, such as Tacuara and Guardia Restauradora Nacionalista in Buenos Aires (both received financial support for their publications), and FEDAN and Montonera in Montevideo. They also tried to mobilize support among Arab-Argentines and Arab-Uruguayans for their anti-Israel campaigns. Articles from the monthly *Nación Árabe* were republished in nationalist periodicals on both shores of the River Plate.[65]

In Buenos Aires, Triki tried to alert the public to Jewish self-defense activities, warning that "some twenty-six training camps for Israeli military personnel are operating on Argentine territory, and even the Spanish language is forbidden there."[66] According to one of the people I interviewed, Triki left Buenos Aires after the Irgún firebombed the Arab League offices, which were located on the first floor, overlooking the street. The Irgún initiated a coordinated action in which one group blocked traffic in the area, another threw Molotov cocktails, and a third conveyed a message to Triki threatening that if he did not leave they would kill him.[67] In fact, Triki was declared persona non grata by President Arturo Illia's government in 1965 and left the country

64 Amado, *Mandato de Sangre*: 268.

65 Michael Goebel, "Von der hispanidad zum Panarabismus," *Geschichte und Gesellschaft* 37, no. 4 (2011): 523–58; Andrés Kilstein, "Vínculo entre sectores del peronismo de derecha y la Liga Árabe en la Argentina entre 1962–1964," ponencia presentada en el 2º Congreso de Estudios sobre el Peronismo (1943–76), UNTreF, November 2010; Leonardo Senkman, *El antisemitismo en la Argentina*, vol. 1 (Buenos Aires: CEAL, 1986), 52–53.

66 Quoted in Yehuda Adín, "Nationalism and Neo-Nazism in Argentina [in Hebrew]," *Betfutzot Hagola* 7, no. 33 (1965): 77. On Triki, his views and his activities, see Hussein Triki, *He aquí Palestina ... El sionismo al desnudo* (Madrid: Afrodisio Aguado 1977).

67 Author's interview with H. P., Buenos Aires, July 2014.

shortly thereafter. Triki was accused of meddling in Argentine internal affairs and violating anti-discrimination laws.[68] Another attack on the Arab League offices in Buenos Aires took place in 1970.[69]

As elsewhere, the Six Day War resonated strongly among many Jews in Uruguay. Very few events since the Holocaust had such a galvanizing effect on the organized Jewish community. Except for the Communist sector, all Jewish organizations rallied to support Israel's cause. Moreover, people who were far removed from Jewish life were attracted to Jewish organizations. Fundraising and political mobilization increased sharply. Some 200 volunteers, many of whom had previous experience in self-defense activities, traveled to Israel as soon as the war broke out. Most returned after six months, but there was an increase in emigration (*aliyah*) to Israel. Chaim Sivan was 17.5 years old when the war broke out and was among the volunteers who traveled to Israel. Upon his return to Montevideo, he was recruited to the self-defense organization. However, Sivan had only a couple of training sessions since he already had plans to "make aliyah," and indeed he moved to Israel in 1969.[70] Notably, according to demographers, in 1967, while the greatest support for Israel came from medium-size Jewish communities in Western Europe, with a combined rate of 8.4 volunteers per thousand Jews, Uruguay followed with 6.9 volunteers per thousand Jews, far ahead of other Latin American countries, including Argentina (1.9), and the U.S. and Canada (only 0.2).[71]

The Uruguayan government expressed solidarity with Israel and the parliament unanimously called for "peace without harm to Israel." Many non-Jews participated in a massive demonstration in support of Israel on June 8. The small groups of anti-Semitic radicals who published the newspaper *El Federal* did not react to the war in the Middle East and seemed almost to disappear from the public eye.[72]

Emanuel Adler, who was born in Montevideo in 1947, joined the self-defense organization. Remembering those times with nostalgia, he declared, "I was a simple soldier." He explained that the volunteers were divided into two groups,

---

68 "Argentina Talks of Ousting Arab," *New York Times*, September 8, 1964, 9; *La Nación*, May 4, 1965; *La Prensa*, July 1, 1965.

69 "Atentado en la oficina de la Liga Arabe," *La Nación*, February 2, 1970.

70 Author's interview with Chaim Sivan, April 2020.

71 Sergio DellaPergola, Uzi Rebhun, and Rosa Perla Raicher, "The Six-Day War and Israel-Diaspora Relations: An Analysis of Quantitative Indicators," in Eli Lederhendler ed., *The Six-Day War and World Jewry* (Bethesda, MD: University Press of Maryland, 2000), 11–50, 34. On Uruguayan and Argentine volunteers, see *El Mundo*, June 9, 10, 1967; *La Razón*, June 9, 10, 1967; *Clarín* and *La Prensa*, June 10, 1967.

72 Egon Friedler, "Uruguay," in *American Jewish Yearbook* (New York: American Jewish Committee, 1968), 426–32.

"I" for intelligence, and "D" for defense. Since he was in the first group, he was never involved in physical or violent actions. His task, in the aftermath of the Six Day War, was to observe the activities of certain Uruguayans of Arab origins. The Israeli victory in the 1967 war evoked great enthusiasm among Diaspora Jews and strengthened their self-confidence, as well as the Zionist identity of many of them.

Emanuel had been active in the Maccabi youth movement, one of the two largest Zionist youth movements in Montevideo at the time. He had spent one year on a kibbutz in Israel and did not receive any significant training for his intelligence duties. He added:

> We were sent to Middle Eastern restaurants to order humus and begin conversing with the people sitting at other tables. Sometimes we followed persons entering or leaving certain offices or private houses and compiled … lists of them.

In the Uruguay Jewish community, he recalled, "it was an honor [he used the Yiddish word *yiches*] to be part of the self-defense organization."[73] Sephardic members of the Bitachon, from the AJISU (Agrupación de la Comunidad Israelita Sefaradí del Uruguay) managed to infiltrate some Arab groups. Like others who testified, Adler noted the involvement of the Israeli embassy in Montevideo in their activities: "In fact, it was all managed by the embassy."[74]

Between 1966 and 1973, the key figure in Jewish self-defense was Julio Ring. Born in Montevideo in 1944 to a lower-middle-class family, his father had been a member of Betar in Poland, before relocating to Uruguay. It was only "natural" for Julio to join Betar in Montevideo a little after his bar mitzvah. He advanced to become a *madrich* (councilor), member of the national board and *mefaked* (commander) of the Continental Camp in 1961, a joint activity of the various branches of Betar in Latin America. By then he had already joined the self-defense network. In 1963 the family immigrated to Israel, where Julio participated in activities of the Revisionist movement and became an assistant to former Irgun member Ben Zion Givony. After completing his military service in the Israeli air force in 1966, he returned to Uruguay to study medicine at the Universidad de la República. Based on his previous activism in Betar and his military service in Israel, Ring was assigned a leading role in the self-defense organization:

---

73 Author's interview with Emanuel Adler, Kineret College, June 2019.
74 Author's interview with Adler, and with A. S., April 2020.

FIGURE 8.2
Julio (Jaim) Ring with Menachem Begin in Montevideo
COURTESY OF JULIO RING'S FAMILY

Immediately upon arriving in Montevideo, my cousin [Mario Julio Yudka], who served as an intelligence officer in the local Misgeret, put me in contact with the Mossad agent who worked undercover in the Israeli embassy. [On the one hand] the local station was falling apart and was completely dependent on Buenos Aires; on the other hand, due to a long period of relative peace, the Misgeret (composed of volunteers) was characterized by a very low level of intelligence and operative activities. My own cousin was about to retire from his job. I was invited to become the contact person with [the] Buenos Aires [station] and … to become the intelligence officer as a paid job …

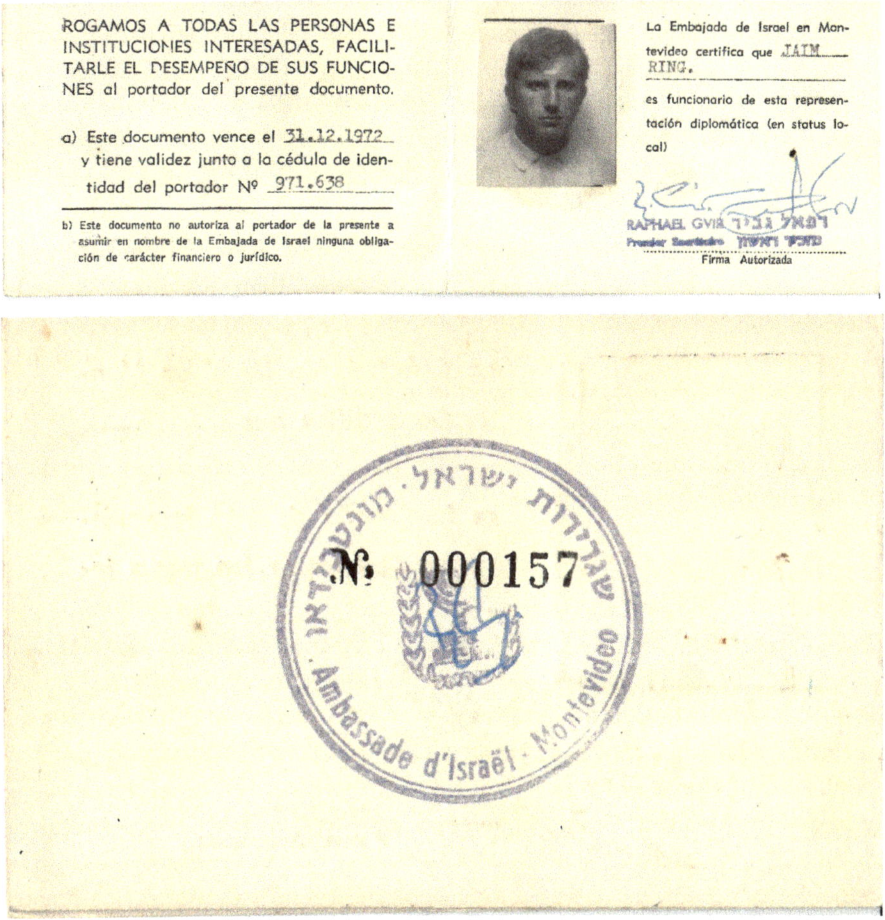

ROGAMOS A TODAS LAS PERSONAS E INSTITUCIONES INTERESADAS, FACILITARLE EL DESEMPEÑO DE SUS FUNCIONES al portador del presente documento.

a) Este documento vence el 31.12.1972 y tiene validez junto a la cédula de identidad del portador Nº 971.638

b) Este documento no autoriza al portador de la presente a asumir en nombre de la Embajada de Israel ninguna obligación de carácter financiero o jurídico.

La Embajada de Israel en Montevideo certifica que JAIM RING. es funcionario de esta representación diplomática (en status local)

RAPHAEL GVIR רפאל גביר

Firma Autorizada

Nº 000157

שגרירות ישראל. מונטבידאו

Ambassade d'Israël · Montevideo

FIGURE 8.3 Ring's staff ID card, issued by the Israeli embassy in Uruguay
COURTESY OF JULIO RING'S FAMILY

Things evolved in such a way that I worked on three levels: a) operational (commando force) by organizing initial [training] courses and consulting regarding the management of the various groups; b) intelligence, in fact, I reorganized the activities according to units and observation and surveillance squads; gathering of information from open sources, infiltrating hostile organizations, running agents, the network of training institutions, courses, etc.; c) guarding the [Israeli] embassy, since there was no security officer, and in this capacity, especially when preparing for [Israeli] visits, I was in contact with the local security forces.[75]

75 Julio Ring's handwritten curriculum vitae to the Mossad in 1974.

Thus, Ring (and his successors) worked from the Israeli embassy, collaborated closely with the organized community, especially through the Comisión de Prensa y Difusión del Comité Central Israelita (COPREDI), and was in touch with the Uruguayan police and secret services in order to avoid unnecessary clashes. Still, "we were very careful in our dealings with the police as we were not always sure about the ideological inclinations of the officers we were in contact with."[76]

Another cousin of Julio's, Miguel Ring, followed a similar trajectory. At age 14, he was a Betar member ("I was sent to Betar, there was no question at all about that," he smiled), and in 1963 arrived in Israel, did his military service at the same time as Julio (1964–66), and by early 1967 was back in Montevideo. The following year he was hired by Julio to work in the Israeli embassy: "Julio was the security officer and in charge of the consular section. Aida Stoler left her position and Julio offered me the job ... Julio's office was on the top floor."[77] It seems that the entire Ring family was involved in these activities: Julio, his wife Irene who sometimes helped him write reports, his brother who was in the defense wing, and his two cousins, Mario Julio Yudka and Miguel.

Although anti-Semitic manifestations began to decline after the campaigns that followed the Eichmann and Cukurs affairs, the regional political unrest which had put students at the center of political activism and radicalization would soon provide an anti-Semitic channel for a portion of them in Uruguay. According to the testimony of people who took part in the Jewish self-defense groups during the early 1970s, one of the main sources of their concern at the time was the Juventud Uruguaya de Pie, known as JUP. The JUP was a rightist group that appeared on the political scene in reaction to the emergence of the Left and the Tupamaros. It was composed mainly of anti-Communist students who wanted to reinvigorate the traditional values of Uruguayan society. Some of them promoted the idea of a national revolution with Falangist characteristics. The organization was active from 1970 and two years later began to take part in the para-military violence that anteceded the military regime. The JUP was dissolved in 1974.[78]

Julio Ring met his future wife, Irene, in March 1971, at a recruiting session of new members for the self-defense organization. She was a Maccabi member. Following training in the woods outside the capital city, she was charged with

76 Author's interview with C. G., April 2020.

77 Author's interview with Miguel Ring, Givataim, July 2019.

78 Gabriel Bucheli, *O se está con la patria o se está contra ella: una historia de la Juventud Uruguaya de Pie* (Montevideo: Fin de Siglo, 2019).

surveillance activities.[79] One of her missions was to go with friends to a rural area where the JUP was organizing a demonstration. The Jewish squad was supposed to photograph the participants. Once they arrived in a village and the police detained them; they had to spend the night in the local police station. She called her father for help. "Grandpa had connections with the local police and he used them more than once," Julio and Irene's son Idan told me.[80] The Misgeret received information on JUP's activities from the medical faculty's security officer, who was closely related to the nationalists. "We knew all about them," affirmed a former member of the Jewish self-defense.

Like other Jewish-Latin Americans involved in self-defense, Julio Ring, too, trained in the Maccabi basement in Buenos Aires, under the guidance of Tata Furmanski. That same year, 1971, he spent six months on a training program in Israel, before returning to Montevideo, self-defense activities, and the Israeli embassy.

In 1970, when the JUP began its activities, R. S. was already a member of the Jewish self-defense organization. As a student of architecture, when he went to construction sites in remote areas he always carried a gun. He took part in three violent operations:

> We burned down several print shops in southern Montevideo. At night, we threw Molotov bombs at them and set them alight. Those shops published anti-Semitic propaganda, like *Mein Kampf* and other pamphlets.

According to another former member: "These activities were well-planned and executed after months of preparations."[81] In one case, "we warned a printing house to stop selling anti-Semitic literature at their stand in one of the city's *ferias*. They did not listen to us, so we had to take 'good care' of them."

R. S. also provided protection for visitors from Israel, such as Shlomo Lahat (a former general in Israel's defense forces and mayor of Tel Aviv, 1974–93) and Pinhas Sapir (Israel's minister of finance, 1963–74), with whom he shared dinners and long conversations in English and Yiddish. He recalled having met Aliza and Menahem Begin in Montevideo as well.

R. S.'s main task, however, was to surveil the surroundings of the Israeli embassy in Montevideo and to escort people (for example, Jews trying to flee Uruguay because of persecution under the military regime) from the embassy to the airport, or vice versa. He would sit in the back seat of a car with people

---

79 Author's telephone interview with Irene Treibich, April 2020.
80 Author's interview with Idan Ring, Tel Aviv, July 2019.
81 Author's interview with C. G., April 2020.

who were "armed to the teeth" and had orders to open fire in the event of confrontation. On one occasion, this nearly happened:

> We were stopped by the military police. Two jeeps came between our two cars and blockaded us. We came out carrying our guns in hand. We did not know who they were. [Ambassador Dov] Shmorak quickly exited his car and shouted to the police not to fire since we were embassy personnel.[82]

At the end of each mission of this kind, the guns were deposited in the Israeli embassy.

In 1973 Julio Ring and his family relocated to Israel and the leadership of the Bitachon was transferred first to Salomón Sapov and then to Jaime G. The latter served as its head for a longer period than any other person, until the return of democracy in the early 1980s. Like Ring, G. was born in Montevideo to a lower-middle-class family, and joined Betar, where he was recruited for self-defense activities. He also studied in the Universidad de la República's medical faculty, received part of his training in Buenos Aires and in Israel, and had an office in the Israeli embassy, "an arrangement that does not exist anymore."[83]

5 Epilogue: The Paladino Affair and a Warning to All Anti-Semites

From 1973 to 1985 a military regime ruled Uruguay, a country that had been viewed as Latin America's most stable, free, and democratic nation. Following the crisis of its democratic institutions, the country was subjected to a series of repressive authoritarian regimes, a sharp economic decline, rising industrial militancy, ideological extremism, and the growth of the urban guerrilla movement, the Tupamaros.[84] According to a Jewish community leader, the Jewish self-defense organization was very active during the dictatorship in trying to prevent anti-Semitic elements from exploiting the over-representation of Jews in left-wing organizations and harming the Jews of Uruguay as a whole. According to one testimony, coded advertisements were published in the Jewish newspaper *Semanario Hebreo*, founded in 1960 by José Jerozolimski, conveying messages to members of the self-defense organization about the

---

82 Author's interview with R. S., Hertzliya, June 30, 2019.

83 Author's Skype interview with J. G., April 2020.

84 Hebert Gatto, *El cielo por as alto: El movimiento de liberación nacional (Tupamaros) y la izquierda uruguaya (1963–1972)* (Montevideo: Taurus, 2004).

execution or cancellation of planned activities.[85] The organization also acted to alert Jews who were potential targets of the security forces. He emphasized that no link existed between the Jewish organization and those forces. On the contrary, a few members of the organization were detained by them for being involved in "suspicious" subversive activities. A former president of the Comité Central Israelita described the defenders as "the most generous Jews, they are people who do a highly dangerous job and remain unknown. It is a very thankless task."[86]

The case of Hector José Paladino Ruvira, during the transition to democracy in Uruguay, taught the Jewish community and the self-defense organization a hard lesson and proved to be a turning point in its activities. Paladino had been watched closely by the organization in order to intimidate him because of his overt anti-Semitic activities. When Paladino noticed this vigilance, the conflict between them escalated. He painted the Nazi abbreviation "SS" on the façade of his house in red, and members of the Bitachon threw tar to cover it. Paladino repainted the letters in white.[87]

Many people in Montevideo had got used to seeing a Nazi flag in front of Paladino's house. On December 21, 1987, he left his home armed with a semi-automatic gun and took the lives of Simón Lazovski, a furniture shop owner, and Delfino Sicco, director of TV's Channel 4. He also injured Horacio Scheck when he tried to kill his father, who was the director of Channel 12 and the newspaper *El País*. When Paladino surrendered to the police, he declared that he had also planned to assassinate José Jerozolimski, director of the most widely read Jewish publication, *Semanario Hebreo*.[88] Paladino was convicted but since he was considered mentally unsound, he served part of his imprisonment in the Vilardebo Hospital.[89] A month after Paladino's crimes, the Uruguayan weekly *Jaque* dedicated the section "Crónicas y Hechos" to the public gathering organized by Jewish institutions and the Catholic Church authorities to condemn the killings. The weekly quoted the interior minister, Antonio Marchesano, who declared that there were "individuals of neo-Nazi

85 Author's conversation with E.N., Givataim, April 2020.
86 Amado, *Mandato de Sangre*, 261–71.
87 Ibid., 270.
88 Author's telephone interview with Ana Jerizolimski, April 2020; Ana Jerozolimski, "El Discurso Del Odio … y Un Recuerdo Personal," *Columnistas.Montevideo.Com.Uy*, 2016, http://columnistas.montevideo.com.uy/ucimprimir_301976_1.html (accessed August 24, 2019).
89 "En 1987 Un Nazi Asesinó a Dos Personas y Se Entregó," *El Observador*, 2016, https://www.elobservador.com.uy/nota/en-1987-un-nazi-asesino-a-dos-personas-y-se-entrego-2016310500 (accessed August 24, 2019).

ideology but [they were] not organized," and that Paladino did not even have friends.[90]

L. was among the persons who had been in charge of watching Paladino's house and following him, and wrote biweekly reports about his activities. Born in Montevideo in 1967, he had been an active member of the Zionist movements Dror, Habonim, and Chazit Ha-Noar since age fourteen. His first training in self-defense was during a summer camp in Colonia, where girls and boys were instructed in separate groups. During the first exercise, each boy had to slam into the one next to him. The instructors were the same ones L. would see at regular activities, but during the training sessions they used different names. In later years, when he would become a *madrich* himself, his codename would be Fernando. At age fifteen or sixteen, L. and his group would train once a week at school in the evening, and every two weeks they worked out and practiced "some punches." He recalled that he felt "like James Bond." There were non-Jewish kids that used to throw objects and shout anti-Semitic insults at them from a roof next to the school. Aware of the situation, their instructors provoked L. and his friends: "How come you allow them to talk to you like that?" The Jewish boys decided to wait for the kids to come down from the roof "and we savaged them with blows." Following this event, "hundreds" of non-Jewish kids gathered to take their revenge. The Jewish youngsters and their instructors, with clubs in hand, were ready to fight back. Fortunately, the directors of the Jewish Integral School managed to calm the two camps down.[91]

One year after the abovementioned episodes, L. was recruited to the intelligence branch of Bitachon and was instructed by the security officer of the Israeli embassy. After training in Israel, L.'s best friend became the commander of the operational branch. L. had a liaison person in the Jewish community who gave him weekly tasks. For example, he would go on Sundays to the Tristan Narvaja market to observe the bookstands and take note of those selling anti-Semitic or neo-Nazi materials. He also watched the local mosque, observing those entering and leaving the building. He took part in the security of Jewish schools and guarded groups of people during synagogue tours. In 1986 he spent one year in Israel as part of the Machon le Madrichim and four months on a kibbutz. During that time, he received firearms training and once back in Uruguay he began to work for the Israeli embassy as part of the local security personnel. L. explained that his self-defense activities made him realize the magnitude of anti-Semitism in Uruguay. He never told his parents about his role in the Bitachon but his father, born in Montevideo in 1937, had told

90 "Paladino: Un Mes de Conjeturas," *Jaque* (Montevideo), August 20, 1988, 7.

91 Author's interview with L., Tel Aviv, March 18, 2019.

him that when he was young and the Jews suffered attacks, they took spontaneous self-defense actions and some used to carry guns. L. characterized self-defense activities as "semi-secret," since Jewish community leaders knew about them, as did the Israeli embassy. He could not say whether the security forces knew "but of course the three policeman assigned to the Israeli embassy [knew]" because they practiced together with L. and his colleagues at the shooting range.[92]

The Uruguayan self-defense organization remained mostly clandestine until the mid-1990s. Following the AMIA Jewish community center bombing in Buenos Aires (1994), however, there was a need for the organization to establish new relations with the local security forces. Nothing changed officially, and there was no mention in any official report or bulletin, but in practice the organization now had a mutual understanding with the state. As of 1994, the Bitachon was visible as part of the security task force organized to receive Israeli personalities, when mass gatherings of Jews took place to mark the New Year, Yom Kippur, or Kristallnacht. They worked in coordination with the police and the armed forces. Such relations were referred to humorously by a Jewish-Uruguayan leader, who recalled a dialogue between the interior minister and the president of the Jewish community during one such mass gathering. The police and the Jewish self-defense were in charge of security and the minister, who wanted to send one of his subordinates to buy cigarettes, pointed to a young person and asked the community leader: "Is that one of yours or mine?"[93]

Acknowledgments

I wish to thank Idan Ring for sharing his family documents with me; Magdalena Broquetas, Gabriel Hojman, Jana Beris, Fernando Butazzoni, and Clara Aldrighi their help in locating several primary sources; and Lelya Stadler, for connecting me with former members of Jewish-Uruguayan self-defense groups. I would also like to thank the Fritz Thyssen Foundation for their support of my research project, "Fighting anti-Semitism with a Club in Hand: Jewish Self-Defense in Latin America."

92 Ibid.
93 Amado, *Mandato de sangre*: 264.

Bibliography

Aldrighi, C. et al. (2000) *Antisemitismo en Uruguay: Raíces, discursos, imágenes (1870–1940)*, Montevideo: Trilce.

Alonso, R., and C. Demasi (1986) *Uruguay 1958–1968: Crisis y estancamiento*, Montevideo: EBO.

Alpini, A. (2015) *La derecha política en Uruguay en la era del fascismo, 1930–1940*, Montevideo: Universidad de la República.

Amado, F. (2012) *Mandato de Sangre*, Montevideo: Sudamericana.

Angrick, A., and P. Klein (2009) *The Final Solution in Riga: Exploitation and Annihilation, 1941–1944*, Oxford: Berghahn Books.

Bar Zohar, M., and N. Mishal (2012) *Mossad: The Greatest Missions of the Israeli Secret Service*, New York: Ecco.

Begin, M. (1951) *La rebelión en Tierra Santa*, Buenos Aires: Santiago Rueda.

Ben Dror Blit, G., and V. (1986) *Radical Socialist Zionism on the Banks of Rio de La Plata: Mordechai Anielevich Movement in Uruguay, 1954–1976* [Hebrew], Rehovot: Havatzelet Group.

Black, I., and B. Morris (1991) *Israel's Secret Wars: A History of Israel's Intelligence Services*, New York: Grove Press.

Bohoslavsky, E. and M. Broquetas (2017) "Vínculos locales y conexiones transnacionales del anticomunismo en Argentina y Uruguay en las décadas de 1950 y 1960," *Nuevo Mundo Mundos Nuevos*, May, http://journals.openedition.org/nuevomundo/70510; DOI: 10.4000/nuevomundo.70510 (accessed October 1, 2019).

Broquetas, M. (2010) "A Propósito de las Repercusiones del 'Caso Eichmann': Antisemitismo y Anticomunismo en Uruguay (1960–1962)," *Encuentros Uruguayos* 3: 47–63.

Broquetas, M. (2012) "'Los frentes del anticomunismo': Las derechas en el Uruguay de los tempranos sesenta," *Contemporánea. Historia y Problemas Del Siglo XX* 3: 11–29.

Broquetas, M. (2014) *La trama autoritaria: Derechas y violencia en Uruguay, 1958–1966*, Montevideo: Banda Oriental.

Bucheli, G. (2019) *O se está con la patria o se está contra ella: una historia de la Juventud Uruguaya de Pie*, Montevideo: Fin de Siglo.

Cherem, I. (2019) *Judíos sin kipa*, n.p., n.e.

DellaPergola, S., Rebhun, U. and Raicher, R.P. (2000) "The Six-Day War and Israel-Diaspora Relations: An Analysis of Quantitative Indicators," in Eli Lederhendler ed., *The Six-Day War and World Jewry*, Bethesda, MD: University Press of Maryland, 11–50.

Feldman, M. (2001) *Tiempos difíciles: Inmigrantes judíos en Uruguay, 1933–1945*, Montevideo: Universidad de la República.

Friedler, E. (1968) "Uruguay," in *American Jewish Yearbook*, New York: American Jewish Committee.

Gatto, H. (2004) *El cielo por asalto: El movimiento de liberación nacional (Tupamaros) y la izquierda uruguaya (1963–1972)*, Montevideo: Taurus.

Goebel, M. (2011) "Von der hispanidad zum Panarabismus," *Geschichte und Gesellschaft 37*, no. 4: 523–58.

Halkin, H. (2014) *Jabotinsky: A Life*, New Haven, CT: Yale University Press.

Harkavi, I. (1989) "En Uruguay durante la época de la captura de Eichmann," in *Historia viva: Memorias del Uruguay y de Israel*, Jerusalem: Hebrew University.

Jung, M.E. (2017) "La Universidad de la República como enemigo interno: La reacción de las derechos uruguayas, 1958–1973," *Anuario IEHS 32*, no. 2: 149–70.

Kaplan, E. (2005) *The Jewish Radical Right: Revisionist Zionism and Its Ideological Legacy*, Madison, WI: University of Wisconsin Press.

Kilstein, A. (2010) "Vínculo entre sectores del peronismo de derecha y la Liga Árabe en la Argentina entre 1962–1964," *ponencia presentada en el 2° Congreso de Estudios sobre el Peronismo (1943–76)*, Buenos Aires: UNTreF.

Künzle, A., and G. Shimron (2004) *The Execution of the Hangman of Riga: The Only Execution of a Nazi War Criminal by the Mossad*, Hertfordshire: Mitchell Vallentine.

Lambroza, S. (1987) "Jewish Responses to Pogroms in Late Imperial Russia," in Jehuda Reinharz ed., *Living with Antisemitism: Modern Jewish Responses*, Hanover and London: Brandeis University Press, 253–74.

Markarian, V. (2017) *Uruguay, 1968: Student Activism from Global Counterculture to Molotov Cocktails*, Oakland, CA: University of California Press.

Monk, A. and J. Isaacson (1966) *Comunidades Judías de Latino América: 1966*, Buenos Aires: Comité Judío Americano.

Nahum, B. et al. (2007) *El fin del Uruguay liberal, 1959–1973*, Montevideo: EBO.

Padrón, J.M. (2017) *Ni yanquis, ni marxistas! Nacionalistas*, La Plata: UNLP.

Pi Hugarte, R. (2005) "Asimilación cultural de los siriolibaneses y sus descendientes en Uruguay," *Anuario de Antropología Social y Cultural en Uruguay*: 53–58.

Pilosof, N. (1964) *"Uruguay," in American Jewish Yearbook*, New York: American Jewish Committee.

Porat, D. (2009) *The Fall of the Sparrow: The Life and Times of Abba Kovner*, Redwood City, CA: Stanford University Press.

Porzecanski, R. (2006) *El Uruguay judío: Demografía e identidad*, Montevideo: Trilce.

Raicher, R.P. (2003) *Uruguay, La Comunidad Israelita y el Pueblo Judio*, Montevideo: Universidad de la Republica-Universidad Hebrea de Jerusalem.

Raviv, D. and Y. Melman (2012) *Spies against Armageddon* [Hebrew], Tel Aviv: Yedioth Aharonot.

Rein, R. (2010) *Argentine Jews or Jewish Argentines? Essays on Ethnicity, Identity and Diaspora*, Boston: Brill.

Rein, R. (forthcoming, 2021) "Defying Traditional Shtadlanut: Jewish Self-Defense in Argentina," in Katalin F. Rac and Lenny U. Valerio eds., *Jewish Experiences in the Americas: Local Histories through Global Lenses*, Gainesville: University Press of Florida.

Schmelz, U.O., and S. DellaPergola (1985) "The Demography of Latin American Jewry," *American Jewish Yearbook 85*: 51–102.

Senkman, L. (1986) *El antisemitismo en la Argentina*, vol. 1, Buenos Aires: CEAL.

Shindler, C. (2009) *The Triumph of Military Zionism: Nationalism and the Origins of the Israeli Right*, London: I.B. Tauris.

Shorer-Kaplan, M. (2016) "Ethnic Migration in Comparative Perspective: A Case Study of Jewish Migration from Uruguay to Israel and Other Countries, 1948–2010," *Hagira—Israel Journal of Migration 6*: 25–56.

Triki, H. (1977) *He aquí Palestina ... El sionismo al desnudo*, Madrid: Afrodisio Aguado.

Tsur, Y. (1981) *Credential No. 4: First Diplomatic Mission in South America* [Hebrew], Tel Aviv.

CHAPTER 9

From Zionist Movements to Guerrilla Groups: David Armando Laniado, Raúl Milberg, and Political Radicalization in Argentina

Adrián Krupnik

The large number of Jews among the illegally detained and disappeared during the *Proceso de Reorganización Nacional* in Argentina have been the focus of several academic works.[1] In addition to the blatant anti-Semitism that permeated the repressive forces, there is no doubt about the presence of Jews in the armed guerrilla movement. Jewish involvement was first evidenced in Argentina during the early experience of the Ejército Guerrillero del Pueblo (people's guerrilla army) in 1963, the focus of Lucas Böckmamnn's chapter in this volume. Böckmamnn employs the biography of Adolfo Rothblat to consider a broader phenomenon: the passage of Jews from the Argentine Communist Party to guerrilla groups, and I follow a similar approach. This chapter illuminates a particular group of Jews—those who had previously been active in Zionism—who decided to join the guerrillas during the 1970s. Drawing on oral testimonies and archival materials, I present the cases of David Laniado, formerly a member of Hashomer Hatzair, and Raúl Milberg, originally from Hechalutz Lamerchav, who died as members of the Liga Comunista Revolucionaria and the Montoneros, respectively. Their biographies are representative of a social phenomenon which involved thousands of Jewish-Argentines.

In her contribution to this volume, Stephanie Pridegeon quotes a fictional character who finds it hard to believe or understand that an Argentine Jew could have joined the Peronist urban guerrilla Montoneros. In reality, for those familiar with the history of the Jews in Argentina, such a question appears to derive from two biases: a negative one regarding the Montoneros and a positive

---

1 Edy Kaufman/Beatriz Cymberknopf, "La dimensión judía en la represión durante el gobierno militar en la Argentina (1976–1983)," in Leonardo Senkman ed., *El Antisemitismo en la Argentina* (Buenos Aires: Centro Editor de América Latina, 1989), 260; Emmanuel Kahan, "'Memories That Lie a Little': New Approaches to the Research into the Jewish Experience during the Last Military Dictatorship in Argentina," in Raanan Rein and Adriana Brodsky eds., *The New Jewish Argentina. Facets of Jewish Experiences in the Southern Cone* (Leiden: Brill, 2013), 302.

 | DOI:10.1163/9789004462540_010

one regarding the Jews. In my view, the presence of Jewish-Argentines within the ranks of armed guerrillas during the 1970s should surprise us no more than the consumption of marihuana by today's youth, or their use of social networks. In other words, as part of Argentine youth, Jews of the 1960s and 1970s were influenced by strong forces that shaped the culture of their time, both locally and globally. In many ways they were no different from their non-Jewish compatriots. However, highlighting their national identity as a result of their integration into Argentine society does not mean that their particular history and cultural heritage should be ignored.

This chapter focuses on the intersection of ethnic and national elements of identity of Jewish-Argentine youth. I explore the biographies of two young Jewish-Argentines whose ethnic and national identities played different roles in their political choices throughout their lives. Approaching identity as a dynamic phenomenon I raise four arguments. First, the incorporation of many young Jews into the guerrilla movement during the 1960s and 1970s reflected, simultaneously, their successful integration into society as well as their conscious desire to deepen such integration. Second, the fascination of revolutionary organizations was so powerful that they were able to recruit followers even from the heart of Jewish organizations in which Zionism was dominant. Third, young people who joined the armed struggle represented the extreme along a continuum of options followed by thousands of persons of their generation, but by no means were they isolated or extraordinary cases. Last, the life experiences of Raúl Milberg and David Armando Laniado reflect not only the broader phenomenon of Jewish-Argentines whose ethnic particularities played a certain role in their path of politicization and radicalization; their cases also call to attention the need to consider the role of ethnic heritage in the lives of Argentine activists of other origins, such as German, Spanish, or Italian.

Upon their arrival as immigrants Jews in Argentina became progressively integrated into every sphere of society. Therefore, by the early 1960s they had been accepted in all main political parties.[2] At the same time, Zionism had become dominant in most Jewish institutions, including the main umbrella organizations. Various youth movements were active in Argentina, instilling in thousands of Jews identification with Israel, the values of kibbutz life, and promoting emigration to Israel as their final goal. Some of these movements, like the Socialist-oriented Hashomer Hatzair, created a network of organizations

2 Raanan Rein, *Argentina, Israel, and the Jews: Perón, the Eichmann Capture and After* (Bethesda: University Press of Maryland, 2002).

which included their own press.[3] Nevertheless, Zionism was not the only political expression present in the life of the Jews. A considerable portion of Jewish-Argentine members of ethnic organizations adhered to Communism.[4] Moreover, most Jews were not affiliated with any ethnic body and their relation to the Argentine political arena was therefore unmediated by a particular Jewish institutional approach.

From the early 1960s, tensions created by the Vietnam War, the emergence of the New Left, Third Worldism, and the Cuban Revolution affected the younger generation of Latin American Jews, who followed various paths in order to accommodate their Jewish and Latin American identities, traditions, and political views. Ethnic and national politics were intertwined in a complex relationship. In Argentina, political radicalization inspired by Fidel Castro and Che Guevara spread quickly among vast youth sectors. The extremely repressive dictatorship which governed the country from 1966, and the proscription of the Peronist Party, whose exiled leader remained the preferred politician of a large portion of the working class, constituted the main reasons for this trend.[5] As historian Valeria Manzano argues, Argentine youth experienced expanded educational and labor opportunities, the increasing cosmopolitanism of the country's cultural life, and the chance of participating in a transnational repertoire of ideas, images, and sounds. Nevertheless, their lives were also shaped by the pervasiveness of authoritarianism in the intimate cultural, and political milieus.[6] The experience of young Jewish-Argentines active in Zionism also included recent memories of the Holocaust and the resort to self-defense as a legitimate response to local anti-Semitism. From the time of the Cordobazo, a violent confrontation of workers and students against military rule in 1969, several new political parties and guerrilla groups appeared on the political scene. They attracted thousands of young Argentines. Many of those who had been active members of the Zionist movement combined or changed their original affiliation to join guerrilla groups.

3 Ana Weinstein and Eliau Toker, *Trayectoría de una Idea. Nueva Sión 50 años de periodismo judeo-argentino* (Buenos Aires: Fundación Mordejai Anilevich, 1999).

4 Silvia Schenkolewski-Kroll, "Ideology and Propaganda in the Collective Memory's Construction: Zionism and Communism in Argentina," in August Grabski ed., *Rebels against Zion: Studies on the Jewish Left Anti-Zionism* (Warsaw: Jewish Institute of History, 2011); Nerina Visacovsky, *Argentinos, judíos y camaradas: Tras la utopía socialista* (Buenos Aires: Biblos, 2015).

5 On political struggles and social tensions which led to the Cordobazo, see James P. Brennan and Mónica B. Gordillo "Working Class Protest, Popular Revolt, and Urban Insurrection in Argentina: The 1969 'Cordobazo'," *Journal of Social History* 27, no. 3 (1994): 477–98.

6 Valeria Manzano, *The Age of Youth in Argentina* (Chapel Hill: The University of North Carolina Press, 2014), 158.

1 The Overwhelming Forces of History

The lives of David Laniado and Raúl Milberg represent just two cases of a social phenomenon that involved thousands of young Jewish-Argentines. The increasing gap between Jewish youth and Zionism due to their attraction to leftist ideologies was discussed by observers even prior to May 1968. For instance, a letter from Buenos Aires sent to the Israeli newspaper *Davar* warned in 1966 of Fidel Castro's appeal among the majority of Jewish youth, who were distancing themselves from Zionist and Jewish affairs.[7] However, this process strengthened noticeably from 1973 and historical evidence is therefore abundant. In May 1973, Héctor José Cámpora, Perón's designated candidate, was elected president of Argentina. Perón had been exiled for almost twenty years and was banned from running in the elections. As a result of a plot within the Peronist movement, Cámpora resigned after forty-nine days and Perón won an overwhelming victory in September.[8] The Cámpora and Perón governments shared an economic program, with José Ber Gelbard as minister of the economy.[9] It was aimed at the expansion of domestic production based on an alliance between unions and national producers, a partnership that represented the core of Peronism. In an attempt to overcome cyclical problems and bottlenecks in Argentina's economy, which were leading to currency shortages and devaluations, Gelbard tried to increase exports.[10] Gelbard's Jewishness was often targeted by his political enemies.[11] The nation was in the

7 "Castroism Attracts Youth," *Davar* [Hebrew], July 22, 1966, 15.

8 Guido Di Tella, *Argentina under Perón, 1973–76* (London: Macmillan, 1983).

9 On Gelbard's economic program and the challenges of Peronism during their third government, see Cecilia Vitto, "Plan Económico Del Tercer Gobierno Peronista: Gestión de Gelbard (1973–1974)," *Problemas Del Desarrollo 171*, no. 43 (2012): 111–34; Julián Blejmar (2020), "El Plan Gelbard y el populismo económico," *Realidad económica 49*, no. 330: 141–154.

10 Marcelo Diamand, "La Estructura Productiva Desequilibrada Argentina y El Tipo de Cambio," *Desarrollo Económico 12*, no. 45 (1972): 25–47; Martín Fiszbein, "La Economía Del Stop and Go: Las Ideas Estructuralistas En La Argentina, 1945–1976," *Desarrollo Económico 55*, 216 (2015): 187–210.

11 On Gelbard, see Leonardo Senkman, "Populismo y Empresarios Judíos: Actuación Pública de Horacio Lafer y José B. Gelbard Durante Vargas y Perón," *Araucaria: Revista Iberoamericana de Filosofía, Política y Humanidades 15* (2006): 46–76; Raanan Rein and Claudio Panella eds., *La Segunda Línea Liderazgo Peronista 1945–1955* (Buenos Aires: Universidad Nacional de Tres de Febrero, 2013); María Seoane, *El Burgués Maldito: Los Secretos de Gelbard, El Último Líder Del Capitalismo Nacional* (Buenos Aires: Editorial Sudamericana, 2002). On Jews and Peronism, see Raanan Rein, *Los Muchachos Peronistas Judíos: Los Argentinos Judíos y El Apoyo Al Justicialismo, Los Argentinos Judíos y El Apoyo Al Justicialismo* (Buenos Aires: Sudamericana, 2015).

midst of an unprecedented upheaval which included actions by leftist guerrillas and right-wing paramilitary groups. Soon after Perón's electoral victory, two political assassinations marked a watershed in the escalation of violence. First, the union leader and Perón ally José Rucci was shot as he was leaving his family home. Two days later a Peronist Youth leader, Enrique Grynberg, was shot in what appeared to be revenge by Rucci's followers. Both crimes remain unsolved.

When Kalman Grossman, a Jewish Agency (JA) emissary based in Buenos Aires to promote immigration to Israel reported on the political situation, he referred to Grynberg as a Peronist leader and a Jew.[12] It was neither the first nor the last time that Jewish leaders reflected Jewish-Argentine youth participation in local politics. Moreover, almost concomitantly with Perón's swearing-in, the Yom Kippur War broke out. Therefore, new considerations would be added to the already complex Argentine political scenario. By 1973 Jewish-Argentines were rooted in Argentine society and were familiar with the entire spectrum of political ideas and movements of their time. This reality was captured by the *American Jewish Year Book*, which referred to two synagogue gatherings organized to demonstrate support for Israel; nevertheless, it also reported that

> it was a source of no little consternation that many leftist and unaffiliated Jews were persuaded not only not to support Israel's cause but also actively voiced opinions and attitudes unfavorable to Israel. A large part of the medical staff at the Hospital Israelita Ezrah refused to back a demonstration in support of Israel, and when funds were finally raised, they were donated to the Red Cross rather than to Israel. Many Jewish teachers in Jewish schools were dismissed for refusing to back the Israeli war effort in the classroom and openly opposing Israel's position. Although no figures were available on the funds raised, it was admitted by some of the foremost communal leaders that the amounts fell short of expectations.[13]

It is clear, then, that wide circles of Jewish-Argentines were influenced by the leftist winds that characterized the period and did not identify with Israel.

---

12 K. Grossman to U. Narkis, October 4, 1973, TLC, 220754. For a comprehensive study of Argentine guerrilla groups, see María Jose Moyano, *Argentina's Lost Patrol: Armed Struggle, 1969–1979* (New Haven, CT: Yale University Press, 2012). On the related general strike and massive demonstration of August 31, 1973, see Norberto Galasso, *Perón: Exilio, Resistencia, Retorno y Muerte, 1955–1974* (Buenos Aires: Ediciones Colihue, 2005), 1231.

13 Naomi F. Meyer, "Latin America Argentina Domestic Affairs," *American Jewish Year Book* (New York: American Jewish Committee, 1976), 262–86, http://www.ajcarchives.org/AJC_DATA/Files/1976_8_LatinAmerica.pdf (accessed May 1, 2019), 363.

Between them and those who decided to relocate to Israel there was a wide spectrum of Jews who, as Kalman Grossman reported, just wanted to live their lives "as usual." Grossman did not hide his disappointment at the relative indifference of Jewish-Argentines to Israeli needs.[14] His Córdoba-based colleague Y. Arazi was clear about the weakening of Zionism as an option for Jewish youth, warning:

> The youth problem continues to worry and disturb: they are divided and separated into different sectors which cannot find common ground; leftist ideas are growing and moving this youth far from all interest in Israeli issues. The lack of an Israeli *sheliach* [emissary] dedicated exclusively to youth education sharpens a problem that it is not only an educational issue but also a very significant and dangerous one.[15]

Reports on the "weakening of Jewish national values, apathy, and assimilation," and the need to strengthen Zionist youth commitment were recurrent at the time.[16] However, while JA officials believed that the lack of an emissary was a problem, *La Luz*, a bi-weekly founded in 1931 by the Elnecavé family, expressed the opposite view. It referred to the large number of emissaries as a "plague" of "horrific" influence on the Jewish-Argentine community, and accused them of introducing to Jewish youth "Third World philosophies," which were contrary to Zionist and Jewish values.[17] Similar allegations were reiterated in 1978 by a committee of Jewish-Argentine leaders in conversation with Prime Minister Menachem Begin. In their view, emissaries from leftist Israeli parties were conducting dangerous activities in Argentina which, "on more than one occasion led the youth under their supervision to incarceration."[18] Since the discussion centered on the small number of Argentine immigrants to Israel, such mutual accusations between Zionist Israeli and Jewish-Argentine leaders lacked a thorough comprehension of the unstoppable sociological forces pushing Jewish-Argentines toward local political trends.

---

14 K. Grossman to U. Narkis, January 3, 1974, TLC, 220756.

15 Y. Arazi to U. Narkis, April 30, 1974, TLC, 220756.

16 K. Grossman to U. Narkis, November 25, 1974, TLC, 220757; Y. Rivkin to U. Narkis, December 5, 1974, TLC, 220757.

17 "La espeluznante plaga del 'shelijismo'," *La Luz*, April 17, 1974, 33.

18 Bar-Mor to Office of the Foreign Ministry, August 23, 1977, ISA, 10907/5. According to the schedule, Argentine leaders also toured an absorption center on August 23. "Delegación de Dirigentes Comunitarios Judíos. República Argentina," August 19–25, 1977, DUDR, meypw 04052.

In addition to Jewish-Argentine youth connections with leftist ideas in general, their prominent participation within the Peronist movement was far from being a secret. Some Jewish youth, Grossman explained, had celebrated Héctor José Cámpora's electoral victory and were embittered after his removal from office.[19] He continued:

> Vast sectors of this youth have adopted leftist positions, which characterize Argentine youth in general, including the vision of Israel as the "aggressor" oppressing Palestinian rights. The rightist orientation of the Peronist government caused frustration and confusion among Argentine youth and the influence of this atmosphere is well entrenched among Jewish youth.[20]

In fact, the Peronist Left adopted the terms "oppressor" and "Third World" defined by the New Left. Therefore, Argentina and the Palestinians were perceived as allies in the global fight against imperialism.[21] Nevertheless, the combination of ideologies and political trends created a spectrum of positions and experiences that were much more complex than any dichotomy. For instance, chants at a Zionist youth gathering in the main synagogue of Buenos Aires during the Yom Kippur War, combined support for Israel with leftist and Peronist terminology, such as: "Fuera los Yankees de America Latina, pero también de Israel y Palestina" (Yankees, get out of Latin America but also out of Israel and Palestine), "Venga, venga, venga, venga compañero, que acá se está luchando por la paz del mundo entero" (Come, come, come, come, comrade, here we are fighting for world peace).[22] There is no doubt that leftist ideas and other Argentine political views were present within Jewish and Zionist organizations. Therefore, both non-affiliated and affiliated Jewish-Argentines were involved equally in a dynamic cultural and political mix.

Moreover, Jewish-Argentine youth involvement in the country's political effervescence, and particularly in Peronism, was not limited to that generation. Older generations were also engaged, either as sympathizers or voters. The Jewishness of Cámpora and Perón's economy minister José Ber Gelbard was well known and many of his associates were Jewish, too. These people were

19 María Cecilia Míguez, "El Regreso del Peronismo al Poder en 1973 y las Relaciones Internacionales: La repercusión regional y mundial de la asunción de Héctor Cámpora," *Postdata: Revista de Reflexión y Análisis Político* 23, no. 1 (2018): 11–47.

20 K. Grossman to U. Narkis, December 6, 1974, TLC, 220757.

21 Manzano, "Argentina Tercer Mundo."

22 "La solidaridad con Israel expresada con un enfoque de juventud progresista y pacífica," *La Opinión*, October 9, 1973, 6.

just the tip of the iceberg of vast sectors of Jewish-Argentines who, since their immigration to the country, had contributed to its industry. The third Peronist government's economic plan was beneficial to their factories and small businesses. These facts, as well as the need for a new institutional perspective on Peronism, were recognized by Nehemias Resnisky, general secretary of the Delegación de Asociaciones Israelitas de la Argentina (DAIA), after Cámpora's electoral triumph. He believed that the vast majority of voters favored "change" and opposed underdevelopment, misery, oppression, dependence, monopolies, and international companies. The envisioned "change," he said, would not be "detrimental to vital and sectoral interests of the Jewish community." In his view, the vast majority of the Jewish community had interests that coincided with those of broad popular sectors.[23] Commenting on his discussions with Jewish leaders from Israel and the United States he said:

> They do not notice that Latin American revolutionary processes are captivating, fascinating because they create ideology, good or bad; they fuel ideology and captivate our youth … For example, they could not understand that there are Peronist Jews.[24]

Nissim Elnecavé, director of *La Luz*, also wrote of the political inclinations of Jewish-Argentines: some felt they were facing an emergency and were leaving for Israel; a number were going to other countries; and another, "non-negligible," group "identified with the ultra-Left."[25] In 1974 the *American Jewish Year Book* stated:

> Writers, journalists, lecturers, and Jewish community leaders seemed to agree that in 1974 the crisis of Argentine Jewish life was more acute than ever before. The inroads by extreme leftist political organizations among Jewish youth were repeatedly decried. This tendency reached its zenith in the latter part of the year when Jewish youth painted on the walls and blackboards of many Jewish schools and organizations such slogans as, "For a Free Palestine," "Freedom for the Palestinians," and "Down with American Imperialistic Zionism."[26]

---

23 "En una reunión de DAIA se evaluó el proceso eleccionario del país," *Informativo DAIA*, April 1973.

24 "Informe del secretario general sobre la reunión del ejecutivo del CJM, en la sesión especial del Consejo Directivo de DAIA," *Informativo DAIA* (Buenos Aires), July 1973.

25 Elnecavé Nissim, "Latin American Jewry," *Jewish Chronicle*, August 10, 1973, 19.

26 Meyer, "Argentina Domestic Affairs": 262.

Toward the end of 1974 and beginning of 1975, Tzvi Rivkin, the JA emissary in Córdoba, continued expressing his apprehensions regarding the risky position of Jewish youth in a context characterized by military victories over the guerrillas.[27] Their involvement with the Left was evident and Zionism, which in the best case scenario would mean immigration to Israel, was considered a safe alternative.

> A number of key people in the Jewish community in Córdoba and local institutions are very worried about the activity of groups of young people who are involved with the radical Left, which many young Jews have joined. It was decided that the presence of a special youth emissary is urgently required to be fully dedicated to the problem and to connect the youth to Zionism and Jewish affairs.[28]

The military regime that ousted María Estela Martínez de Perón, who had succeeded her late husband as president, lasted from March 24, 1976, to December 10, 1983. During this period, the nation experienced some of the darkest moments in its history. In a climate dominated by the Cold War doctrine of national security and the struggle against leftist guerrillas, the military government systematically violated human rights in order to annihilate political opponents. State terrorism and an economic program aimed at the destruction of national industry led to rapidly declining living conditions for vast sectors of the population and forced many into exile. As a result, a JA representative in Argentina referred to an increase in the number of immigrants to Israel in early 1977, explaining:

> In 1976, there were approximately 1,900 *olim* [immigrants] compared with approximately 1,000 the previous year. It is thought that the [problem with] this aliyah [immigration] stems from the fact that the majority are young people fleeing regime persecution because they were leftists or relatives of leftists.[29]

---

27 Y. Rivkin to U. Narkis, January 16, 1975, TLC, 220757.
28 Y Rivkin to U. Narkis, February 4, 1975, TLC, 220757.
29 "Report on the Situation in Argentina" [no details], January 12, 1977, TLC, 471090.

2 David Armando Laniado: A Little Sephardic Monster among Marxist Guerrillas

One day during the early 1960s Carlos Wolpowicz, a member of the Sephardic Zionist group Baderech, received an unusual telephone call from a fellow activist. A mutual friend, David Armando Laniado, had been engaged in a fight with members of the anti-Semitic organization Tacuara and was hospitalized. Carlos and other members of Baderech rushed to the hospital fearing for David's condition. They had good reason to worry. Tacuara had been responsible for many attacks against Jewish students and buildings.[30] For instance, fifteen-year-old Edgardo Trilnik was wounded by gunfire in 1960 and university student Graciela Sirota was forcibly tattooed with a swastika on her breast in 1962.[31] However, this time the young Jew's injuries resulted from different circumstances. When his friends arrived at the hospital David was being treated for injuries on his knuckles: "You have no idea how much I pummeled them," he said. His cheerful attitude and good health relieved his friends, who consequently nicknamed him "Monstruito" (Little Monster).[32] Carlos explained:

> Tacuara was active in his secondary school and the day of the fight they had called him a "Communist Jew." "Monstruito" suited him because he was like a beast due to his disproportionate reactions. He seemed to us … like [the Incredible] Hulk or a golem … He had an outstanding intellectual capacity but when the monster came out of him everything else was overwhelmed.[33]

In those days fights between young Jewish-Argentines and members of Tacuara were not unusual. Many Jewish boys took up self-defense and Zionist

30 On Tacuara, see Juan Manuel Padrón, *Ni yanquis, ni marxistas! Nacionalistas: Nacionalismo, militancia y violencia política: el caso del Movimiento Nacionalista Tacuara en la Argentina, 1955–1966* (La Plata: UNLP, 2017).

31 In May 1960 the Nazi war criminal Adolf Eichmann was kidnapped by Mossad agents, triggering a wave of anti-Semitic attacks. Tacuara was the main offender and the campaign lasted two years. Haim Avni, "Jewish Leadership in Times of Crisis: Argentina during the Eichmann Affair (1960–1962)," in *Studies in Contemporary Jewry: XI: Values, Interests, and Identity* (Oxford University Press, 1996), II, 117–32; Raanan Rein, "The Eichmann Kidnapping: Its Effects on Argentine-Israeli Relations and the Local Jewish Community," *Jewish Social Studies* 7, no. 3 (2001): 101.

32 Author's interview with Carlos Wolpowicz, November 25, 2020.

33 Ibid.

organizations played a key role in their training.[34] Moreover, some members of Zionist organizations were tasked with obtaining information from within anti-Semitic organizations by joining their events. Through these activities they engaged in the protection of their community. Néstor B. took part in a self-defense action that occurred during the first years of the 1960s at the main entrance to Sociedad Hebraica Argentina (SHA), a prominent Jewish cultural and sport organization:

> We knew they were going to come. We had obtained the information in advance. Then, we waited for them inside the club. When they came we went out and we were ready. During the fight I saw a classmate. I beat him up and the next day at school I saw the marks on his face. I smiled and asked him, "How was the fight with the Jews?"[35]

David's fight against Tacuara, as well as other episodes and experiences during his life, were far from being unique. In fact, his story is representative of a wide sector of Jewish-Argentine youth during the 1960s and 1970s. Successive military dictatorships which interrupted democratic rule created the conditions for political violence to be accepted in Argentine society. A decade later such violence would reach unprecedented heights. The emergence of armed guerrillas and state-sponsored paramilitary groups would take the lives of thousands. At the time of David's youthful exploit, his family and friends would never have believed that a decade later he would be riddled by police bullets as a member of the armed Liga Comunista Revolucionaria (LCR). The LCR was a Trotskyist group created by former members of the Partido Revolucionario de los Trabajadores–Ejército Revolucionario del Pueblo (Workers' Revolutionary Party–People's Revolutionary Army; PRT-ERP).[36] How did David, who began his life as an activist within the Sephardic Zionist movement, come to finish it as a member of a Marxist guerrilla group?

An incident that occurred in 1969 is important for understanding the background to David's passage from Sephardic Zionism to the Argentine armed struggle. Salomón "Moni" Balé, another member of Baderech, played a key role

34 Raanan Rein and Ilan Diner, "Unfounded Fears, Inflated Hopes, Passionate Memories: Jewish Self-Defence in 1960s Argentina," *Journal of Modern Jewish Studies* 11, no. 3 (2012), 357–76; Raanan Rein, "Jewish Self Defense in 1960s and Early 1970s Argentina: A Mirror Image of Anti-Semitic Violence?" *Israel* (Studies in Zionism and the State of Israel) [Hebrew], no. 25 (2018): 227–55.

35 Author's interview with Néstor B., May 3, 2004.

36 Federico Cormick, *Fracción Roja: Debate y Ruptura en el PRT-ERP* (Buenos Aires: El Topo Blindado, 2012).

in his transition. Moni was older than David, and therefore had little contact with him. Nevertheless, he remembered him well enough to relate that "he was a lovely kid."[37] Moni had been a member of the Zionist Sephardic movement since its inception; he was active in Jewish-Argentine self-defense during the early 1960s and by 1969 was the secretary of Baderech. The group had become part of Hashomer Hatzair.[38] Moni and other members of the movement, such as Néstor B., Osvaldo S., and Alex H., had been planning a student front sponsored by Hashomer to take part in Argentine politics.[39] While such participation was already taking place individually, the idea of channeling it through Zionism was resisted by the adult leadership. They were focused on immigration to Israel and saw Argentine politics as a dangerous activity that frowned upon emigration from the country. The violence around them precipitated these young people's decisions. When New York governor Nelson Rockefeller visited Argentina as President Richard Nixon's envoy, there were massive demonstrations against him. Violent police repression caused the demonstrators to leave in disorder and one of the organizers, journalist Emilio Jáuregui, was killed. According to the official version, Jáuregui had a gun and threatened a policeman.[40] However, two young Zionists knew this to be untrue. They had witnessed the murder through the window of the Maccabi sport club where they were working as instructors. They were Moni and Clara C. and the scene was immortalized in a novel penned by the prize-winning writer Bernardo Verbitsky.[41]

Osvaldo S. had already begun his university studies and was therefore in touch with students engaged in Argentine politics. He established a connection between Moni and the General Federation of Argentine Workers (CGTA), and a press conference in which witnesses rebutted the police version, was organized. A few days later, Peronist union leader Augusto Vandor was assassinated; the government immediately declared a state of siege and carried out

37 Author's interview with Salomón Balé, September 19, 2019.

38 On Baderech, see Adriana Brodsky, "Argentine Sephardi Youth: Between Aliyah and Activism, 1960–1970," *Journal of Jewish Identities* 8, no. 2 (2015): 113–35.

39 Author's interviews with Néstor B., May 3, 2004, Alejandro H., April 17, 2004, and Osvaldo S., October 7, 2004; Herbert Waserstein to the Office of Communication at Kibbutz Gazit, August 30, 1969. Waserstein read the document to a gathering at Kibbutz Gazit in 2016. It was the first time the movement had welcomed Moni since the 1969 incident. A copy of the letter was contributed by Salomón "Moni" Balé to the author.

40 On Jáuregui, see Juan Sebastián Califa, (2018) "El '69 porteño. El ascenso de masas en la Universidad de Buenos Aires y su contexto," *Argumentos: Revista de Crítica Social*, no. 20, https://publicaciones.sociales.uba.ar/index.php/argumentos/article/view/3093 (accessed January 4, 2021).

41 Bernardo Verbizky, *Etiquetas a Los Hombres* (Buenos Aires: Editorial Planeta, 1972).

massive arrests and incarcerations.[42] Moni was imprisoned for three weeks,[43] producing an earthquake within Hashomer in both Argentina and Israel. Almost fifty years later, Hashomer members are still licking their wounds. For young Zionists who were willing to take part in Argentine politics, including Moni, it was a question of ethics and civic responsibility. Those who believed that immigration to Israel was the only goal of the movement considered Moni and his friends irresponsible. Moreover, they claimed, they put the Jewish community in peril.[44] Thereafter, those who wanted to be active in both Argentine politics and Zionism became the Amos Movement of Leftist Zionism (Amos).[45] This short-lived experience, touched on in a novel by Mauricio Goldberg, demonstrated that there was fluid communication between Zionist activists and Argentine political parties.[46] Moreover, Argentina-oriented political groups endeavored to recruit new members among Zionist activists and vice versa. While David "Monstruito" Laniado's attitude is unclear during Amos's rupture with Hashomer, there is no doubt that, as a member of Baderech, he witnessed the process and knew about Moni and his friends' ideas. In fact, after Moni's incarceration those seen as conflictive members were excluded from Hashomer and special activities were held to explain why.[47]

According to one testimony, during the creation of Amos, David was active in both Argentine politics and Zionism. Carlos became David's friend since, due to their physical strength both were assigned to be the "vanguard" in charge of organizing one of Baderech's camps. Their friendship continued from that time when they carried the tents that would shelter 300 members, until the early 1970s, when Carlos began to feel that frank conversations with David had become impossible. According to Carlos, David was active in both Baderech and Argentine politics through his secondary school in the Caballito

42 Herbert Waserstein to the Office of Communication at Kibbutz Gazit, August 30, 1969. "En los episodios ocurridos en cercanías de Plaza Once fueron detenidas 102 personas," *Clarín*, June 29, 1969.

43 Herbert Waserstein to the Office of Communication at Kibbutz Gazit, August 30, 1969; "El Caso Jáuregui," talk delivered by Salomón "Moni" Balé to the gathering at Kibbutz Gazit in 2016.

44 Author's interview with Alejandro H., 2004; Beatrice D. Gurwitz, *Argentine Jews in the Age of Revolt between the New World and the Third World* (Boston and Leiden: Brill, 2016).

45 Adrián Krupnik, "Cuando Camino Al Kibbutz Vieron Pasar Al Che: Radicalización Política y Juventud Judía: Argentina 1966–1976," in Emmanuel Kahan et al. eds., *Marginados y Consagrados. Nuevos Estudios Sobre La Vida Judía En Argentina* (Buenos Aires: Lumiere, 2011), 311–27.

46 Mauricio Goldberg, *Donde Sopla La Nostalgia* (Buenos Aires: Grupo Editor Latinoamericano, 1985).

47 Herbert Waserstein to the Office of Communication at Kibbutz Gazit, August 30, 1969.

neighborhood. His activism led him to link up with Hugo S., a student at the Carlos Pellegrini School who was also a Jewish-Argentine of Sephardic origin. Hugo was a member of the Tendencia Estudiantil Revolucionaria Socialista (TERS), a branch of the Trotskyist party's Política Obrera (PO), led by José Saúl Wermus, known as Jorge Altamira.[48] Hugo introduced David to texts that "spellbound" him. According to Carlos, "when he read Althusser, for instance, it was as if his life had changed." Nevertheless, his influences also came from the Jewish community. Another source of David's intellectual formation was SHA's school of leaders directed by Janán Nudel. There, group dynamics was particularly inspiring.[49]

Carlos and David shared many experiences connected to street demonstrations and politics, particularly after the Cordobazo. On one occasion, at his grandfather's luxurious house in the Recoleta neighborhood, David said to his friend: "I am in the perfect place; no one would suspect that a person like me in a place like this is a revolutionary." Something about being part of these two different and even contradictory realities seemed to attract him. He enjoyed strolling around downtown wearing a suit and tie, pretending to be a businessman, and then suddenly joining a demonstration, after which he would leave calmly, just like any other passerby. David underwent the experience of "proletarization," as it was called then, and took a job in a factory around 1970. The idea was to become more closely acquainted with the Argentine masses. However, he was also enjoying his youth. He made friends among recruits to the Argentine army in Plaza Italia and frequented working-class bars outside the capital. Like many Argentines of his generation, political activities offered him a lively social life. Nonetheless, the situation was becoming riskier every day. "I remember him," Carlos related, "talking about how it would feel to be shot." That discussion took place while they were collecting Marxist books from David's house in order to burn them so as to avoid trouble in the event of a police raid.[50]

While politics gave activists the opportunity to get to know new people and to acquire a wide range of intense experiences, the radicalization process also created splits between old friends. In the case of Carlos and David, the former felt that David's reserved attitude, justified allegedly on security grounds, served as an excuse to thwart discussions which would challenge his point

48 On the PO and their Jewish leaders, see Maximiliano Jozami, (2019) "Argentine Left Parties and the 1967 Six-Day War through the Prism of Global Networks and South-South Connections," *Jahrbuch für Geschichte Lateinamerikas Anuario de Historia de América Latina*, no. 56: 15–41, 23, https://doi.org/10.15460/jbla.56.125 (accessed January 4, 2021).

49 Author's interview with Carlos Wolpowicz, November 25, 2020.

50 Ibid.

of view.[51] Their friendship waned. In 1974, Carlos took part in a large demonstration which included hooded people among the crowd. This was their way to avoid being identified by security forces. At a certain point, one of the cloaked demonstrators moved close to Carlos and whispered in his ear: "Don't you recognize me? I'm David." It was the last time Carlos saw David, whose preventive measures he judged to be unsophisticated: "They concealed themselves in the demonstrations, but we used to meet at popular places such as Café La Paz and to stroll along Corrientes Avenue."[52]

Isabel Steinberg, who was David's girlfriend from 1973 until his death, also believed, in retrospect, that David's notion of a clandestine life was not entirely realistic. More than once he told her: "We can't meet for the next three months because I need to hide." Nevertheless, "once he returned to visit his family; we met and he even walked me to my house." His intermittent periods of hiding were not strange to her. As a student of psychology, she had known of several acts of violence, including the assassination of Silvio Frondizi, a Marxist intellectual and brother of a former president, in September 1974. Frondizi's daughter studied with Isabel, who was horrified by the crime committed by the paramilitary Triple A (Alianza Anticomunista Argentina). "I was very young," she explained, "and violence was internalized. They could kill someone just for talking at a university assembly." She knew that David used guns and on one occasion he took her to an isolated location and taught her how to shoot "in case I needed to defend myself ... We were intellectually precocious but at the same time very naive."

Isabel was three years younger than David and had also been a member of Baderech, but they did not meet there. While she had never heard of his nickname "Monstruito," she said that he was big and strong and the kind of person who would defend a stranger in the street out of his dedication to justice. In fact, a friend of hers used to call him Robin Hood. She remembered the time in Baderech as "being in paradise, singing 'The Internationale' in Yiddish, Hebrew and Spanish." Afterward, from age twenty to thirty, she felt she was just "surviving." Isabel also left Baderech and joined the Marxist PO first and the Peronist movement later, when she was a student of psychology at the University of Buenos Aires. She was aware that David had been a PRT member, and had left to join the LCR, but little else: "He took care to avoid letting me know anything about the people he was with." For two years Isabel joined David's family during the Jewish high holidays, enjoying the warm atmosphere, Sephardic style, and Ladino songs. Like many traditional parents, she believed

51 Ibid.

52 Ibid.

David's father and mother expected to see the young couple married soon: "I think they expected me to 'save him,' to get him out of politics."

David told Isabel about his desire to join the Fourth International. He also told her that he had visited France to that end, as well as Germany, where he met members of the Baader Meinhof group. With some reservations regarding secrecy but wishing to help her support herself financially, he offered Isabel a job at a print shop for long-playing record covers, which was the organization's facade for their actual printing house. However, Isabel got scared and did not take the job.

According to her, David's activities did not affect his identification as a Jew or his connection with Israel. For instance, at the end of 1974 his grandfather died, leaving him an airline ticket to Israel. David had been chosen to honor his grandfather in accordance with Jewish-Sephardic tradition. Therefore, he took the ticket printed for "David Armando Laniado" and visited Israel for a week.

> He went to the Wall ... He went to ... Jerusalem. I remember his stories and he showed me pictures ... I was ready to marry him even though he was not going to leave politics. We were going to do half what his family wanted and half what we wanted ... He was a man of impressive creativity and artistic sensibility ... He was full of vitality. [However] three months before his assassination he told me ... "I know that my life is ending" ... The night before his assassination he rang the bell to my house at 10 pm. My father said I was not home, locked me in my room, and threw him out. He went to the printing house and was murdered the next morning ... He wanted to stay at our home. I don't know what would have happened.[53]

David's tragic death on May 1975 appeared in the press as the result of a confrontation with the police. Such reports provided abundant information on the LCR and its activities. Accordingly, following a meeting held by the organization in March 1975, its leadership was identified, tracked down, and murdered, along with most of its members.[54] Isabel felt endangered and changed her jobs many times until she relocated to Israel in 1977, convinced by her brother that she should go into exile. She volunteered in a kibbutz and traveled around the country. She also met exiled Montoneros in Tel Aviv. They were living in a building whose walls were decorated with the organization's flag. In Israel she read the foreign press and became aware of the real scale of Argentine state terrorism. Moreover, her cousin disappeared while he was a conscript in the

53 Author's interview with Isabel Steinberg, February 25, 2020.

54 Ibid.; see also Cormick *Fracción Roja*: 161–75.

Argentine army. Despite her knowledge of the situation, she returned to her native homeland "out of guilt." She worked in SHA as a secretary and there she found an atmosphere where her life and previous political activism were not questioned. She had no photos, but showed me a caricature done by an artist at a Buenos Aires restaurant. In it, David has a prominent moustache and holds a wooden caveman's club, drawn by the artist at his client's request.

3 Raúl "Rafi" Milberg: Bearing the Holocaust Legacy with the Montoneros

In April 1972, central Jewish-Argentine organizations commemorated the Warsaw Ghetto uprising. Raúl Milberg, known by his Hebrew name Rafi within Zionist circles, delivered a speech. He did so as a member of the Jewish-Argentine Youth Federation to which he was affiliated through the Zionist movement Hechalutz Lamerchav. Emphasizing that most of the ghetto fighters were Zionists, "the vanguard of the Jewish people," he continued:

> These young heroes commanded by Mordechai Anilevich decided to wield the rifle, demonstrating that the Jewish people does not dwell on the margins of history ... and took part in the fight for dignity and human freedom [along] with anti-Nazi fighters. They occupy a golden page in the history of the peoples and ... of those who at any given time preferred to die on their feet rather than live on their knees. We young Jews ... of the ... national liberation and social normalization movement of our people, Zionism, do not say we will remember them as long as we live. Instead, we say that we will live as long as we remember them. We, who carry the flag of those who knew how to die on their feet, cannot do less than they did in those adverse circumstances. We should not only [see ourselves] as an oppressed and dispersed minority, but as the vanguard [of] ... *aliyah* [immigration].[55]

Raúl also urged his young audience to combat local anti-Semitism and anti-Zionism. Undoubtedly, no one among his listeners imagined that he would die as a member of the armed Peronist Montoneros. "I thought he was going to end up in Israel," said fellow student Esparis,[56] who met him when both attended ORT Secondary School. "He was very committed [to Zionism], intellectually

55 "Palabras de Rafi Milberg," *Informativo DAIA*, April 15, 1972, 7.
56 Author's interview with Esparis, November 22, 2004.

superb and awe-inspiring." Since Raúl was a year older than Esparis and they were not close friends, Esparis lost track of him when he finished school in 1972. However, they met by chance in early 1975 at the well-known corner of Corrientes and Callao streets in downtown Buenos Aires.

> He asked me which organization I had joined, because at the time if you were not in politics you were [considered] a kind of faggot … If you were on the left, [you joined] the People's Revolutionary Army, if you were a Peronist sympathizer, [you joined] the Montoneros … I also asked and he said: "I am now with *los muchachos* [the guys], I am with Peronism. Did you know about the Peronist march?" … It was the last time I saw him and I thought "this guy is crazy."

Esparis also recalled that Milberg, like other Zionist activists, had been very active at the school students center, established in 1972. Secondary school students in general had been taking part in massive protests and demonstrations for several years but one particular conflict pushed many ORT students toward politics. In September journalistic sources announced that the provincial government of Buenos Aires was considering enacting a decree which would downgrade the professional relevance of technical degrees. Although there was no precise information about the act, which became known as "the phantom law," secondary school students, along with their teachers and parents, launched a series of mass demonstrations.[57] Argentine technical schools were considered an alternative for the middle or lower-middle-class population, which aspired to get a job without pursuing university studies. Therefore, new limitations to their professional possibilities were seen as an attack.

In the early 1970s, ORT Argentina was not the prestigious, large educational organization which it is today. It was a small institution whose student body was predominantly male and was divided into three specializations: chemistry, mechanics, and electronics. Many of the students were assisted by generous scholarships, and some were even given lunch. The school was headed by Eva Staif, and her husband, Tobías Kamenszain, was a Zionist activist who occupied several posts in Jewish organizations. In fact, as some graduates recall, Zionism was dominant among the students, although a minority were affiliated with the Communist Youth Federation. Nevertheless, students were aware of national politics. For instance, following the execution of sixteen leftist and

57 Pablo Augusto Bonavena, "La Lucha de Los Estudiantes de Las Escuelas Técnicas Contra La Ley Fantasma," *X Jornadas de Sociologia* (Buenos Aires: Universidad de Buenos Aires, 2013).

Peronist political prisoners, mimeographed copies of related press bulletins appeared on the walls of public toilets denouncing what became known as the Trelew Massacre.[58] Discussions and criticism among the students regarding the military government were not new; however, social reaction against the "phantom law," also known as "the struggle of the technical schools," convinced them to create a student committee at the school.

As part of their struggle, ORT students took part in mass street marches and demonstrations at Plaza de Mayo, during which they began to coordinate their actions with other schools. Outstanding students were appointed as committee delegates, who participated in meetings with their colleagues from other institutions and quickly learned that being involved in politics outside their classrooms was tougher than they expected and not exactly based on free speech. As Esparis recalled:

> These general gatherings turned out to be managed by older students who studied at night schools because they worked during the day. They were linked to right-wing Peronist unions ... They met with the authorities and student mobilization declined. When at the assembly someone said, "Hey, you are betraying us! You are negotiating!" one of them would stand right next to you [in order to intimidate you]. [These older students] labeled some schools *los zurdos* [lit., the left-handed, meaning leftist]. This is what they were called if they were democratic and pluralistic.[59]

Additional testimonies provide details about Raúl's experiences as a member of the ORT student committee. Secondary school activism in 1972, however, was one of the factors contributing to his engagement in Argentine politics. In 1973, ORT students, led by their committee, marched again to Plaza de Mayo, this time in order to demonstrate against the assassination of Chilean president Salvador Allende. Such participation in Latin American politics was compatible with Zionist solidarity with Israel during the Yom Kippur War. In fact, during this emergency, one student's parents who were chemists coordinated the creation of a blood bank at the school. By then, political activism

58 Author's interviews with Esparis, November 22, 2004, Enrique Fukman, October 13, 2004, and Enrique P., September 6, 2005. On the Trelew massacre, see Roberto Pittaluga, (2006) "La memoria según Trelew," *Sociohistórica*, nos. 19–20: 81–111, http://www.memoria.fahce.unlp.edu.ar/art_revistas/pr.3610/pr.3610.pdf (accessed January 4, 2021); Pablo Augusto Bonavena, (2011), "El movimiento estudiantil frente a la masacre en Trelew del 22 de agosto de 1972," *Conflicto Social* 4, no. 5, https://publicaciones.sociales.uba.ar/index.php/CS/article/view/377 (accessed January 4, 2021).

59 Author's interview with Esparis, November 22, 2004.

had become common among the students and there were high expectations regarding the Peronist government led by Héctor Cámpora, who was linked by Peronist left-wing organizations to the idea of *la patria socialista*. These sectors received a hard blow when the exiled leader Juan Domingo Perón returned to Argentina and his right-wing followers fired on left-wingers near the airport.[60] Apparently, the so-called Ezeiza massacre was a warning to the school authorities about increased violence and the dangers to come. The student committee was disbanded and the counselors, who had been university students of psychology, sociology, and philosophy, and were themselves protagonists of youth politics, were replaced by instructors. Disciplinary rules, such as lining up before entering the classroom, were imposed. The year 1973 marked the zenith of the student committee at ORT.

Raúl completed secondary school in 1972 and continued his path to radicalization as a member of the Zionist movement Hechalutz LaMerchav (Hechalutz). This organization, created by Jewish-Argentines after the establishment of the State of Israel, was not formally aligned with any Israeli political party. Moreover, unlike Hashomer, Hechalutz did not explicitly ban their members from participating in local politics. Therefore, their political activity intensified during the years 1972–74. Hechalutz member Misha Senderovsky explained:

> The revolutionary option ... was vast and attractive due to [our] willingness to act, to be rebellious, and to search for an identity ... The challenge was to channel [our] Jewish identity together with all that. Tacuara was not [as] active as it had been in the 1960s and therefore, one's questions were related to daily life from the point of view of youth of the world rather than [that] of a young Jew from a Jewish community and a particular location.[61]

According to Miguel Benasayag, Raúl started abandoning Zionist activism for the Argentine armed struggle during the Yom Kippur War. Miguel had been his close friend and fellow Zionist in Hechalutz from age thirteen, even though they attended different neighborhood centers, or "nests" (*ken*). Miguel was part of

60 On the Ezeiza massacre, see Sebastián Carassai, (2015), "The Dark Side of Social Desire: Violence as Metaphor, Fantasy and Satire in Argentina, 1969–1975," *Journal of Latin American Studies* 47, no. 1: 31–63; Daniel Feierstein and Douglas Andrew Town, "Toward a Periodization of Genocide in Argentina," in *Genocide as Social Practice: Reorganizing Society under the Nazis and Argentina's Military Juntas*, 161–85 (New Brunswick, NJ: Rutgers University Press), 166–67.

61 *Hejalutz Lamerjav 50 Años En Movimiento* (Buenos Aires: Hejalutz Lamerjav, 1998), 13.

the *ken* that functioned in the Maccabi sport club, in Abasto, and Raúl attended the Flores one. They participated together in many street demonstrations after the Cordobazo and got used to witnessing violent clashes with the police and escaping from the repressive forces. For their own amusement, they would print leaflets containing slogans against the military government and deliver them in the streets while lighting the wick of a fake bomb made out of carton boxes. In 1970, Miguel left Hechalutz and joined the Marxist guerrilla ERP-22.[62] During the Yom Kippur War he went to look for his friend at Natan Gezang, a Jewish school which shared a building with the Jewish Agency. Raúl was there volunteering in the solidarity campaign and Miguel showed up in order to convince his friend to follow his path. It did not take much effort because the same day, as he left Natan Gezang, Raúl decided to join Miguel, whose main argument had been that the kibbutz was indeed an option for liberation but they were living in Argentina not in Israel. In the Marxist guerrilla movement of that time, Miguel explained, "we felt no contradiction between being Zionist and leftist. I don't know how it was for Raúl later in the Montoneros." Miguel remembered that Raúl had always been a bold person: "When he volunteered to fight in our self-defense training sessions at Hechalutz no one wanted to fight him. He did not fool around. He would pummel you really hard."[63]

Miguel was imprisoned in 1974 and liberated in 1978 due to his French citizenship. Once in Paris he got in touch with guerrilla members. Through such connections he knew that Raúl had been training in Lebanon when Israeli jet fighters bombed the training camp. Even today Miguel continues asking himself what his friend would have felt at that moment: "He could have joined the Israeli army but he was running to the shelters because of the Israeli bombs." About his friend's attitude as a member of the Montoneros, Miguel concluded:

> Rafi was always angry with me because I used to criticize the organizations I joined. I criticized the ERP-22, in which we were both active, and later I criticized the ERP ... I tried to explain to him that as members we should be the first to criticize. He never understood that. In his view we had to believe.[64]

---

62 ERP-22 de Agosto was an Argentinian guerrilla faction that split from the People's Revolutionary Army (ERP). The group's name refers to the date of the Trelew Massacre, August 22, 1972.
63 Author's interview with Miguel Benasayag, November 20, 2020.
64 Ibid.

Raúl was not the only Zionist from Hechalutz drawn to Argentine politics, and those like him had formed a faction within the movement. Moreover, later, impressed by Raúl's abandonment of Zionism, some members of this group launched the magazine *Aporte para el Análisis* (Contribution to the analysis), which opposed the organization's official publication *Debate*.[65] These activists promoted participation in Argentine politics and a Latin American perspective. They were known as *La Tendencia* because many of them had joined the leftist Peronist Youth. Until 1975 they were active in both Hechalutz and Argentine politics. Later, they immigrated to Israel, left Zionism, or were expelled from the movement. Enrique Fukman stayed in Hechalutz while remaining active in the Peronist Youth movement. His path was similar to Raúl's. His brother was shot dead in the street and Enrique was abducted from his home where he had hidden a Montoneros printing machine. Asked about his reasons for joining the organization, he told me, in Hebrew, that as a Jew he had been taught to combat oppression, "since the time we were slaves in Egypt." He also told me that his grandparents had been politically active in Europe and named Spartacus as one of the most celebrated stories in his family home.[66] Fukman was a disappeared detainee at the infamous Army Mechanics School (ESMA). He survived illegal detention and became active in the teachers union until his death in 2016.

Journalist Elio Brat has blurred memories of a visit to Raúl's house in 1974. Raúl had left Hechalutz and Elio was one of the editors of *Aportes*. In fact, the reason for the visit was to bring mimeographed copies of the publication to his former Zionist colleague. On the one hand, Rafi's views were important to Elio and his friends; on the other, they were trying to draw him back to their group. "Rafi looked at the publication and laughed a lot," he recalled. Later, said Elio, a Zionist activist called Vicky T. saw Raúl in the street and called out to him by his Hebrew name, Rafi, without getting a reaction. At the time Milberg was answering only to "Raúl": "It seems he had left that aspect of his life behind."[67] While still within the Zionist movement, Elio, Fukman, and Rafi's friends continued their politicization, which pushed them gradually closer to Argentina and further from Israel. When they traveled to Israel in 1975, these members of Hechalutz, who were also engaged in Argentine politics, decided to stay in a Hashomer Hatzair kibbutz. They found this option to be congruent with their Socialist inclinations. While becoming more distant from their

---

65 Author's interview with Elio Brat, November 25, 2020; see also *Hejalutz Lamerjav 50*: 27–28.

66 Author's interview with Enrique Fukman, October 13, 2004.

67 Author's interview with Elio Brat, November 25, 2020.

Zionist colleagues in Hechalutz, they continued to adhere to Zionism. Back in Argentina, however, they were not given positions as *madrichim* (instructors) and left the movement. It was the end of their dual activism.

Peronist researcher Roberto Baschetti created a website to commemorate activists of the movement killed by the military. According to the entry on Raúl Milberg, he joined the Peronist Youth movement around 1972 and left Argentina as a member of the Montoneros with the rank of second lieutenant in October 1977. In Spain, he acted as the organization's spokesperson regarding the actions it carried out during the 1978 FIFA World Cup in Argentina. His clandestine activity abroad continued in Mexico and Lebanon, where he was trained between April and May 1979. Back in Argentina, as part of a "special unit," he took part in an attempt to assassinate Walter Klein, an economic adviser of the military government. The bomb in Klein's house killed two policemen. According to Baschetti, Milberg then left the country until his final return in 1980.[68] According to Elio Brat, he had heard that Raúl, the person he had admired in the Zionist movement, had joined guerrilla activities in the province of Córdoba. However, he is not certain whether this is true or not.[69]

According to Luciana Milberg, a human rights activist and lawyer who researched the case of her uncle and testified in court, Raúl began to live clandestinely in 1975. After that, his family never knew where he was even though there was sporadic communication. Andrea Bello, who survived illegal detention as a disappeared, was Raúl's friend from the Peronist Youth. During her captivity at ESMA she was forced to eavesdrop on his telephone conversations with his family using the technology of military intelligence. In 1980 the military conducted "Operation Storage," aimed at capturing a group of Montoneros who had entered Argentina to launch what the organization's leaders bombastically called *Contraofensiva* (counterattack).[70] After the military identified storage facilities for hiding arms, they put them under surveillance and waited for guerrilla members to show up. Each capture led to the next. Raúl Milberg was taken on February 27, 1980. According to military documents, he was apprehended alive, and tortured and killed at the Campo de Mayo military base.[71]

68 "Militantes del Peronismo Revolucionario Uno por Uno," http://www.robertobaschetti.com/biografia/m/202.html (accessed December 11, 2020).

69 Author's interview with Elio Brat, November 25, 2020.

70 On the *contraofensiva*, see Hernán Eduardo Confino, "La contraofensiva estratégica montonera en las memorias de sus participantes: Crónica de un objeto polémico," *Aletheia* 6, no. 11 (2015), http://www.memoria.fahce.unlp.edu.ar/art_revistas/pr.6969/pr.6969.pdf (accessed January 4, 2021).

71 Author's interview with Luciana Milberg, December 8, 2019.

Based on testimonies, Raúl's motivation for joining the armed struggle might have had elements in common with Fukman's argument based on Jewish history. However, in Raúl's case his reasons did not go back to biblical times but to family memories intertwined with history. Raúl's parents were Jewish-Ukrainians who arrived in Argentina after escaping from the Nazis. Part of their family stayed in Europe and fought against the Nazis either as partisans in the East or as resistance members in the Warsaw Ghetto. One day at his parents' home Raúl reproached them for not having fought the Nazis, infuriating his typically calm father to the point that he slammed his fist on the table. The scene was witnessed by Raul's girlfriend Patricia Roisinblit, who also disappeared during military rule.[72]

4 Conclusions

Jewish-Argentine youth were fully integrated into Argentine society by the 1960s. Therefore, when the entire region experienced a rise in politicization and radicalization inspired by liberation movements, Third Worldism, and the New Left, Jewish-Argentines and their organizations did not remain indifferent. Moreover, most Zionist movements such as Hashomer and Hechalutz had instilled Socialist values in their members in order to prepare them for their expected future life on a kibbutz. When their members began to engage in Argentine politics and gave up the idea of immigration to Israel, their Zionist education served as fertile soil for new revolutionary ideas. In addition, they were aware of anti-Semitic attacks and many were trained in self-defense. As part of their Jewish education and family memories, they had inherited a visceral rejection of Fascism and great respect for those who fought against Nazism. Leftist organizations and Peronism offered Jewish-Argentine youth the opportunity to channel their desire for social transformation as never before and thousands of them grasped this opportunity. In doing so, their Jewish ethnic values and history were not necessarily abandoned; rather, their actions were guided by a universalistic, regional, or national perspective. The biographies of David Armando Laniado and Raúl Milberg suggest that there was an elective affinity between the young generation of Jewish-Argentines educated in the Zionist movement and thousands of non-Jews with whom they joined in armed guerrilla groups such as the ERP and the Montoneros. Once a young Zionist Argentine decided to postpone or relinquish their immigration to Israel, it was possible to replace the idea of the new Jew built

72 Ibid.

through agricultural work in Israel with admiration for Che Guevara. Ancestral memories of oppression and still fresh family stories regarding the Holocaust provided additional inspiration for "wielding the rifle" and "dying on their feet" rather than "living on their knees."

The biographies explored in this article are representative of a social phenomenon which involved thousands of Jewish-Argentines who abandoned Zionism to become revolutionary activists in Argentina. Nevertheless, the opposite pathway also existed. The experience of those who in the midst of increasing violence decided to leave Argentina for Israel or to abandon their political activism deserves further research. In addition, while those who joined the armed struggle in Argentina chose to follow a political program in harmony with their national identity, it is impossible to know what future choices they would have taken. Their assassination put a violent end to the dynamic process of identity building. What the evidence indeed makes clear is that for these people, their incorporation into Argentine revolutionary groups reflected their integration into society, and their conscious desire to strengthen it. They decided that their political fight was in Argentina and not in Israel. This was so because the appeal of the guerrillas reached broad sectors of Argentine youth, including those who, like the Zionists, were affiliated with an ethnic political movement. The lives of Raúl Milberg and David Armando Laniado represented the extreme along a continuum of similar experiences, as reflected by the testimonies and archival evidence quoted in this article. However, their cases were not extraordinary; many like them followed the same radical option. Their experience was part of the broader phenomenon of Jewish-Argentines whose ethnic particularities played a certain role in their path of politicization and radicalization. Additional research might compare this Jewish-Argentine chapter of history with that of German-Argentines, Spanish-Argentines, or Italian-Argentines.

Bibliography

Avni, H. (1996) "Jewish Leadership in Times of Crisis: Argentina During the Eichmann Affair (1960–1962)," in *Studies in Contemporary Jewry: XI: Values, Interests, and Identity*, Oxford University Press, 11, 117–32, https://doi.org/10.1093/acprof:oso/9780195103311.003.0008.

Blejmar, J. (2020) "El Plan Gelbard y el populismo económico," *Realidad económica* 49, no. 330: 141–54.

Bonavena, P.A. (2011) "El movimiento estudiantil frente a la masacre en Trelew del 22 de Agosto de 1972," *Conflicto Social 4*, no. 5, https://publicaciones.sociales.uba.ar/index.php/CS/article/view/377 (accessed January 4, 2021).

Bonavena, P.A. "La Lucha de Los Estudiantes de Las Escuelas Técnicas Contra La Ley Fantasma," *X Jornadas de Sociologia*, Buenos Aires: Universidad de Buenos Aires.

Brennan, J.P., and M.B. Gordillo (1994) "Working Class Protest, Popular Revolt, and Urban Insurrection in Argentina: The 1969 'Cordobazo'," *Journal of Social History 27*, no. 3: 477–98.

Brodsky, A. (2015) "Argentine Sephardi Youth: Between Aliyah and Activism, 1960–1970," *Journal of Jewish Identities 8*, no. 2: 113–35.

Califa, J.S. (2018) "El '69 porteño. El ascenso de masas en la Universidad de Buenos Aires y su contexto," *Argumentos: Revista de Crítica Social*, no. 20, https://publicaciones.sociales.uba.ar/index.php/argumentos/article/view/3093 (accessed January 4, 2021).

Carassai, S. (2015) "The Dark Side of Social Desire: Violence as Metaphor, Fantasy and Satire in Argentina, 1969–1975," *Journal of Latin American Studies 47*, no. 1: 31–63.

Confino, H.E. (2015) "La contraofensiva estratégica montonera en las memorias de sus participantes: Crónica de un objeto polémico," *Aletheia 6*, no. 11, http://www.memoria.fahce.unlp.edu.ar/art_revistas/pr.6969/pr.6969.pdf (accessed January 4, 2021).

Cormick, F. (2012) *Fracción Roja: Debate y Ruptura En El PRT-ERP*, Buenos Aires: El Topo Blindado.

Diamand, M. (1972) "La Estructura Productiva Desequilibrada Argentina y El Tipo de Cambio," *Desarrollo Económico 12*, no. 45 (1972): 25–47.

Feierstein, D., and D. Town (2014) "Toward a Periodization of Genocide in Argentina," in *Genocide as Social Practice: Reorganizing Society under the Nazis and Argentina's Military Juntas* (New Brunswick, New Jersey and London: Rutgers University Press), 161–85.

Fiszbein, M. (2015) "La Economía Del Stop and Go: Las Ideas Estructuralistas En La Argentina, 1945–1976," *Desarrollo Económico 55*, no. 216: 187–210.

Galasso, N. (2005) *Perón: Exilio, Resistencia, Retorno y Muerte, 1955–1974*, Buenos Aires: Ediciones Colihue.

Goldberg, M. (1985) *Donde Sopla La Nostalgia*, Buenos Aires: Grupo Editor Latinoamericano.

Gurwitz, B.D. (2016) *Argentine Jews in the Age of Revolt Between the New World and the Third World*, Boston and Leiden: Brill.

Hejalutz Lamerjav (1998) *Hejalutz Lamerjav 50 Años En Movimiento*, Buenos Aires.

Jozami, M. (2019) "Argentine Left Parties and the 1967 Six-Day War through the Prism of Global Networks and South-South Connections," *Jahrbuch für Geschichte*

Lateinamerikas Anuario de Historia de América Latina, no. 56: 15–41, https://doi.org/10.15460/jbla.56.125 (accessed January 4, 2021).

Kahan, E. (2013) "'Memories That Lie a Little': New Approaches to the Research into the Jewish Experience during the Last Military Dictatorship in Argentina," in Raanan Rein and Adriana Brodsky eds., *The New Jewish Argentina: Facets of Jewish Experiences in the Southern Cone* (Leiden: Brill), 293–314.

Kaufman, E., and B. Cymberknopf (1989) "La dimensión judía en la represión durante el gobierno militar en la Argentina (1976–1983)," in Leonardo Senkman ed., *El Antisemitismo en la Argentina* (Buenos Aires: Centro Editor de América Latina), 235–273.

Krupnik, A. (2011) "Cuando Camino Al Kibbutz Vieron Pasar Al Che. Radicalización Política y Juventud Judía: Argentina 1966–1976," in Emmanuel Kahan, Laura Schenquer, Damian Setton, and Alejandro Dujovne eds., *Marginados y Consagrados. Nuevos Estudios Sobre La Vida Judía En Argentina*, Buenos Aires: Lumiere, 311–27.

Manzano, V. (2014) *The Age of Youth in Argentina*, Chapel Hill: The University of North Carolina Press.

Meyer, N.F. (1976) "Latin America Argentina Domestic Affairs," *American Jewish Year Book*, New York: American Jewish Committee, 262–86, http://www.ajcarchives.org/AJC_DATA/Files/1976_8_LatinAmerica.pdf (accessed May 1, 2019), 363.

Moyano, M.J. (2012) *Argentina's Lost Patrol: Armed Struggle, 1969–1979*, New Haven, CT: Yale University Press.

Padrón, J.M. (2017) *Ni yanquis, ni marxistas! Nacionalistas: Nacionalismo, militancia y violencia política: el caso del Movimiento Nacionalista Tacuara en la Argentina, 1955–1966*, La Plata: UNLP.

Pittaluga, R. (2006) La memoria según Trelew. *Sociohistórica*, nos. 19–20 (2006): 81–111, http://www.memoria.fahce.unlp.edu.ar/art_revistas/pr.3610/pr.3610.pdf (accessed January 4, 2021).

Rein, R. (2001) "The Eichmann Kidnapping: Its Effects on Argentine-Israeli Relations and the Local Jewish Community," *Jewish Social Studies* 7, no. 3: 101–30, https://doi.org/10.1353/jss.2001.0016.

Rein, R. (2002) *Argentina, Israel, and the Jews. Perón, the Eichmann Capture and After*, Bethesda: University Press of Maryland.

Rein, R. (2015) *Los Muchachos Peronistas Judíos: Los Argentinos Judíos y El Apoyo Al Justicialismo, Los Argentinos Judíos y El Apoyo Al Justicialismo*, Buenos Aires: Sudamericana.

Rein, R. (2018) "Jewish Self Defense in 1960s and Early 1970s Argentina: A Mirror Image of Anti-Semitic Violence?" *Israel* (Studies in Zionism and the State of Israel) [Hebrew], no. 25: 227–55.

Rein, R., and I. Diner (2012) "Unfounded Fears, Inflated Hopes, Passionate Memories: Jewish Self-Defence in 1960s Argentina," *Journal of Modern Jewish Studies 3*: 357–76, https://doi.org/10.1080/14725886.2012.722766.

Rein, R., and C. Panella, eds., (2013) *La Segunda Línea Liderazgo Peronista 1945–1955*, Buenos Aires: Universidad Nacional de Tres de Febrero.

Schenkolewski-Kroll, S. (2011) "Ideology and Propaganda in the Collective Memory's Construction: Zionism and Communism in Argentina," in August Grabski ed., *Rebels against Zion: Studies on the Jewish Left Anti-Zionism*, Warsaw: Jewish Institute of History.

Senkman, L. (2006) "Populismo y Empresarios Judíos: Actuación Pública de Horacio Lafer y José B. Gelbard Durante Vargas y Perón," *Araucaria: Revista Iberoamericana de Filosofía, Política y Humanidades 15*: 46–76.

Seoane, M. (2002) *El Burgués Maldito: Los Secretos de Gelbard, El Último Líder Del Capitalismo Nacional*, Buenos Aires: Editorial Sudamericana.

Verbizky, B. (1972) *Etiquetas a Los Hombres*, Buenos Aires: Editorial Planeta.

Visacovsky, N. (2015) *Argentinos, judíos y camaradas: Tras la utopía socialista*, Buenos Aires: Biblos.

Vitto, C. (2012) "Plan Económico Del Tercer Gobierno Peronista: Gestión de Gelbard (1973–1974)," *Problemas Del Desarrollo 171*, no. 43, 111–34.

Weinstein, A.E., and E. Toker (1999) *Trayectoría de una Idea: Nueva Sión 50 años de periodismo judeo-argentino*, Buenos Aires: Fundación Mordejai Anilevich.

CHAPTER 10

Revolutionary Eschatology: The Argentine Ejército Guerrillero del Pueblo and the Secularization of Religious Traditions

Lukas Böckmann

On November 4, 1963, a young Argentine film student, Adolfo Rotblat stood trial in a revolutionary tribunal. Only a few weeks earlier he had left his parental home in the Argentine capital where he bade farewell to his sister Martha Schapiro with an announcement that was intended to signal his unassailable resolve: "We shall return triumphant—or we shall not return at all!"[1] Together with three other young men, he spent the following days traveling to the extreme north of Argentina. In Aguas Blancas, he crossed the border to Bolivia, where he became one of the first recruits to join the emerging Ejército Guerrillero del Pueblo (people's guerrilla army, EGP).[2] Under the command of the Argentine journalist Jorge Ricardo Masetti (1929–64), the group was preparing for guerrilla operations in the Argentine Yungas. Conceived by Masetti and his close friend Ernesto Guevara (1928–67), the EGP was the first attempt to ignite an armed revolution in both men's homeland, following the Cuban example, with direct support from Havana. With Peronism proscribed since 1955, the Socialist and Communist parties became marginalized and, following the overthrow of Arturo Frondizi by the military in 1962, the objective conditions for a revolution seemed ideal.[3]

In September 1963, the EGP managed to cross the border undetected and over the following weeks set up camps, arms depots, and clandestine transportation routes near the small town of Colonia Santa Rosa. At the same time, the young guerrilla group began to develop an urban support network in Córdoba and Buenos Aires among Communist dissidents, who supplied the group with

1 Author's interview with Martha Schapiro by correspondence, December 2017; Daniel Ulanovsky Sack, "La olivdada guerrilla del Che en la Argentina," *Viva*, March 22, 1998.

2 Causa Penal, "Asociación ilícita, intimidación pública, delitos contra la seguridad de la nación, atentado y resistencia a la autoridad con muerte misma y homocidio," Cuerpo V, folio 914, 924–26.

3 Ciro Bustos, *Che Wants to See You: The Untold Story of Che Guevara* (New York: Verso, 2013), 48–51; Jon Lee Anderson, *Che: A Revolutionary Life* (London/New York: Bantam Books, 1997), 537.

 | DOI:10.1163/9789004462540_011

provisions, equipment, and above all recruits. By the end of the year, the group had grown to about thirty fighters. However, this development in the jungle of Salta increasingly attracted the attention of the Argentine security services.[4] In March, the Gendarmería began to cordon off large parts of the mountainous area. The brief skirmishes that ensued between the guerrilla group and the security forces claimed two lives on each side. Another six guerrillas succumbed to accidents or undernourishment over the following weeks as they were pursued through the jungle. Only a few were able to escape being captured by the government forces. Eighteen survivors were apprehended and sentenced to several years of imprisonment. All traces of Masetti and a companion were lost in the mountains of Salta, where they apparently died.[5]

Hopes for a Socialist revolution implied in Rotblat's parting words were dashed. However, Rotblat's fate differed from that of his comrades. A few weeks after his enlistment in the EGP, it was clear that the young film student was not cut out for the pressures of guerrilla life. By October, Rotblat was physically enervated and descending into a state of apathy. Desertion by a member at the end of the month led to a marked rise in Masetti's fear of further absconding and to mistrust of Rotblat specifically. The group's military code, as he declared shortly thereafter, would now be applied with full force to cowards and the weak.[6] Over the following days, Masetti was unwilling to remove Rotblat from combat operations, despite suggestions to that effect from other guerrilla leaders.[7] Rather, he reacted to Rotblat's physical and emotional decline with authoritarian severity, declaring that an example needed to be made in order to combat the moral decline of the group and to liberate them from such corrosive influence.[8] On November 4, 1963, Masetti formed a revolutionary tribunal and accused Rotblat of planning to desert. According to the military statutes of the group, this amounted to a capital break with the guerrillas' moral and

4 "Más detalles de la detención de 7 detenidos," *Clarín*, March 3, 1964, 19; "Guerrilleros: sigue la búsqueda," *Clarín*, March 17, 1964, 17.

5 Causa Penal, Cuerpo II, folios 306, 310; Cuerpo III, folio 480; Sergio Bufano and Gabriel Rot, "La guerrilla salteña: Entrevista a Héctor Jouvé," *Lucha Armada 1*, no. 2 (March–May 2005): 46–61; Alberto Oscar Fernández, "Crónica de una guerrilla, I Parte," *Revista de Gendarmería Nacional*, no. 8 (1965): 56–69, and "II Parte," *Revista de Gendarmería Nacional*, no. 9, (1965): 19–29; Gabriel Rot, *Los orígenes perdidos de la guerrilla en la argentina* (Buenos Aires: Waldhuter, 2010), 287–94.

6 Bustos, *Che Wants to See You*: 144; Causa Penal, Cuerpo V, folio 925–7.

7 Bustos, *Che Wants to See You*: 152; "Testimonio de Héctor Jouve," *La Intemperie* 15–16 (October–November 2004), reprinted in Pablo René Belzagui ed., *No matar. Sobre la responsabilidad* (Córdoba: Universidad Nacional de Córdoba, 2008), 14; author's interview with Alberto Castellanos, Havana, June 24, 2017.

8 Bustos, *Che Wants to See You*: 153; Anderson, *Che*: 578.

military code.[9] The tribunal found Rotblat guilty and, although he could hardly have been accused of more than weakness, he was sentenced to death. Bound up in his hammock, Adolfo Rotblat was shot to death by his comrades in the early morning of November 5.[10] Masetti justified the killing as a measure "that sanitizes the group's psychology, that liberates it from this thing that is corroding it."[11]

Due to the clandestine nature of the group and the loss of its written sources, including notes or diaries,[12] the Rotblat case lacks evidence needed for a detailed reconstruction. Nonetheless it is part of a broader picture. On the very night that Rotblat was executed, another group of recruits arrived at a camp nearby. Without knowledge of the shooting, they integrated into the ranks of the EGP during the next weeks. One of their number, Henry Lerner, soon felt singled out and mistrusted by Masetti. On Christmas Eve, the tensions between them culminated, with Masetti accusing Lerner of planning to desert. Two months later, Masetti ordered Lerner to supervise two other members of the group selected for punishment in order to prove his revolutionary morality and reliability. One of them, Federico Frontini, had recruited his friend Adolfo Rotblat the previous August and then played a major role in organizing the EGP's support structure in Buenos Aires. Masetti accused Frontini of anti-revolutionary behavior and embezzlement. He ordered him to return to Salta from Buenos Aires, and separated him from the rest of the fighters. The second man under surveillance was an eighteen-year-old bank employee, Bernardo Groswald, who had joined the guerrillas at the beginning of the year. Like Rotblat, he could hardly have been reproached for more than having broken down emotionally as well as physically.[13]

After a week, Lerner reported to Masetti that Frontini was recoverable but had nothing positive to say about Groswald, who had spent days cowering in an improvised bunker.[14] On February 18, 1964, Masetti held a second trial for Groswald who was found guilty of intentional desertion, insubordination, neglecting military equipment, and lacking revolutionary morale. Under pressure from Masetti, Groswald was sentenced to death. In the early morning of

---

9 Causa Penal, Cuerpo III, folio 422.

10 Causa Penal, Cuerpo III, folio 429; Bustos, *Che Wants to See You*: 153.

11 Anderson, *Che*: 578.

12 One exception is the *Diario de Guerra* included in the Causa Penal, which mentions the execution only very briefly. Causa Penal, Cuerpo III, folio 429: "On day 5 [of November] Cordobés [Héctor Jouve] left with the newcomers, and we shot Pupi [Adolfo Rotblat]."

13 Causa Penal, Cuerpo III, folio 431; Author's interviews with Lázaro Henry Lerner, Carabaña, April 2, 2016, and Ana María Kaufmann, Buenos Aires, November 11, 2016.

14 Interview with Lerner.

February 19, the commander ordered all guerrillas to convene in a glade near the camp and had Bernardo Groswald executed by firing squad.[15] To Lerner it was clear that Masetti wanted to set an example to improve group morale.[16]

Years later, after the EGP's dismantlement by the Gendarmería, Henry Lerner—who by then was living in exile in Spain—realized that Rotblat, Groswald, and he himself shared a biographical element: they all stemmed from Jewish families.[17] Furthermore, they had grown up in families that had immigrated to Argentina from Eastern Europe and had first settled in one of the agricultural settlements of the Jewish Colonization Association (JCA). Each had aligned with the Partido Comunista (PCA) and then joined the EGP. Six of Masetti's guerrillas came from Jewish families and shared similar backgrounds. Though a rather small empirical sample, it should be noted that Jews comprised less than 2 percent of the total population, but about 20 percent of the EGP.[18] Their histories reveal much about Jewish experience in Argentina, Jewish-Argentine ties to the labor movement, their complex links with the PCA, the affinity of later generations for the revolutionary movements, and the hidden religious basis of its ideology. The trajectories of Adolfo Rotblat and Henry Lerner reflect the biographies of many Jewish-Argentines who joined the revolutionary Left in the late 1960s and early 1970s.

1 A New Dawn: The Attraction of the Cuban Revolution in the Early 1960s

In mid-1958, the Buenos Aires-based station Radio El Mundo broadcast a feature that caused a stir among the politicized youth of the country. In its four episodes a little-known journalist named Jorge Ricardo Masetti recounted how he had traveled to Cuba earlier that year and visited Fidel Castro and Ernesto Guevara in their hideout, while they were still fighting against Cuban dictator Fulgencio Batista.[19] Henry Lerner, who had just started studying medicine

---

15 Causa Penal Cuerpo III, folio 431, 435, 438; "Guerrilleros fusilados," *Clarín*, May 14, 1964, 20; Interview with Lerner; Bufano and Rot, "La guerrilla salteña": 46–61.

16 Interview with Lerner.

17 Anderson, *Che*: 591; Interview with Lerner.

18 Of about thirty fighters in the EGP, at least six were of Jewish origin. In 1964, there were slightly more than 300,000 Jews living in Argentina, out of a total population of 22,045,000, meaning a proportion of about 1.4 percent. See Sebastian Klor, *Between Exile and Exodus: Argentinian Jewish Immigration to Israel, 1948–1967* (Detroit: Wayne State University Press, 2017), 24; United Nations, *Demographic Yearbook 1964* (New York, 1965), 123.

19 Juan Martín Guevara and Armelle Vincent, *Mein Bruder Che* (Stuttgart: Tropen, 2018), 149.

at Córdoba University, was electrified. A member of the Féderación Juvenil Comunista (FEDE), he was so impressed that he immediately acquired a copy of the report *Los que luchan y los que lloran* (Those who fight and those who cry), when it appeared in print in October of that year,[20] and elevated it, as he later recalled, to his "Bible."[21]

Lerner, like so many of his peers, followed the subsequent triumph of the Cuban Revolution with overwhelming enthusiasm. Fidel Castro's victorious entry into Havana in January 1959 and the regime's subsequent consolidation of power had proven before the eyes of the entire world that revolution was still possible despite the circumstances of the Cold War. In contrast to Cuba, the Argentine workers' movement seemed fragmented and increasingly leaderless.[22] Following the ousting of Juan Domingo Perón in 1955 and the proscription of Peronism, which had hitherto succeeded in uniting the great mass of the lower segments of society, a political vacuum had emerged left of center.[23] The PCA, as the self-declared representative of the Argentine workers' movement, was unable to fill this void.[24] While young intellectuals and students were beginning to look toward Cuba, the PCA was somewhat dismissive. Nevertheless, in 1960, after Lerner had been sent to the countryside to organize and politicize rural communities, the party assigned him to organize "Brigades for the Defense of Cuba" and to establish contacts with young Communists in support of the revolution.[25] But the party continued to orient itself less toward pressing developments in Latin America and clung to the general policy of peaceful coexistence decreed by Moscow. It insisted that the path to Socialism could be brought about only by a mass movement that extended into parliament and not through a violent uprising, even when the democratically elected president Arturo Frondizi was deposed by yet another

20 Jorge Ricardo Masetti, *Los que luchan y los que lloran: El Fidel Castro que yo ví* (Buenos Aires: Freeland, 1958).

21 Interview with Lerner.

22 Carlos Altamirano, *Peronismo y cultura de izquierda* (Buenos Aires: Siglo XXI, 2011), 63–65; María Cristina Tortti, "Izquierda y 'nueva izquierda' en la Argentina: El caso del Partido Comunista," *Sociohistórica*, no. 6 (1999): 221–32.

23 David Rock, *Argentina 1516–1987: From Spanish Colonization to Alfonsín* (Berkeley: University of California Press, 1987), 334–36; Monica Gordillo, "Protesta, rebelión y movilización: De la resistencia a la lucha armada, 1955–1973," in Daniel James ed., *Nueva historia argentina: Violencia, proscripción y autoritarismo* (Buenos Aires: Sudamericana, 2003), 333.

24 Ernesto Giudici, "Neocapitalismo, neosocialismo, neomarxismo," *Cuadernos de Cultura*, no. 50 (November–December 1960): 8–44.

25 Interview with Lerner.

military coup in March 1962.[26] Particularly following the thwarted Bay of Pigs invasion in April 1961, the Cuban espousal of Marxism-Leninism and Castro's unshakable certainty that revolution was imminent across all of Latin America turned the Cuban leadership into a beacon for the younger generation.[27] Castro and his revolutionaries were epitomized as the alternative to the orthodoxy of the PCA.[28]

Catalyzed by Castro's revolution, numerous party members turned their backs on the PCA in the early 1960s.[29] Probably the most paradigmatic split occurred in Lerner's hometown of Córdoba. A circle of young intellectuals from the FEDE concentrated around José María Aricó, Oscar Del Barco, and Héctor Schmucler, had founded the theoretical and cultural journal *Pasado y Presente* (Past and Present) in spring 1963. Following the philosophy of Antonio Gramsci, they questioned the PCA's policy of peaceful coexistence, which the authors viewed as idle stagnation.[30] Instead, they called for an action strategy based on the opportunities presented by contemporary events. When the leadership of the PCA reacted to the publication of the first edition of *Pasado y Presente* by expelling the editors,[31] the conflict that had been brewing for some time between the orthodox party leadership and the younger generation came to a head. Shortly afterward, almost the entire university section of Córdoba, which numbered some 140 members, quit the party.[32]

By then, Adolfo Rotblat was living in Buenos Aires and supported the FEDE. PCA cadres responsible for the youth organization accused him of being an

---

26 Victorio Codovilla, "Por la acción de masas hacia la conquista del poder: Informe rendido en nombre del Comité Central ante el XII° Congreso del Partido Comunista de la Argentina, 22 de Febrero de 1963," Buenos Aires, 1963, 10, 24.

27 Raúl Burgos, *Los gramscianos argentinos: Cultura y política en la experiencia de* Pasado y Presente (Buenos Aires: Siglo XXI, 2004), 70.

28 Boris Goldenberg and Klaus Eßer, *Zehn Jahre kubanische Revolution* (Hanover: Verlag für Literatur und Zeitgeschehen, 1969), 71.

29 Tortti, "Izquierda y nueva izquierda": 230–32; Claudia Hilb and Daniel Lutzky, *La nueva izquierda argentina, 1960–1980: Política y violencia*, (Buenos Aires: Centro Editor de América Latina, 1984).

30 Raúl Olivieri, "El problema del determinismo en el materialismo dialéctico," *Cuadernos de Cultura*, no. 58 (July–August 1962): 11–30; Oscar del Barco, "Notas sobre Antonio Gramsci y el problema de la 'objetividad'," *Cuadernos de Cultura*, no. 59 (September–October 1962): 29–41; Raúl Olivieri, "El materialismo dialéctico y la objetividad," *Cuadernos de Cultura*, no. 60 (November–December 1962): 23–39; Oscar del Barco, "Respuesta a una crítica dogmática," *Cuadernos de Cultura*, no. 63 (May–June 1963): 34–57.

31 Laura Prado Acosta, "El Partido Comunista argentino y la ruptura con los 'muchachos' de la revista *Pasado y Presente*," *Prismas: Revista de historia intellectual*, no. 18 (2014): 185–88.

32 Author's interview with Héctor Schmucler and Oscar Del Barco, Córdoba, November 3, 2016; Burgos, *Los gramscianos argentinos*: 77.

undercover agent of the security services sent to spy on it, and this became the pretext for revoking his party membership. Rotblat's girlfriend at the time considered it inconceivable that there was any basis for the claims against him.[33] However, Rotblat had become alienated from the party even before the procedures against him began. He had been increasingly fascinated by the Cuban Revolution, as a result of which he had been drawing closer to the dissident circle around the young intellectual Juan Carlos Portantiero, who had joined the party's youth organization in 1952 and had since then risen to an editorial position in the party journal *Cuadernos de Cultura* (notebooks of culture).[34] When Portantiero published an article in the first edition of *Pasado y Presente* openly positing the existence of a "revolutionary situation" in Argentina and demanding that theory needed to be combined with practice,[35] he was also expelled from the party along with some of his close friends.[36]

This was around the time that Jorge Ricardo Masetti and the core group of what would become the EGP relocated to the Salta wilderness, following military training in Cuba and visits to Czechoslovakia and Algeria.[37] The commander of the group sent one of his confidants, the painter Ciro Bustos, to Córdoba in order to establish contact with the circle around *Pasado y Presente* with the aim of setting up an urban support network and of recruiting new fighters for this first guerrilla group in Argentina.[38] Although the editors remained reserved regarding any involvement in warfare, they gladly established contact between him and the group Henry Lerner was part of which was already preparing for armed struggle. During the next months those surrounding the journal editorship in Córdoba and Portantiero in Buenos Aires became central links between the guerrilla group and the milieu of dissident Communists, providing the vast majority of recruits who joined the EGP in Salta. When in August 1963 Ciro Bustos proclaimed that "the guerrilla has begun," his words, as Héctor Schmucler summarized aptly, had a "metaphysical effect, as though he had said 'an angel has been born'."[39]

33 Author's interview with Lucila Geralnik, December 21, 2017.
34 Interview with Kaufmann.
35 Juan Carlos Portantiero, "Politica y clases sociales en la Argentina actual," *Pasado y Presente*, no. 1 (April–June 1963): 18–23.
36 Estela Morales Campos, "Juan Carlos Portantiero," *Archipélago*, no. 57 (2007): 23; Hernán Camarero, "Juan Carlos Portantiero, 1934–2007," *Prismas 11*, no. 2 (2007): 305; Burgos, *Los gramscianos argentinos*: 52, 70, 78.
37 Rot, *Orígenes perdidos*: 158–61.
38 Bustos, *Che Wants to See You*: 128.
39 Interview with Schmucler and Del Barco.

2 Transformed Hope: Jewish Experience between Emigration, Party Communism, and Guerrilla Warfare

Henry Lerner was born in 1938 in Cosquín, a small town near Córdoba; Adolfo Rotblat was born two years later in Buenos Aires, in the capital's most emblematic Jewish neighborhood, Villa Crespo. Like other Jewish EGP members, their grandparents had originated in the traditional world of the Eastern European shtetl and had immigrated to the Rio de la Plata at the turn of the century.[40] With only a few exceptions, they settled in the agrarian colonies established by the JCA, and not in the country's capital. In 1905, Rotblat's grandfather, Hilel Simkin, became one of the founding members of the Jewish colony Rivera.[41] But shortly afterward, he relocated his family to Buenos Aires.

Rotblat's paternal grandfather, Boris Rotblat, had settled immediately in Buenos Aires on his arrival from the Russian Empire. He contributed significantly to the establishment of the local Ashkenazi community, intermittently running Jewish educational institutions and serving as a founding member of the Gran Templo Paso congregation.[42] Over the years, his religious adherence abated, but his family continued to observe Jewish holidays and practices. Also, the Russian language and Yiddish continued to be present in the family household.[43] In accordance with traits characterizing the second generation, Adolfo spent the mornings during his childhood attending a state primary school in Villa Crespo and his afternoons learning English, Hebrew, and Yiddish at the district's Escuela Scholem Aleijem, which had links to Bundism and the Jewish workers' movement.[44] In the 1950s he frequently visited the Circulo Israelita cultural club in Villa Crespo where he met his close friend Marcelo Cosín. In 1955, Cosín recalls, they had cheered the overthrow of Juan Domingo

40 Interviews with Lerner and Shapiro; Israel Werthein, "85 años: Toda una vida," unpublished manuscript (Harvard College Library, 1990); Haim Avni, *Argentina y la historia de la immigración judía, 1810–1950* (Jerusalem: Magnes, 1983).

41 Gregorio Verbitsky, *Rivera. Afán de medio siglo* (Buenos Aires: Comisión del Cincuentenario de Rivera y sus Colonias, 1955), 44, 81.

42 As late as 1943, a commemorative publication praising his far-reaching activities within the Jewish community appeared: Mauricio Silbert and Isaac Stambler, *A Idisher klal tuer in Argentine, h. Boris Rotblat. Baerung far zayn frukhtbare aktivitet in dem groysen templ fun Idishn hilfs un kult farayn "Talmud Torah harishonah ve-Dr. Hertsel"* (Buenos Aires: Fraynt, 1943); interview with Shapiro.

43 Interview with Schapiro.

44 Nerina Visacovsky, *Argentinos, judíos y camaradas: Tras la utopía socialista* (Buenos Aires: Biblos, 2015), 95–96; Klor, *Between Exile and Exodus*: 33–34; Frank Wolff, *Neue Welten in der Neuen Welt: Die Transnationale Geschichte des Allgemeinen Jüdischen Arbeiterbunds 1897–1947* (Cologne: Böhlau 2014), 388.

Perón, whom they considered an authoritarian caudillo in the tradition of Italian Fascism. A year after the start of de facto president Pedro Aramburu's administration, and influenced by a group of exiled Spanish intellectuals and poets, the two friends joined the Communist youth organization's branch in Villa Crespo.[45]

As in Rotblat's family, Jewish belonging and traditions were preserved throughout the various stages of migration in Henry Lerner's family as well, clearly distinguishing them from the surrounding culture. Lerner's father Jacobo, born in 1908 in Colonia Mauricio, lived a childhood defined by privation, agricultural labor, provincial isolation, and anti-Semitic discrimination, about which Henry learned only much later when the family moved to Cosquín.[46] After his parents settled in the latter, his grandparents continued to read the Jewish community periodicals *Di Presse* and *Die Jidishe Tsaitung*, spoke Yiddish, and prepared traditional dishes on festive occasions. Lerner's parents opted for a secular education for their children and, like his Jewish contemporaries at primary school, Lerner and his brother were not allowed to attend religious classes. They did not go to the local synagogue or have a religious wedding. At the beginning of the 1920s, Lerner's father joined the PCA's Jewish section and soon gained publicity as one of its local officials.[47] The establishment of the Idsektie, the Jewish section of the PCA, which emerged from a schism in the Bund in 1920, highlighted the striking affinity between Jewish Argentines and working class politics at the beginning of the twentieth century.[48] Despite the marginal proportion of Jews within the total population of the country, the Idsektie would, for the duration of its existence, constitute the second-largest idiomatic section of the Communist Party, following the Italian one.[49] However, it emerged less in recognition of Jewish particularism than as a means of reaching out to the Yiddish-speaking sector of the working masses. Although the PCA, at least in its early years, did not demand any immediate or comprehensive acculturation from its members, it

45 Author's phone interview with Marcelo Cosín, July 19, 2020.

46 Interview with Lerner.

47 Ibid.

48 Daniel Kersfeld, *Rusos y rojos: Judíos comunistas en los tiempos de la comintern* (Buenos Aires: Capital Intelectual, 2012), 104–5; Silvia Schenkolewski-Kroll, "The Jewish Communists in Argentina and the Soviet Settlement of Jews on Land in the USSR," *Jews in Eastern Europe* 49 (2002): 81–82.

49 According to Schenkolewski-Kroll, 54 percent of party members were foreigners, of whom 28.4 percent belonged to the Italian and 13.7 percent to the Yiddish sectors. There were also German, Yugoslav, Czechoslovak, Armenian, Hungarian, and Polish sections, though these made up altogether less than 12 percent. Schenkolewski-Kroll, "The Jewish Communists": 82.

was nevertheless grounded in the conviction that the final emancipation of the Jews would only be possible through their absorption into the workers' movement, whose members in turn were to be liberated through Socialism.[50] As Jewish affiliation to Argentine society was increasingly questioned by Catholic nationalist sectors during the first decades of the century, many Jewish immigrants who had already encountered Bundism, Socialism, or anarchism in Eastern Europe, felt attracted to the promise of participation on an equal footing in a community transcending their former sphere of belonging.

World War II, and especially the destruction of European Jews by National Socialist Germany, emphasized those ties. Henry Lerner remembers celebrating the fall of Berlin with other Socialists and Communists in May 1945. At the climax of the festivities a group of party officials held a parade during which they sang the "Marsellaise" and waved Allied flags. The red banner of the Soviet Union was held by his father Jacobo.[51] But in the 1950s, the affinity between the party and the descendants of Jewish immigrants became increasingly fragile. The anti-Jewish show trials in the Soviet Union were condemned by the umbrella organization of Argentine Jewish organizations, the Delegación de Asociaciones Israelitas Argentinas (DAIA). The PCA, represented on the democratically elected board of the DAIA by a large number of delegates, refused to support the declaration agreed to by all the other affiliated organizations and denounced the public condemnations of the show trials as anti-Soviet propaganda directed by the United States.[52] The party's loyalty to Moscow had now become an issue of open conflict, which culminated in the de facto exclusion of all Communist representatives from DAIA in 1953.[53]

That same year Henry Lerner, together with José María Aricó, who later became editor of *Pasado y Presente*, founded the Cordobá branch of the FEDE. For Lerner, the image of the Soviet Union as savior of the world from National Socialism during World War II and as a protector of the Jews was still valid. That image proved equally true for Héctor Schmucler and Alberto Szpunberg, who became part of the urban network of the EGP in 1963; they hailed from

50 Claudia Bacci, "Las políticas culturales del progresismo judío argentino. La revista Aporte y el ICUF en la década de 1950," *Politicas de la Memoria*, no. 5 (2004/2005): 160–162; Edgardo Bilsky, "Etnicidad y clase obrera: La presencia judía en el movimiento obrero argentino," *Estudios Migratorios Latinoamericanos*, no. 4: 41–42.

51 Interview with Lerner.

52 Lawrence D. Bell, "Bitter Conquest: Zionists against Progressive Jews and the Making of Post-War Jewish Politics in Argentina," *Jewish History* 17, no. 3 (2003): 294–302.

53 Silvia Schenkolewski-Kroll, "Ideology and Propaganda in the Collective Memory's Construction: Zionism and Communism in Argentina," in August Grabski ed., *Rebels against Zion: Studies on the Jewish Left Anti-Zionism* (Warsaw: Jewish Institute of History, 2011), 136.

Eastern European Jewish families and joined the Communist youth organization during the 1950s.[54] It was only at the end of the decade, after the Communist Party of the Soviet Union had begun to deal with criticism of the recent past, as outlined in Khrushchev's speech "On the Cult of Personality and Its Consequences," that Lerner, Schmucler, and Szpunberg slowly began to question the policies of the PCA. But the Argentine party refused to address Stalinism in its own ranks. At the time that the guerrilla group was first being formed, the renowned Jewish psychoanalyst José Bleger was publicly attacked by the Argentine party leadership and eventually expelled from the PCA following a lengthy article he wrote detailing his 1962 trip to the Soviet Union and expressing fundamental criticism of the Communist regime over its treatment of the Jewish population.[55] Both, Szpunberg and Lerner later cited the orthodox refusal of the PCA's leadership to openly discuss the Prague trials as one of the reasons for their increasing disappointment with the party.[56]

In the early 1960s, anti-Jewish prejudices in Argentina were on the rise. Immediately following reports of Adolf Eichmann's arrest in May 1960, the Jewish population was confronted with a violent wave of anti-Semitic attacks by right-wing nationalists. Adolfo Rotblat was studying at the University of Buenos Aires' Law School at the time and experienced the threats at close range. One of his fellow students was briefly kidnapped and tortured by members of Tacuara, a group of nationalist Catholics inspired by German Fascism and anti-Semitism. The victim was the descendant of a Jewish mother and a Catholic father. Since he considered himself Jewish, the group had decided to make an example of him as a betrayer of the Catholic faith.[57] Rotblat and his close friend Marcelo Cosín were distressed by the event. Although they were neither religious nor connected to Jewish community institutions, they still adhered to a mainly cultural Jewish self-understanding, as Cosín recalls.[58]

---

54 Interview with Héctor Schmucler and Oscar del Barco; author's interview with Alberto Szpunberg, Buenos Aires, October 25, 2016.

55 José Bleger, "Los judíos en le Unión Soviética," *Nueva Sión* (January 17, 1963), reprinted in Ana E. Weinstein and Eliahu Toker, *Trayectoría de una idea: Nueva Sión 50 años de periodismo judeo-argentino* (Buenos Aires: Fundación Mordejai Anilevich, 1999), 91–93.

56 Interviews with Szpunberg and Lerner.

57 Interview with Cosín; Raanan Rein, *Argentina, Israel, and the Jews: Perón, the Eichmann Capture and After*, (Bethesda: University Press of Maryland, 2002), 184–228; Rein, "Argentine Jews and the Accusation of 'Dual Loyalty'," in Rein, *Argentine Jews or Jewish Argentines? Essays on Ethnicity, Identity, and Diaspora* (Leiden: Brill, 2010), 169–93; Daniel Gutman, *Tacuara: Historia de la primera guerrilla urbana argentina* (Buenos Aires: Vergara, 2003).

58 Interview with Cosín.

Despite Rotblat's consternation over these events, the extent to which they influenced his distancing from the Communist Party remains unclear. Instead of clinging to a party that vehemently rejected any reference to Soviet missteps,[59] he turned, like Henry Lerner, in a direction in which salvation was cast as national liberation. As a former medical student suffering from asthma, Rotblat began identifying with Ernesto "Che" Guevara as the ideal of the New Man he sought to become himself.[60] According to Lerner, they joined the EGP "because we believed in [the revolution], we believed in the New Man, we believed in the example of Che … The image of Che is like the crucifix, or like the Virgin Mary, it is the banner of the Revolution."[61]

3 Converging Traditions: Religious Roots and Revolutionary Eschatology

Lerner's mindset was common among adherents of the emerging Argentine New Left. The traditional parties and labor groups, including the PCA with its rigid commitment to the Soviet policy of peaceful coexistence, appeared toothless. In contrast, Castro and Guevara managed to reignite their revolutionary faith. Their specific interpretation of the ideal of the New Man emphasized subjective will and devotion as central elements in the reshaping of society. While events following Eichmann's capture had demonstrated that being of Jewish origin in Argentina still made one vulnerable to attacks, developments in the Soviet Union gave the impression that the Communist regime had failed to spawn a New Man disengaged from former bondages. Moreover, the trials in Prague had demonstrated that the burden of origin had been tragically resurrected. In contrast, at the same time Ernesto Guevara had become not only the political icon of the emerging revolution but the Latin American fulfillment of the utopian New Man.

In the early 1960s Guevara began developing the notion that if societal change neglected the people's consciousness, it would serve only as a measure for redistribution; he sought an all-encompassing revolutionary mindset.[62] In the first edition of *Pasado y Presente*, in 1964, implicitly dedicated to the EGP's struggle in Salta which José María Aricó had visited in January, Héctor

59 Rubén Sinay, *La invención del 'Antisemitismo Sovietico'* (Buenos Aires: Ediciones Tribuna, 1963).

60 Interview with Cosín.

61 Interview with Lerner.

62 Jean Daniel, "Une affaire de famille, où en est Cuba? Che Guevara a réspondu à Jean Daniel," *L'Express*, July 25, 1963, 9.

Schmucler summarized: "The revolution (as Che Guevara reminded us in a report about Algeria) is not aimed at redistribution, but at the creation of a new, unalienated, free man."[63] Guevara himself expanded on this idea only later, in 1965, in an open letter to the Uruguayan weekly *Marcha*, titled "El socialismo y el hombre en Cuba." He wrote that the revolution had to uproot "bad seeds" in order to create "a new human being who would represent neither the ideas of the 19th century nor those of our own decadent and morbid century. What we must create is the human being of the 21st century."[64]

The motif of the New Man invokes historical semantics that can be traced back to early Christian texts. After his conversion, Paul the Apostle wrote in a letter to the Ephesians "that you put off, concerning your former conduct, the old man who grows corrupt according to the deceitful lusts, and be renewed in the spirit of your mind, and that you put on the new man which was treated according to God, in true righteousness and holiness."[65] For Paul, turning to Christ in conversion and baptism entailed the elimination of the Old Man and his world of darkness and replacing it with Christian enlightenment. As the renewed man already experienced fulfillment thanks to his faith in Christ—though eschatological salvation was still pending until the coming of the Kingdom of God—individual and universal refinement continued to be separated and subordinated to the will of God.[66]

In his methodological writings *Guerrilla Warfare*, Guevara conceptualized a guerrilla-fighter tied to Paul's writings. Prototypically embodied by himself, this heroic and ascetic fighter was willing to sacrifice himself while serving as the revolutionary vanguard of societal change.[67] His individual willpower, self-discipline, and obedience were likewise a driving force behind his political transformation and proof of the metamorphosis of the Old into the New Man. The *guerrillero heroico* was at the same time bearer of the revolution and forestaller of its telos. It is unclear whether Guevara was aware that the motif of the New Man originated in Paulinian thought. However, he tended to use Christian metaphors in private correspondence and deemed himself a "little wandering prophet, who announces with a resounding voice the coming of the Last Judgement."[68] Likewise, he signed his open letter to *Marcha*: "Receive our

63 Héctor Schmucler, "Problemas del tercer mundo," *Pasado y Presente*, no. 4 (January–March 1964): 288.

64 Ernesto Guevara, "El socialismo y el hombre en Cuba," *Marcha*, March 12, 1965, 14–15, 20.

65 Eph. 4, 22–24.

66 Gottfried Küenzlen, *Der Neue Mensch: Eine Untersuchung zur säkularen Religionsgeschichte der Moderne* (Frankfurt: Suhrkamp, 1997), 54–56.

67 Ernesto Guevara, *Guerrilla: Theorie und Methode* (Berlin: Wagenbach, 1968), 52.

68 Ernesto Guevara, *Aquí va un soldado de América*, (Buenos Aires: Planeta, 1987), 101–02.

ritual greeting—which is like a handshake or an 'Ave María Puríssima'. Patria o muerte."[69]

When Guevara began instructing Jorge Ricardo Masetti to be the commander of the EGP in 1961, the latter had only recently experienced a Pauline conversion. Born in a suburb of Buenos Aires to Italian immigrants in 1929, Masetti was raised in a Catholic household.[70] In the mid-1940s, at age fifteen, he joined the Unión Nacional de Estudiantes Secundarios (UNES), the youth organization of the proto-fascist Catholic Alianza Libertadora Nacionalista (ALN), which at the time engaged in violent confrontations with its political opponents on the streets of Buenos Aires.[71] The ALN constituted a decisive point of intersection between nationalism, Peronism, and a Catholicism that repeatedly employed anti-Jewish tropes. This ideological mixture was condensed in two subtly interrelated slogans introduced by members of the ALN to the streets of the capital, "Death to the Jews" and "Long live Jesus Christ!"[72] As it began aligning itself with Peronism, and particularly following its departure from excessively right-wing nationalist positions and the break with the Catholic authorities that this entailed, Masetti turned his back on the group. He began a journalistic career with the *Tribuna*, a daily newspaper that emerged from that milieu, and worked until the 1950s for publications which, like *Cabildo*, had either an anti-Semitic profile or, like *Pregón*, a radical right-wing nationalist one, but which above all were organs of political Catholicism.[73] Although Masetti increasingly gained a journalistic foothold outside of this spectrum, he would retain his deeply-felt Catholic convictions until his departure for Havana, visiting Sunday mass every week without exception. On his return from Cuba, however, he confided to his sister-in-law, "I don't know whether I still believe in God."[74]

His 1958 report, *Los que luchan y los que lloran*, was not only an account of his Cuba journey, but also a testament to his political metamorphosis. As Masetti admitted in the preface, he had left for Havana filled with doubt. He

69 Guevara, "El socialismo": 20.

70 Rot, *Orígenes perdidos*: 31.

71 Hernán Vaca Narvaja, *Masetti: El periodista de la revolución* (Buenos Aires: Sudamericana, 2017), pos. 146–95; Rubén Furman, *Puños y pistolas: La extraña historia de la Alianza Libertadora Nacionalista, el grupo de choque de Perón* (Buenos Aires: Sudamericana, 2014), 11–12, 279.

72 Michael Goebel, *Argentina's Partisan Past: Nationalism and the Politics of History* (Liverpool: Liverpool University Press, 2011), 73; Graciela Ben-Dror, *The Catholic Church and the Jews: Argentina, 1933–1945* (Lincoln: University of Nebraska Press., 2008), 103, 164 and 227ff.

73 On *Cabildo*, see Ben-Dror, The Catholic Church: 103, 164, and 227ff; Furman, *Puños y pistolas*: 279; Rot, *Los orígenes perdidos*: 36–38.

74 Rot, *Orígenes perdidos*: 57.

had wanted to find out who the men were who were trying to topple Batista, whether they were receiving help from abroad, or "if Latin America [was witnessing] the intriguing exception of a revolution underway toward its triumph financed by its own people."[75] Yet with every step toward the rebels, his questions gave way to euphoria, especially when Guevara revealed to him that Castro was no Communist and that his movement was more in the way of a "revolutionary nationalist" one.[76] On his way to see Fidel Castro in March 1958, he spent a night in a provisional rebel camp. In the early morning, Castro appeared out of the darkness with a dim lamp to wake him.[77] Aside from the vision of the revolutionary leader appearing before him in this way, Masetti's portrayal of the encounter is layered with religious, symbolically-charged iconography. The semi-dilapidated hut in which he was sleeping turned out to have been an abandoned chapel, full of traces of its former religious use. Aside from a large, white wooden cross there was a small altar in one of the corners on which an image of the Virgin Mary as well as that of a mustached man shone forth beneath the wax of long extinguished candles.[78]

Regardless of whether this first encounter with Castro took place as described or was apocryphal, the scene had far-reaching significance. Masetti returned to Buenos Aires a committed supporter of the Cuban Revolution. He had also made so strong an impression on Castro and Guevara that they invited him to return to Havana in January 1959 to work for the new regime.[79] Together with Guevara, Masetti founded the Cuban news agency Prensa Latina, and served as its first director.[80] In 1961, after he had to leave his post due to the pressure of high ranking Cuban Communists, who wanted the agency to be directed by a Cuban, Guevara entrusted him with preparations for an independent guerrilla operation in Argentina. In keeping with the imagery of his report, Castro had guided Masetti out of the darkness and into the light of revolution.

Although the revolutionary regime regarded itself at least as anticlerical, the religious motifs present in both Lerner's and Masetti's accounts are not surprising. From the outset the Cuban Revolution was infused by religious iconography. When Masetti interviewed Castro initially, the rebel leader conveyed the myth of an apostolic number of twelve guerrilla survivors of the

75 Masetti, *Los que luchan*: 32–3.
76 Ibid.: 83.
77 Ibid.: 101.
78 Ibid.: 101.
79 Furman, *Puños y pistolas*: 281; Rot, *Orígenes perdidos*: 84; Anderson, *Che*: 385; Vaca Narvaja, *Masetti*: pos. 979–93.
80 José Bodes Gómez, *Los años precursores: Memorias de Prensa Latina (1959–1962)* (Havana: Prensa Latina, 2009), 34.

Granma landing, clinging to a biblical seven guns.[81] This widely disseminated claim turned out to have been allegorical.[82] Even so, the notion of a massively outnumbered group prevailing due to their indomitable faith remained. Soon after his victory the national and international media romanticized Castro as a Christ-like savior.[83]

Charged with the religious force of the Cuban Revolution, the EGP's attraction derived less from its program than from the eschatological charisma of ostensibly self-explanatory tropes of a new social order. In fact, it was precisely faith in the abstract ideals of National Liberation and the New Man mentioned by Lerner that guaranteed the inner cohesion of the EGP. In order to safeguard this faith belief from all doubt, Masetti demanded that his guerrillas swear an oath to him on their lives. In an article written before the establishment of the EGP, he had claimed that true revolutionaries were those who sacrificed their lives on the altar of the fatherland.[84] As though intended to demonstrate their willingness to make this sacrifice, all recruits were required to break immediately with their former civilian lives before joining the guerrillas. Donning the olive-green uniforms of the guerrilla, they handed their IDs over to a member of the leading cadre and adopted pseudonyms in place of their civilian names. The recruits' visual adaptation to the other members of the group and the extinction of all traces of their former lives may have reflected the conventional security measures of a subversive organization. At the same time, the related act of initiation also marked the final break with their former lives. This symbolic death of the civilian individual was followed—if a probationary phase lasting several weeks was passed successfully—by the recruits' resurrection as guerrillas who, with the ceremoniously declared oath "Revolution or Death," swore their allegiance to the EGP and its statutes on their lives.[85]

Compared with their share in the total population, the EGP exerted a disproportionate attraction to young men of Jewish origin. At the same time, the guerrilla group's commander, Jorge Ricardo Masetti, had in the early stages of his political activity, when he adhered rigorously to Catholicism, declared Jews and Communists to be Argentina's primary enemies. Thus, the guerrilla

81 Masetti, *Los que luchan*: 145.

82 Maurice Halperin, *The Rise and Decline of Fidel Castro: An Essay in Contemporary History* (Berkeley: University of California Press, 1972), 17; Lillian Guerra, *Visions of Power in Cuba: Revolution, Redemption, and Resistance, 1959–1971* (Chapel Hill: University of North Carolina Press, 2012), 16; Anderson, *Che*: 213.

83 Mario Kuchilan Sol, "Historia de un retrato," *Bohemia*, August 30, 1959, 50–51; "Fidel Castro: An die Wand," *Der Spiegel*, no. 27 (1960): 41.

84 Jorge Ricardo Masetti, "Que es un revolucionario?" [n.d.], Masetti, *Los que luchan*: 220–22.

85 Causa Penal, Cuerpo III, folio 436; Bufano and Rot, "Entrevista a Héctor Jouvé": 53.

group he headed combined traditions which in previous decades had still been dissonantly opposed to one another. The disproportionate presence of Jews and Catholics, whose religious ties increasingly disappeared from view, suggests that these bonds did not simply disappear but were rather suspended in a traditional sense in order to be preserved in a philosophy of history oriented toward the future.[86] Both the group commander's Catholic influences and Jewish experience appeared to coalesce in a new form through the image of the ascetic as well as virtuous New Man, and his embodiment in the heroic guerrilla. Due to the conversion experiences of its members, the group mechanisms, which were reminiscent of religious formations, and the utopian political promise of humankind's salvation through National Liberation to which the guerrillas had sworn their lives, the crystallization of the EGP bears the traits of a process of secularization through which formerly religious articles of faith were transferred to the realm of the profane. Indeed, the surviving members of the group would later point to this affinity themselves. The former political commissar of the EGP, Héctor Jouve, mentioned that members had discussed in prison the differences between their revolutionary aspirations and Christianity. These, according to him, were far fewer than the unrecognized similarities. After all, the overarching goal of the guerrilla had been the creation of "paradise on earth."[87]

Existing sources suggest that the Jewish origins of Adolfo Rotblat and Bernardo Groswald were not part of the revolutionary trials against them. Rather, their psychological and physical weaknesses eventually led to their executions. However, Masetti had demonstrated his willingness in other cases to allow exhausted recruits or those found to be unfit to return to the city. By contrast, he had repeatedly and pointedly bullied Rotblat and Groswald even before the tribunals were organized against them. Their consequent breakdown and disintegration exposed the fiction underlying the central narrative of the guerrilla as the harbinger of the New Man, which in turn raises doubts about the eschatological ideology of the guerrilla. The affinity between religious and revolutionary eschatology may have allowed Masetti to strengthen the guerrilla group morally by making an example of members whose Jewish origins marked them a history apart from the collective of the Catholic imagination.

---

86 Jan Gerber, *Karl Marx in Paris. Die Entdeckung des Kommunismus*, (Munich: Piper, 2018), 181; Dan Diner, *Aufklärungen. Wege in die Moderne*, (Stuttgart: Reclam, 2017), 42–43.

87 Daniel Ávalos, *La guerrilla del Che y Masetti en Salta, 1964: Ideología y mito en el Ejército Guerrillero del Pueblo—Testimonio de Héctor Jouve* (Córdoba: La Intemperie, 2005), 156, 199.

Bibliography

Interviews with Alberto Castellanos (Havana, June 24, 2017), Marcelo Cosín (July 19, 2020) Lucila Geralnik (December 21, 2017), Ana María Kaufmann (Buenos Aires, November 11, 2016), Henry Lázaro Lerner (Carabaña, April 2, 2016), Martha Schapiro (December 2017), Héctor Schmucler and Oscar del Barco (Córdoba, 3. November 2016), Alberto Szpunberg (Buenos Aires, October 25, 2016).

Newspapers: *Bohemia, Clarín, Der Spiegel, Hoy, L'Express, New York Times.*

Causa Penal, "Asociación ilícita, intimidación pública, delitos contra la seguridad de la nación, atentado y resistencia a la autoridad con muerte misma y homocidio," archived at Biblioteca J. Armando Caro, Cerrillos, Salta.

Altamirano, C. (2011) *Peronismo y cultura de izquierda*, Buenos Aires: Siglo XXI.

Anderson, J.L. (1997) *Che. A Revolutionary Life*, London: Bantam Books.

Ávalos, D. (2005) *La guerrilla del Che y Masetti en Salta, 1964: Ideología y mito en el Ejército Guerrillero del Pueblo—Testimonio de Héctor Jouve*, Córdoba: La Intemperie.

Avni, H. (1983) *Argentina y la historia de la immigración judía, 1810–1950*, Jerusalem: Magnes.

Bacci, C. (2004/2005) "Las políticas culturales del progresismo judío argentino. La revista Aporte y el ICUF en la década de 1950," *Politicas de la Memoria*, no. 5: 159–68.

Bell, L.D. (2003) "Bitter Conquest: Zionists against Progressive Jews and the Making of Post-War Jewish Politics in Argentina," *Jewish History 17*, no. 3: 285–308.

Belzagui, P.R., ed. (2008) *No Matar: Sobre la responsabilidad*, Córdoba: Universidad Nacional de Córdoba.

Ben-Dror, G. (2008) *The Catholic Church and the Jews: Argentina, 1933–1945*, Lincoln: University of Nebraska Press.

Bilsky, E. (1989) "Etnicidad y clase obrera: La presencia judía en el movimiento obrero argentino," *Estudios Migratorios Latinoamericanos*, no. 4: 27–47.

Bodes Gómez, J. (2009) *Los años precursores: Memorias de Prensa Latina (1959–1962)*, Havana: Prensa Latina.

Bufano, S., and G. Rot (2005) "La guerrilla salteña. Entrevista a Héctor Jouvé," *Lucha Armada 1*, no. 2 (March–May): 46–61.

Burgos, R. (2004) *Los Gramscianos argentinos Cultura y política en la experiencia de* Pasado y Presente, Buenos Aires: Siglo XXI.

Bustos, C. (2013) *Che Wants to See You: The Untold Story of Che*, Guevara, New York: Verso.

Camarero, H. (2007) "Juan Carlos Portantiero, 1934–2007," *Prismas 11*, no. 2: 305–308.

Codovilla, V. (1963) "Por la acción de masas hacia la conquista del Poder: Informe rendido en nombre del Comité Central ante el XII° Congreso del Partido Comunista de la Argentina, 22 de Febrero de 1963," Buenos Aires.

Del Barco, O. (1962) "Notas sobre Antonio Gramsci y el problema de la 'objetividad,'" *Cuadernos de Cultura*, no. 59, (September–October): 29–41.

Del Barco, O. (May–June 1963) "Respuesta a una crítica dogmática," *Cuadernos de Cultura*, no. 63: 34–57.

Diner, D. (2017) *Aufklärungen. Wege in die Moderne*, Stuttgart: Reclam.

Fernández, A.O. (1965) "Crónica de una guerrilla: I Parte," *Revista de Gendarmería Nacional*, no. 8: 56–69; "II Parte," *Revista de Gendarmería Nacional*, no. 9: 19–29.

Furman, R. (2014) *Puños y pistolas. La extraña historia de la Alianza Libertadora Nacionalista, el grupo de choque de Perón*, Buenos Aires: Sudamericana.

Gerber, J. (2018) *Karl Marx in Paris. Die Entdeckung des Kommunismus*, Munich: Piper.

Giudici, E. (November/December 1960) "Neocapitalismo, neosocialismo, neomarxismo," *Cuadernos de Cultura*, no. 50: 8–44.

Goebel, M. (2011) *Argentina's Partisan Past: Nationalism and the Politics of History*, Liverpool: University Press.

Goldenberg, B., and K. Eßer (1969) *Zehn Jahre kubanische Revolution*, Hannover: Verlag für Literatur und Zeitgeschehen.

Gordillo, M. (2003) "Protesta, rebelión y movilización: De la Resistencia a la lucha armada, 1955–1973," in Daniel James ed., *Nueva historia argentina: Violencia, proscripción y autoritarismo*, Buenos Aires: Sudamericana, 329–80.

Guerra, L. (2012) *Visions of Power in Cuba: Revolution, Redemption, and Resistance, 1959–1971*, Chapel Hill: The University of North Carolina Press.

Guevara, E. (1965) "El socialismo y el hombre en Cuba," *Marcha* (March 12): 14–15, 20.

Guevara, E. (1968) *Guerrilla. Theorie und Methode*, Berlin: Wagenbach.

Guevara, E. (1987) *Aquí va un soldado de América*, Buenos Aires: Planeta.

Guevara, J.M., and A. Vincent (2018) *Mein Bruder Che*, Stuttgart: Tropen.

Gutman, D. (2003) *Tacuara: Historia de la primera guerrilla urbana argentina*, Buenos Aires: Vergara.

Halperin, M. (1972) *The Rise and Decline of Fidel Castro: An Essay in Contemporary History*, Berkeley: University of California Press.

Hilb, C., and D. Lutzky (1984) *La nueva izquierda argentina, 1960–1980: Política y violencia*, Buenos Aires: Centro Editor de América Latina.

Kersfeld, D. (2012) *Rusos y rojos: Judíos comunistas en los tiempos de la comintern*, Buenos Aires: Capital Intelectual.

Klor, S. (2017) *Between Exile and Exodus: Argentinian Jewish Immigration to Israel, 1948–1967*, Detroit: Wayne State University Press.

Kuchilan Sol, M. (1959) "Historia de un retrato," *Bohemia* (August 30): 50–51.

Küenzlen, G. (1997) *Der Neue Mensch. Eine Untersuchung zur säkularen Religionsgeschichte der Moderne*, Frankfurt: Suhrkamp.

Masetti, J.R. (1958) *Los que luchan y los que lloran: El Fidel Castro que yo ví*, Buenos Aires: Freeland.

Masetti, J.R. (2006) *Los que luchan y los que lloran. Y otros escritos inéditos*, Buenos Aires: Nuestra América.

Masetti, J.R. (2006) Que es un revolucionario? [n.d.], in Masetti, *Los que luchan y los que lloran. Y otros escritos inéditos*, Buenos Aires: Nuestra América, 220–22.

Morales Campos, E. (2007) "Juan Carlos Portantiero," *Archipélago*, no. 57: 23.

Olivieri, R. (1962) "El problema del determinismo en el materialismo dialéctico," *Cuadernos de Cultura*, no. 58 (July–August): 11–30.

Olivieri, R. (1962) "El materialismo dialéctico y la objetividad," *Cuadernos de Cultura*, no. 60 (November–December): 23–39.

Portantiero, J.C. (1963) "Politica y clases sociales en la Argentina actual," *Pasado y Presente*, no. 1 (April–June): 18–23.

Prado Acosta, L. (2014) "El Partido Comunista argentino y la ruptura con los 'muchachos' de la revista *Pasado y Presente*," *Prismas: Revista de historia intelectual*, no. 18: 185–88.

Rein, R. (2002) *Argentina, Israel, and the Jews: Perón, the Eichmann Capture and After*, Bethesda: University Press of Maryland.

Rein, R. (2010) "Argentine Jews and the Accusation of 'Dual Loyalty'." In Rein, *Argentine Jews or Jewish Argentines? Essays on Ethnicity, Identity, and Diaspora*, Leiden: Brill, 169–93.

Rock, D. (1987) *Argentina 1516–1987: From Spanish Colonization to Alfonsín*, Berkeley: University of California Press.

Rot, G. (2010) *Los orígenes perdidos de la guerrilla en la argentina*, Buenos Aires: Waldhuter.

Schenkolewski-Kroll, S. (2002) "The Jewish Communists in Argentina and the Soviet Settlement of Jews on Land in the USSR," *Jews in Eastern Europe*, no. 49: 79–98.

Schenkolewski-Kroll, S. (2011) "Ideology and Propaganda in the Collective Memory's Construction. Zionism and Communism in Argentina," in August Grabski ed., *Rebels against Zion. Studies on the Jewish Left Anti-Zionism*, Warsaw: Jewish Institute of History.

Schmucler, H. (1964) "Problemas del tercer mundo," *Pasado y Presente*, no. 4 (January–March): 234–90.

Senkman, L. (1989) "El antisemitismo y las crisis argentinas, 1959–1975," in Senkman, *El antisemitismo en la Argentina*, Buenos Aires: Centro Editor de América Latina.

Silbert, M., and I. Stambler (1943) [*A Idisher klal tuer in Argentine, h. Boris Rotblat: Baerung far zayn frukhtbare aktivitet in dem groysen templ fun Idishn hilfs un kult farayn "Talmud Torah harishonah ve-Dr. Hertsel"*]. Buenos Aires: Fraynt.

Sinay, R. (1963) *La invención del 'Antisemitismo Sovietico,'* Buenos Aires: Ediciones Tribuna.

Tortti, M.C. (1999) "Izquierda y 'nueva izquierda' en la Argentina: El caso del Partido Comunista," *Sociohistórica*, no. 6: 221–232.

Ulanovsky Sack, D. (1998) "La olivdada guerrilla del Che en la Argentina," *Viva* (March 22): n.p.

United Nations (1965) *Demographic Yearbook 1964*, New York, 123.

Vaca Narvaja, H. (2017) *Masetti: El periodista de la revolución*, Buenos Aires: Sudamericana.

Verbitsky, G. (1955) *Rivera: Afán de medio siglo*, Buenos Aires: Comisión del Cincuentenario de Rivera y sus Colonias.

Visacovsky, N. (2015) *Argentinos, judíos y camaradas: Tras la utopía socialista*, Buenos Aires: Biblos.

Weinstein, A.E., and E. Toker (1999) *Trayectoría de una idea: Nueva Sión 50 años de periodismo judeo-argentino*, Buenos Aires: Fundación Mordejai Anilevich.

Werthein, I. (1990) "85 años: Toda una vida," unpubl. manuscript, Harvard College Library.

Wolff, F. (2014) *Neue Welten in der Neuen Welt: Die Transnationale Geschichte des Allgemeinen Jüdischen Arbeiterbunds 1897–1947*, Köln: Böhlau.

CHAPTER 11

The Montoneros and the Jewish Question in Argentine Fiction

Stephanie M. Pridgeon

"How can a Jew belong to the Montoneros?" (the leftist guerrilla group that emerged in Argentina in 1970), asks a character in Marcelo Birmajer's 2001 novel *Tres mosqueteros* (Three Musketeers).[1] The character is referring to the fact that, contrary to Jewish principles, the armed group engaged in the taking of human life. Moreover, due to the latent Catholic tenor of the Montoneros Jews occupied an ambiguous position in the group. On the other hand, one could also ask, "How could a Jew *not* belong to the Montoneros?" in light of the connections that abounded in mid-twentieth-century Argentina between long-standing Jewish involvement in Socialist movements and Jewish participation in the Montoneros. In many ways, Jewishness facilitated identification with the Montoneros and was therefore not incompatible with the group. Manuela Fingueret's 1999 novel *Hija del silencio* (Daughter of Silence) emphasizes this phenomenon of the connection between Jewishness and the Montoneros, but also includes its protagonist-narrator's reflection: "Unlike others I know in the Movement, I've never denied my Jewish condition."[2] As both novelists show, participation in the Montoneros was both contrary to and consistent with Jewishness. That is, while for some Jewish Argentines being Jewish directly enabled identification with the Montoneros, for others, being Jewish was in opposition to the group. Some twenty years after the publication of *Tres mosqueteros* and *Hija del silencio*, this contradiction remains underexplored within literary and cultural analysis. Recent fiction, whether it vindicates or vilifies the Montoneros as a group, has indeed sought to inquire *how* a Jew could belong to the Montoneros. Regardless of whether the question should be phrased, "How could a Jew belong to the Montoneros?" or "How could a Jew not belong to the Montoneros?" literary explorations of Jewish experiences with the organization consider *how* Jews could—and could not—belong to the group.

1 Marcelo Birmajer, *Tres Mosqueteros* (Buenos Aires: Debate, 2001), 84.

2 Manuela Fingueret, *Daughter of Silence*, trans. Darrell B. Lockhart (Lubbock, TX: Texas Tech University Press, 2012), 134.

 | DOI:10.1163/9789004462540_012

As I underscore throughout my discussion of these two novels, the cultural imaginary makes clear that Jews vacillated, on the one hand, between being drawn to the principles embodied by the Montoneros and, on the other, being deterred by the group's possible anti-Semitism, latent Catholic ideals, and embrace of armed violence. Within this framework, Judaism and Jewishness both facilitated and impeded identification with the Montoneros, a paradox that, as noted, remains critically underexplored.[3] Nonetheless, *Hija del silencio* and *Tres mosqueteros* both present identification and non-identification with the Montoneros in line with their characters' Jewish beliefs. Here, I fill this lacuna through close readings of the two novels after first taking account of the historical context of the Montoneros, as well as a brief consideration of other Jewish-Argentine fiction that references Jewish involvement in the Montoneros.

Beyond strictly religious concerns (as I discuss below), the armed violence in which the Montoneros and other revolutionary groups engaged was hotly debated in Argentina, throughout Latin America, across other formerly colonized regions, and globally. Che Guevara, from whom the Montoneros drew much of their inspiration and whose ideas the characters in both of these novels discuss, advocated violence and bloodshed as a means to reach revolutionary ends. Guevara declared in his address to the 1966 Tricontinental conference:

> Each spilt drop of blood, in any country under whose flag one has not been born, is an experience passed on to those who survive, to be added to the liberation struggle of his own country. And each nation liberated is a phase won in the battle for the liberation of one's own country.[4]

For his part, Frantz Fanon also famously exhorted violence among colonized peoples in *The Wretched of the Earth*. Specifically, he notes: "The mobilization of the masses, when it arises out of the war of liberation, introduces into each man's consciousness the ideas of a common cause, of a national destiny, and

---

3 With the exception of Giffney's book chapter on *Tres mosqueteros*, while existing criticism of these two novels has focused on Jewish identity, national belonging, and militant commitment in them, previous analyses have not explored how they address involvement with the Montoneros as a particularly Jewish issue. Sarah Giffney, "Argentina's Wandering Jews: Judaism, Loyalty, Text, and Homeland in Marcelo Birmajer's *Tres Mosqueteros*," in David William Foster ed., *Latin American Jewish Cultural* (Production, Nashville: Vanderbilt UP, 2009), 97–116.

4 Che Guevara, "Message to the Tricontinental," 1966, https://www.marxists.org/archive/guevara/1967/04/16.htm (accessed October 12, 2020).

of a collective history."[5] Guevara and Fanon, as we see, considered violence in relation to both national and global liberation, an idea that is particularly important in the case of many young Jews' identification with movements that called for violence as a part of identifying fully with the national project. In other words, the capacity of the armed struggle to cultivate a sense of national identity was an aspect of revolutionary culture that was often particularly important for Jews whose parents or (great-)grandparents had immigrated to Argentina, akin to Fanon's focus on the consciousness of a national destiny.

Bearing in mind these Argentine, Latin American, and global ideas on violent revolutionary movements, identification—or the experience of contending with possible identification—with armed guerrilla movements in late 1960s' and 1970s' Argentina was an important facet of national identity. For Jews in Argentina, this aspect of life resonated with events that were affecting Jewish youth around the world. As Brodsky, Gurwitz, and Kranson have argued:

> Young Jews of the period ... grappled with particularly Jewish issues ... they also engaged in the cultural and political rebellions that animated so many others of their age group, joining in struggles against racism, the Vietnam War, sexism, and imperialism.[6]

Involvement in social movements was both an important part of the international zeitgeist and a way of belonging within the national project. In this regard, I liken Fanon's emphasis on the consciousness of a national destiny to Amalia Ran's notion that "historical alternatives offered by fiction can be viewed as equivalent and interchangeable versions of what Jewishness and Argentineanness are all about."[7] In line with Ran's position, I argue throughout my readings of these two novels that fiction is a site through which Jewish Argentine authors navigate the complex points of contact between Jewishness and Argentine national identity, underlined by the particular experience of encounters with armed revolutionary struggles. However, Catholicism and Judaism both played complex roles in such movements.

---

5 Frantz Fanon, *The Wretched of the Earth* (New York: Grove Press, 1963), 93.
6 Adriana Brodsky, Beatrice Gurwitz, and Rachel Kranson, "Editors' Introduction: Jewish Youth in the Global 1960s," *Journal of Jewish Identities* 8, no. 2: 1–2.
7 Amalia Ran, *Made of Shores: Judeo-Argentinean Fiction Revisited* (Bethlehem, PA: Lehigh University Press, 2011), 6.

1 Religion and the Montoneros

How do we make sense, decades later, of the role of religion in a historical moment in which armed guerrilla movements coalesced around grassroots movements that staunchly avowed Catholic values, while the military operatives who later persecuted suspected members of those groups did so on the basis of maintaining Catholic values? As I argue in my book *Revolutionary Visions*:

> Elements of religious culture have entered the political sphere in such a way that political action is made more intelligible and compelling through familiarity with these religious elements … As cultural depictions of these movements continue to remind us, Jews were situated ambiguously vis-à-vis revolutionary practices in Latin America.[8]

Decades later, Jewishness would continue to have an ambiguous role in Peronist politics in Argentina, even in its less militant variants. As Raanan Rein notes, for example, two Jewish Peronist politicians, "Kelly" Olmos and Ana Kessler, recently stated that after meeting in the mid-1980s they shared with one another their mutual experiences of having people react to meeting a Peronist Jew as if such sympathies were unheard of.[9]

The Montoneros were the most well-known organization among a broader swath of revolutionary youth groups in 1960s' and 1970s' Argentina. Avowing Guevarist (after Che Guevara) and Peronist beliefs, the group at first coalesced largely around the principles of class equality and preferential treatment for the poor, as espoused by the Second Vatican Council. The group's latent Catholicism also lay in its ties to the Guardia de Hierro, which disbanded shortly before the Montoneros' formation in 1970, with many of its members proceeding subsequently to join the latter. Unlike the Montoneros, Guardia de Hierro opted against armed violence and perceived Justicialism as the "political expression of faith."[10] While, according to many, the group became much less avowedly Catholic over the course of the 1970s, the Montoneros' bases in Catholic thought nonetheless lingered in its ethos, placing Jews who came to

8 Stephanie M. Pridgeon, *Revolutionary Visions: Jewish Life and Politics in Latin American Film* (Toronto: University of Toronto Press, 2020), 30.

9 Raanan Rein, *Los muchachos peronistas judíos: Los argentinos judíos y el apoyo al Justicialismo* (Buenos Aires: Sudamericana, 2015), 14.

10 Claudia Peiro, "Murió Alejandro Álvarez, Mítico Fundador de Guardia de Hierro—Infobae," June 6, 2006, https://www.infobae.com/politica/2016/06/06/murio-alejandro-alvarez-mitico-fundador-de-guardia-de-hierro/ (accessed October 12, 2020).

identify with them (or who struggled with possible affiliation) in an ambiguous position. In his watershed 2009 study *Sobre la violencia revolucionaria* (On revolutionary violence), Argentine historian Hugo Vezzetti discusses "revolutionary eschatology," a term he borrows from religious historian Norman Cohn, in order to expound on revolutionary violence in 1960s' and 1970s' Argentina.[11] Vezzetti's framework furthers the religious tenor that the Montoneros' legacy takes on in twenty-first century memory of 1970s' politics. In Lucas Lanusse's account, the group moved from "the preferential option for the poor to a preferential option for arms."[12] Within this schema, religion provides a pathway not only for identification with the Montoneros, but specifically (and provocatively) with the armed violence that the group avowed.

Indeed, the major religious issue that complicated Jewish identification with the organization was the taking of human life. To be sure, killing others is no more a Catholic value than a Jewish one. Around the same time that *Hija del silencio* and *Tres mosqueteros* were published, former armed revolutionary Oscar del Barco prompted a heated debate surrounding the legacy of armed violence within the patently religious framework of the biblical commandment not to kill. A series of open letters in response to Oscar del Barco's *No matarás* ("Thou shalt not kill") made clear that the taking of human life and the commandment not to kill took on particular meaning for Jewish Argentines in the context of the Montoneros of the 1970s and memory of that era. For León Rozitchner, a Jewish Argentine revolutionary and writer, this biblical commandment was eclipsed by that of *Vivirás ("Thou shalt live")*, a distinction that he presented as part and parcel of discrepancies between Christian and Jewish thought. Rozitchner asserts in his response to del Barco that the interpretation of "Thou shalt not kill," on which the former revolutionary bases his letter, is rooted in a Christian understanding of the Old Testament Jewish god, a paternal interpretation in contrast to the maternal maxim of "Thou shalt live." Rozitchner concludes:

> Prior to the maternal and the paternal *no matar* was the murmur of life that cries out for the absolute respect of Life as absolute. I believe that the call or plea for *no matar* is life itself and the possibility of life presenting itself and persevering as life.[13]

---

11 Hugo Vezzetti, *Sobre la violencia revolucionaria: memorias y olvidos* (Buenos Aires: Siglo Veintiuno, 2009), 166.

12 Lucas Lanusse, *Montoneros: El mito de sus 12 fundadores* (Buenos Aires: Vergara, 2005), 170.

13 Léon Rozitchner, "Primero Hay Que Saber Vivir: Del Vivirás Materno al No Matarás Patriarcal" and "Intercambio Rozitchner-Del Barco," 2014, https://laempresadevivir

Violence and the taking of life, concepts that were integral to the Montoneros and to similar groups, have remained a point of contention in the discourse surrounding the legacy of revolutionary movements, and, as observed here, have been presented in the context of Jewish and Catholic beliefs.

In terms of global geopolitics, the Montoneros aligned with the liberation of Palestine, a position that often required a moment of reckoning for Jewish Argentines as they grappled with their own identification with the group and with Jewish politics (an issue addressed directly at length in Birmajer's novel). This is not to say that all Jews were Zionists, but that avowing solidarity with an anti-Israel organization was a significant matter for Jewish individuals. Returning briefly to Rozitchner, the philosopher posited in his monograph *Ser judío*, written in the wake of the 1967 war, that Jewish Argentines' preoccupation with the State of Israel was merely a distraction from the politics going on at home in Argentina.[14] This assertion is presented within the broader framework of an essay in which Rozitchner ponders whether, in order to be truly revolutionary, one must cease to be Jewish. The dichotomy that Rozitchner presents is important not because it is necessarily true but because it indicates that, for Jewish Argentines of that generation, there was a perceived tension between being Jewish and being revolutionary. Similarly, Birmajer himself, in his 2002 essay "Ser judío en el siglo XXI," references members of the Uruguayan Tupamaros who, exiled in Israel, began to fight for Israel. Birmajer describes this trajectory as "one of those fast and absurd pirouettes that only two identities as convoluted as Jewish and Latin American could allow for."[15] While not the main focus of my discussion, this tension between revolutionary commitment and allegiance to Israel is also present in *Tres mosqueteros*.[16]

As fictional and historical accounts of Jewish political prisoners during the 1976–83 dictatorship demonstrate, Jews were often persecuted in particularly anti-Semitic ways. When the paramilitary Anti-Communist Alliance (AAA) and so-called *grupos de tareas* (special task forces) proliferated in the years immediately before and during the dictatorship, the military junta and its forces often equated Jewishness with "subversion," or organizing against the state. According to Emmanuel Kahan, an estimated 17 percent of political

.wordpress.com/2010/04/25/intercambio-rozitchner-del-barco/ (accessed October 13, 2020) [my translation].

14 Birmajer, *Tres Mosqueteros*.

15 Birmajer, *Ser judío en el siglo XXI* (Buenos Aires: AMIA/Editorial Milá, 2004), 9.

16 Ran and Giffney both discuss the role of Israel in the novel at greater length. Amalia Ran, "'Israel': An Abstract Concept or Concrete Reality in Recent Judeo-Argentinean Narrative?" in David William Foster ed., *Latin American Jewish Cultural Production* (Nashville: Vanderbilt UP, 2009); Giffney, "Argentina's Wandering Jews."

prisoners during the military dictatorship were Jewish, a figure that is staggeringly disproportionate to the 1–2 percent of the country's Jewish population.[17] Equally troubling is the phenomenon of physical and psychological torture enacted against Jews *as Jews*. That is, Jewish political prisoners were notoriously interrogated and persecuted using anti-Semitic language and tactics. The military's anti-Semitic persecution of Jewish political prisoners on the basis of the assumption that Jews were somehow more likely to engage in "subversive" activities or armed violence remains one of the more insidious aspects of a notoriously ruthless regime.[18]

2 Jewish-Argentine Fiction and the Montoneros

Within the context of the transition to democracy and the post-dictatorship period, narratives about political events became integral to Jewish Argentine fiction. As Amalia Ran has shown, post-dictatorship Jewish Argentine novels "study the void left after the renunciation of the military junta and give a protagonist place to individual actors who played a part in the restoration of a national identity by filling the social gaps left in the collective psyche with personal anecdotes."[19] In other words, recent Jewish Argentine novels have shown that for many authors to be Jewish and to carve out a sense of belonging in twentieth-century Argentina meant—almost necessarily—to grapple with the complicated political landscape of those times. As many of the novels that Ran explores indicate, this co-constitutive relationship between Jewish belonging and the political sphere is particularly evident in the case of Jewish experiences with the Montoneros. Yet existing analyses of Jewish Argentine fiction have not focused as closely on the role of Peronist politics in general as historical

17 Kahan, Emmanuel, *Recuerdos que mienten un poco: vida y memoria de la experiencia judía durante la última dictadura militar* (Buenos Aires: Prometeo, 2014).

18 While I discuss characters who are Montoneros taken as political prisoners, as well as characters that grapple with the ethics of armed violence, I certainly do not mean to suggest that the violence committed by the revolutionary groups was in any way comparable with that of the dictatorship. This argument would perpetuate the so-called *dos demonios* (two demons) theory that became widespread in the post-dictatorship period: the idea that, in the years leading up to the dictatorship, the nation found itself between two forces of evil—the revolutionary militants and the anti-Communist (or anti-"subversive" forces). For more on this, see Marina Franco, "La 'teoría de los dos demonios': un símbolo de la posdictadura en la Argentina," *A Contracorriente: Una Revista de Estudios Latinoamericanos 11*, no. 2 (2014): 22–52.

19 Ran, *Made of Shores*: 4.

discussions have. As Rein shows, Peronism was, for many Jews in Argentina, the first and strongest instance of participating in the political sphere:

> It was Peronism, partly under Socialist influence, that notably accelerated the processes that would give way to a new social, political, and cultural meaning of citizenship ... as in the case of Arabs and Jews, Peronism transformed many of these 'imaginary citizens' into an integral part of society.[20]

Involvement in the Montoneros—the most extreme (and complicated) iteration of Peronist politics—then, is particularly significant for Jews, as these retrospective, fictional stories underscore.

Birmajer and Fingueret are not, strictly speaking, alone in their considerations of participation in the Montoneros from a Jewish perspective. On the contrary, novels and testimonials both before and after these two publications have also waded into the murky waters of what it meant to be Jewish and identified with the Montoneros. To situate these two novels within a broader panorama of Jewish Argentine fiction that has taken up this subject matter in the last twenty-five years, I pause here to take account of works that deal more obliquely than *Tres mosqueteros* and *Hija del silencio* with Jewish experiences with the Montoneros. As I show from these examples, the subject has been of wide concern for Argentine Jewish authors. These other novels' explicit references to Jewish experiences with the Montoneros—albeit relatively few in comparison with *Tres mosqueteros* and *Hija del silencio*—nonetheless lay bare the paradoxical role of Jewishness in both facilitating and impeding identification with the Montoneros and also illustrate the tendency within the military regime to equate Jewishness with subversion.

Novelist Liliana Heker focused on the Montoneros in her controversial 1996 novel *El fin de la historia*. It tells of a Montonera woman, Leonora, who is taken political prisoner and collaborates with the regime after falling in love with one of her captors. Through the narration of Leonora's Jewish Montonero husband, Fernando—filtered through dialogue recapitulated to the main narrator by another narrator—she speaks of the ambiguity of Catholicism and Jewishness vis-à-vis the armed revolutionary group.[21] The novel also tells of

20 Rein, *Los muchachos peronistas judíos*: 22 [my translation].

21 Many critical discussions of *El fin de la historia* have mistaken its more critical narrator (Diana Glass) for its main narrator (Hertha Bechofen), and that misreading has led to further unfavorable interpretations of the novel.

the capture of Fernando, shortly before his death at the hand of military forces, comparing him with Che Guevara and Christ:

> Strangely, he envisioned himself in Higueras [*sic*], in someone else's story that he had dreamed so often—"Don't shoot, I'm Che Guevara"—the man with the naked torso and the face of Christ believing that something in those words would stay the hand of the men aiming at him, or in spite of everything, trusting in his seed and in the unconquerable lineage of the New Man.[22]

After his capture, he dies "with his faith … riddled with bullets from the gunfight and trusting fully in the triumph of the revolution."[23] Yet, elsewhere, we learn that Leonora's mother had wished for her daughter "a good Jewish husband who worries about the household and is married to her and not to politics."[24] But, after Fernando's death and Leonora's capture, her mother quickly becomes disenchanted with nice Jewish husbands: "Life has taught her not to trust even Jewish husbands."[25] In Heker's provocative style, this perspective of the character's mother has changed because her son-in-law "*por algo* ended up shot full of holes."[26] Leonora's mother echoes the "shot full of holes" descriptor but, through the *por algo* modifier implies that "it must be for a reason," in order to make sense of her son-in-law's death.[27] More importantly, she couples his Jewish identity with his militancy and his death in an attempt to understand her daughter's relationship with her captor, one that is coded in hegemonic Catholic terms (up to and including the moment of their wedding). Thus, the character of Leonora's mother registers the complicity of a wide swath of Argentine society (distilled through the three simple words *por algo será*) in religious terms.

Published the following year, Nora Strejilevich's fictionalized testimonial, *Una sola muerte numerosa* (1997), deals explicitly with the military regime's anti-Semitism in its interrogation and torture of Jewish political prisoners.

---

22 Liliana Heker and Andrea Labinger, *The End of the Story* (Emeryville, ON: Biblioasis, 2012), 114–15.

23 Ibid.: 117.

24 Ibid.: 129.

25 Ibid.

26 Ibid.

27 *Por algo será* ("it must be for a reason") became a common refrain among Argentine civilians to suggest that those who were being kidnapped and whose homes were being ransacked were targeted because they had been involved in subversive activities.

Strejilevich recalls being told, "Even if you don't know a thing, you're going to pay just for being a Jew," and goes on to state that the military operatives "centered the interrogation around Jewish matters ... they assured me that they were primarily concerned with 'the problem of subversion', but the 'Jewish problem' was next in importance, and they were gathering information for their files."[28] Hence, the literary recreation of this moment reproduces the ways in which military operatives equated Jewishness with "subversion," going so far as to say that she would be tortured for being Jewish even if she was unable to offer them military information about subversive activities. Both Strejilevich's and Heker's novels mention the "New Man," a concept linked intimately to both Che Guevara and New Testament Christian thought.

Appearing a few years later, in 2003, Andrés Neuman's *Una vez argentina* includes several references to the narrator-protagonist's understanding as a child of his Jewish immigrant family's relationship to Perón and to the Montoneros. From this perspective, Neuman speaks of Jewish-Argentine experiences with the Montoneros spanning the course of four generations. Early on, he relates that his great-grandfather Jacobo was taken to hospital "the same day Perón gave his last speech and disparaged the Montoneros."[29] The narrator equates his family's story with the political landscape of Argentina—namely, Perón and the Montoneros. Similarly, he recounts his mother playing her violin in the orchestra in the presence of Perón at Ezeiza in 1973:

> In Ezeiza ... were the Peronist youth, the Montoneros, people from the ERP [People's Revolutionary Army], and other small revolutionary armies along with thousands of sympathizers, entire families, children of militants who went to lay eyes for the first time on the forbidden idol of their parents.[30]

Neuman's emphasis on "children of militants" suggests an intergenerational political connection between militants and their offspring (a connection further underscored by the fact that the narrator is recounting his mother's journey to Ezeiza in honor of Perón's return from exile). Elsewhere, he relates that his aunt Silvia was taken captive during the dictatorship and interrogated

---

28 Nora Strejilevich and Cristina De la Torre, *A Single, Numberless Death* (Charlottesville: University Press of Virginia, 2002), 19.

29 Andrés Neuman, *Una Vez Argentina*, 2nd ed. Narrativas Hispánicas (Barcelona: Editorial Anagrama, 2004), 27.

30 Ibid.: 51 [my translation].

about the Montoneros, an account that dovetails with Strejilevich's experience of the military regime's linkage of Jewishness with subversion. The novel's final mention of the Montoneros comes in the form of the narrator-protagonist's dream as he is on the verge of leaving the country with his family to immigrate to Spain in 1989 in the midst of the *carapintadas* conflicts.[31] In this context, the narrator recounts:

> The night after amnesty was granted to the military juntas, I had a dream … I dreamt I was a subversive. A Montonero, a militant of the ERP, or something like that. I was imprisoned. There were two others there. Half-closing my eyes, this is the story I remember I dreamt that warm October night: "Talk, you fucking Jew."[32]

The pre-teen narrator dreams that he is a Montonero being tortured by his captors who use language that is patently anti-Semitic.[33]

While Heker, Strejilevich, and Neuman mention the convergence of Jewishness and revolutionary militancy, Fingueret's and Birmajer's novels emphasize in greater detail the complex relationship between the two throughout their narratives. Fingueret and Birmajer explore the ways that Jewishness both facilitated and impeded identification with revolutionary Peronist politics in 1970s' Argentina. In *Hija del silencio* and *Tres mosqueteros*, this fraught relationship comes to the fore in order to determine the respective protagonists' connection to Judaism, Argentina, and the politics of global Jewish life—specifically, memory of the Holocaust, in the case of Fingueret, and Israel, in that of Birmajer.

---

31 The *carapintadas* conflict was a series of uprisings by members of the Argentine army between 1987 and 1990 to express military dissatisfaction with the democratic governments of Raúl Alfonsín and Carlos Menem after the country's return to democracy in 1983.

32 Neuman, *Una Vez Argentina*: 195–96. Amalia Ran also discusses this passage from *Una vez argentina*: "According to Neuman's novel, the main significance of the past lies in the effect remaining at the end; therefore, the protagonist's dream of being abducted and imprisoned during the dictatorship years, a dream provoked by the news of liberating the responsible leaders of the junta, becomes part of the history told by the narrator and constitutes an experience inscribed by its real impression," Ran, *Made of Shores*: 100.

33 The late 1980s in Argentina saw the passing of two amnesty laws—in 1985 and 1987—that effectively granted impunity to those responsible for the military dictatorship and its acts of brutality.

3 *Hija del silencio*

Published in 1999, Fingueret's novel uses silences as a way of bridging intergenerational experiences—namely, those of the narrator-protagonist, political prisoner Rivke (whom her fellow Argentines call Rita), and of her mother, Holocaust survivor Tínkele. The novel recounts Tínkele's experiences in Terezin prior to her immigration to Argentina, where she married and gave birth to Rivke. As the title suggests, silences pervade the novel, evoking Holocaust memory and state repression. Silences also prompt an intergenerational and diachronic consideration of militant commitment. For Daniela Goldfine,

> the legacy of silence is a weight that overwhelms the search for answers, but it is also an opportunity for those of us situated in the (relatively comfortable) place of readers: an opportunity to delve into the elusive spaces, to reflect on the cracks in our own stories and in others' stories, and to contemplate the ellipses in our identity.[34]

Both what is said and what is not said explicitly in *Hija del silencio* prompt reflection on 1970s' politics and the role of Jews within revolutionary movements. Indeed, as I show, the narrator refers to the "movement" and to her *compañeros*—her militant peers, whom she sometimes names "Montoneros"—yet she never explicitly states that she herself is or was a Montonera. Nonetheless, she describes her militancy and references the Montoneros as the group to which she belongs if not by name (as I elaborate further). On the contrary, she often avows her Jewish identity much more directly. Nevertheless, despite this apparent tension, from an early age the narrator finds an affinity between Jewishness and Peronist militancy.

Of her militant peers, Fingueret's narrator states:

> They consider me a tolerable Jew. Why do I put it like that? Is it my mother that once again places herself between others and myself? Or something I sense behind the ironies? Because, unlike others I know in the movement, I never denied my Jewish condition. I never said one of those ridiculous things that you hear so often like, "I have Jewish roots."[35]

---

34 Daniela Goldfine, "La postmemoria del silencio: Transmision truncada y elipsis tangibles en la obra de Manuela Fingueret y Sergio Chejfec," *Estudios Interdisciplinarios de America Latina y el Caribe* 26, no. 1 (2015): 97.

35 Fingueret, *Daughter of Silence*: 189.

As Rita shows here, it was entirely possible to participate in the Montoneros while outwardly avowing Jewishness. Nonetheless, as she makes clear through the references to her reputation as a "tolerable Jew" and to the fact that others in the group denied their Jewishness, participating in the movement did compel some Jews to downplay it as part of their integration into the group.

For Rita, Jewishness and identification with revolutionary Peronism are not mutually exclusive; on the contrary, they are intimately linked. Indeed, Rita recounts her own life in such a way that explicitly couples her intergenerational connection to her mother's Holocaust experiences with empathy for Eva Perón from a young age. Thus, Fingueret ties Holocaust memory and Jewish experiences with Peronist political affinities, a conceptual link that facilitates the young woman's later identification with the Montoneros. Narrated in beautiful vignettes, *Hija del silencio* is a heartbreaking life story which ends with its narrator's tragic (probable) death at the hands of the military regime. (I say "probable" here because, at the novel's end, one of the men holding her captive tells her that she is going to be transferred; at the Naval Mechanics School, the clandestine torture center where she is being held, prisoners were told they were being "transferred" before being drugged and tossed from military helicopters into the Río de la Plata.[36]) Throughout Rita's short life, her familial connections to Jewishness and her affinity for revolutionary politics come together in multiple and sometimes contradictory ways.

Fingueret, who was known better for her poetry than her prose before publishing *Hija del silencio*, at times makes quite explicit the links between Jewish identity and Peronist politics and, at others, relies on creative omissions and juxtapositions to do so. For example, she beautifully interlays the narrator's militant peers' chants of: "If Evita (Perón) were alive, she'd be a Montonera," with her father's "Adonai Eloheinu, Adonai Echad."[37] After recounting her father's prayer on Yom Kippur, she wonders to herself, "Would she be a Montonera?" suggesting an inconsistency with the Montoneros group. This discrepancy becomes more pronounced, as I show, through her questioning of the group's stance on violence. Moreover, as noted, she never proclaims explicitly, "I am a Montonera." Rather, she associates herself laterally with the

36 Amy Kaminsky also interprets the novel's abrupt ending as suggesting Rita's death: "The novel ends inconclusively, but the implication is that Rita does not survive. The first-person narrative simply stops, as does the story she has been reconstructing of her mother." (116). Rita's end as she is being transferred also resonates with her mother's transfer from Terezin to Auschwitz. Amy Kaminsky, "Memory, Postmemory, Prosthetic Memory: Reflections on the Holocaust and the Dirty War in Argentine Narrative," *Hispanic Issues Online* 14 (2014): 104–17.

37 Fingueret, *Daughter of Silence*: 95.

organization through her friendships and love relationships with members. She does, however, count herself among the Montoneros by referring to the moment when "the General threw us out of Plaza de Mayo."[38] Here, Fingueret is referring to the instance—also mentioned in Neuman's novel—when Perón infamously called the Montoneros *imberbes* and ousted them from the Plaza de Mayo. As this moment in history implies, Perón's relationship to the Montoneros was in fact quite ambivalent, another facet of 1970s' Argentine politics that Fingueret explores through *Hija del silencio*.

For the narrator, Juan and Eva Perón are as much a personal pantheon in line with her Jewish beliefs as they are national icons. Indeed, she positions Peronism in opposition to Nazism. She asserts, "For me, Peronism continues to be the line of thinking that will recover the honor and meaning of existence in an Argentina that was destroyed by Nazi-capitalism of his century," extending the particularity of the Nazi atrocities that her mother escaped to the general condition of Argentina in the twentieth century and tacitly equating Peronism with anti-Nazi positions.[39] Thus, the novel evokes Perón's ambivalent stance toward Nazism. The leader infamously supported the Axis before his rise to the presidency, and allowed Nazis to enter the country after World War II; however, he keenly supported the advancement and inclusion of Jews in the workforce and social organizations. While Uki Goñi focused on Perón's strategies to bring Nazis to Argentina, Raanan Rein recently noted that the Jewish community within Argentina was divided regarding its stand on Perón, concluding: "Neither the collective memory of Argentine Jews nor the history books seem to retain much recall of the fact that many Jews did in fact support Perón and the Justicialist movement in its early years."[40] Likewise, in Rita's case, not only did her family support Perón, but she understands Peronism as the antidote to Nazism. Yet, as I will discuss below, different generations had differing stances on Perón and Peronism due to the increasingly militant practices of the Peronist Left in the 1970s.

In addition to her evocation of Perón's relationship to Nazism, Fingueret's hyphenation of "Nazi-capitalism" resonates with Zygmunt Bauman's considerations of the Holocaust and late market capitalism in terms of the Holocaust's "uniqueness and normality" in *Modernity in the Holocaust*:

---

38 Ibid.: p. 102.

39 Ibid.

40 Uki Goñi, *The Real Odessa: How Perón Brought the Nazi War Criminals to Argentina* (London: Granta Books, 2003); Raanan Rein, *Argentine Jews or Jewish Argentines? Essays on Ethnicity, Identity, and Diaspora* (Leiden, Netherlands: Brill, 2010), 167.

> The *possibility* of the Holocaust was rooted in certain universal features of modern civilization; its *implementation* on the other hand, was connected with a specific and not at all universal relationship between state and society.[41]

Put differently, the Holocaust was simultaneously a unique phenomenon of atrocities visited upon the Jews and a broader part of the development of modernity in the twentieth century. The Montoneros' struggles against a capitalist world order resonate with Rita's identity as a Jew and as the daughter of a Holocaust survivor who opposes "Nazi-capitalism."

Furthering her emphasis on the connections between Nazism/the Holocaust and Peronist politics, Rita finds links between her mother, the Holocaust survivor, and Eva Perón. She posits the two as interchangeable and draws strength from their examples. Specifically, she ties her mother's survival in the concentration camps to the militancy she later pursues on the basis of the principles she identifies in Eva Perón. Yet, elsewhere, she suggests that she has exchanged her devotion to Jewish beliefs for militant ones.

> Around the time of Tom and Jerry I stopped believing in that God who dwelled inside the synagogue doors and who people kissed on the holidays. An obvious God who ceased to interest me. I exchanged the harshness of Jehovah for the harshness of militancy. I was active in everything: in politics, in my literary convictions, in atheism, in my relationships with others, in sex.[42]

Yet, even when disclosing that she had exchanged her Judaism for political militancy, she equates the two by underscoring their shared "harshness."

Despite the many silences and elisions that pervade the novel, Fingueret's narrator does not evade the topic of the Montoneros' use of armed violence as a means to effect revolution and enact Peronist ideals. Rita demonstrates that the violence in which the Peronist Left engaged was counter to her parents' values. She relates that her father, who had previously been loyal to Perón, had "fallen in love with Frondizi, violence went against his nature, and he detested the triumphant attitude he perceived in my so-called friends."[43] Almost immediately afterward, she references her father's Yom Kippur prayer, suggesting

41 Zygmunt Bauman, *Modernity and the Holocaust* (Ithaca, NY: Cornell University Press, 2002), 82 [emphasis in original].

42 Fingueret, *Daughter of Silence*: 32.

43 Ibid.: 95.

through juxtaposition that his opposition to violence is somehow related to his Judaism. Elsewhere, Rita shows that she believes violence is a necessary means to an end despite her individual opposition to violence. She states that she personally disagrees with the ways that the Peronist Left sought to reach their objectives: "It was not the third way advocated by our comrades that I longed for, but the only position possible for a country and a world that mercilessly condemns those who live on the margins."[44] Thus, while disclaiming that she personally disavows violence as a means to reach revolutionary ends, the narrator also espouses a position that recalls Fanon's and Guevara's embrace of violence as a necessary tactic for the liberation of peoples in nations such as Argentina. The phenomenon of personally disavowing violence while also advocating for its place in the broader schema of revolutionary praxis was common among the Peronist Left and was, at times, presented in tandem with religious preoccupations. For example, Guardia de Hierro, mentioned at the beginning of this chapter as a group that understood Justicialism as the "political expression of faith," stated that Perón himself envisioned them as the "rearguard" which should not engage in armed violence.[45]

In keeping with the leitmotif throughout the novel, silence characterizes the bond between Rita and her mother, on whom she models herself and bases her militancy. In this regard, silence and elisions are also mechanisms for political militancy. Yet, she and her mother remain silent on the topic of militancy, just as her mother has not spoken to her directly about her experiences as a Holocaust survivor, leaving Rita to learn about her time in the concentration camps by reading Tinkele's writings that she secretly discovers. D. Jan Mennell notes:

> The narrator's inability to penetrate the accusation inherent in Tínkele's silence becomes one of the primary motivating factors in her life, and ultimately induces her to attempt to fill in the gap and emptiness by political militancy.[46]

Similarly, for Amy Kaminsky, "Rita's political activity is 'the space of permanent discussion with [her] mother', but she never tells Tinkele the key fact of that militancy, the fact that her mother is the source of what she is doing and

44 Ibid.: 102.

45 Humberto Cuchetti, *Combatientes de Perón, Herederos de Cristo: Peronismo, religión secular y organizaciones de cuadros* (Buenos Aires: Prometeo, 2010), 99.

46 D. Jan Mennell, "(Im)Penetrable Silence: The Language of the Unspeakable in Manuela Fingueret's *Hija Del Silencio*," *Revista Canadiense de Estudios Hispánicos* 27, no. 3 (2003): 493.

is included in her actions."[47] Kaminsky's reading of the novel concludes that it presents a memory that has been proscribed by the dictatorship—Peronist militancy—as a moral imperative.

Indeed, the intergenerational silencing of memory—Tinkele's silencing of her Holocaust memories and Rita's silencing of her militant actions—becomes a moral enjoinder to remember, but questions the morality of the armed violence in which the Montoneros engaged. Thus, the meaning and substance that political militancy provide as an antidote to the void of Holocaust loss and silence—as a moral imperative—necessarily call for a reckoning with the difficult truths of the legacy of armed violence. Peronist militancy becomes a way for Rita to understand herself as an Argentine citizen, as well as a way to ensure that the horrors of Nazi-capitalism are not repeated.

4 *Tres mosqueteros*

Birmajer's Tres mosqueteros is much more critical of the Montoneros than Fingueret's novel. In essence, it is a conversation between its narrator, journalist Javier Mossen, and Elías Traúm, who has been living in Israel for decades, but has returned to Argentina in order to say *kaddish* following the deaths of his disappeared friends Benja and Guidi. Javier, the narrator, asks, "How can a Jew belong to the Montoneros organization?" Over the course of conversations between Javier and Traúm, readers learn of a complex story of love triangles and secrecy between these *tres mosqueteros*—Traúm, Benja, and Guidi—that tells of the complicated convergences and divergences between Jewishness and identification with the Montoneros. While this question is posited as such only once, the conundrum is at the heart of the novel's central explorations: indeed, how could Jews belong to the Montoneros?

Through these conversations, the novel presents an intergenerational dialogue that is difficult yet necessary for the nation, in order to make sense of its turbulent recent past. For Daniela Goldfine,

> Traúm's visit and the discussions about Argentine 1970s' politics, especially within the Jewish community, covers an aspect of history that resonates with conversations held between that militant generation and

47 Kaminsky, "Memory, Postmemory, Prosthetic Memory": 109–10.

> those born in the 1970s—or, perhaps, it manages to salvage those conversations that needed to take place.[48]

The novel unlocks conversations that should have been held for decades. For Birmajer and his characters, at its heart is the unresolved moral issue of armed violence. In the context of his own mourning for his friends, Traúm engages in conversations about the Montoneros and the moral question of the taking of human life. The novel's retrospective consideration of 1970s' politics is bound up in Jewish beliefs and culture. Like many other works of Argentine fiction from the early twenty-first century, the novel fills in silences left by the gaps in historical memory following the amnesty laws and politics of forgetting that characterized the 1980s and 1990s. Yet, as I outline below, compared with Fingueret's novel, *Tres mosqueteros* broaches the subject of militant action in the 1970s from the foregone conclusion that armed violence is wrong.

Javier introduces Traúm to readers by indicating that the story he has been placed in charge of writing ties together two themes: Judaism and the Montoneros. As the narrator describes them:

> One was vast, the other unfathomable. The fate of the guerrilla fighters of the 1970s—basically the Montoneros, the organization to which Traúm and two of his now deceased friends had belonged—and Judaism. Or the Jews. Or the infinite gamut of Jewish Judaisms.[49]

This idea of a multiplicity of Jewish identities is echoed later by Traúm, who asserts, "For each person, being Jewish is a different thing."[50] For the characters in *Tres mosqueteros*, being Jewish can mean identifying with the Montoneros, as was the case for Benja and Guidi, or it can mean disavowing the group, as Traúm and the narrator both do. As I show in my discussion of Birmajer's novel, his characters mostly adopt a critical stance toward the Montoneros, placing the group in tension with Jewishness. Nonetheless, the coupling of these two themes—"one vast and the other unfathomable"—signals that the complex points of contact between Jewishness and Montoneros continue to be significant for literature and politics in early twenty-first century Argentina.

48 Daniela Goldfine, "Birmajer, Burman, Winograd: Tres mosqueteros que no esperan al Mesías: (Trans)formaciones de la producción cultural judeo-argentina en el siglo XXI," in *Actas Del Congreso de La Asociación Internacional de Hispanistas: Hispanismos Del Mundo: Diálogos y Debates en (y Desde) El Sur VI*, Literatura Latinoamericana (2013): 264.

49 Birmajer, *Tres Mosqueteros*: 9 [all quotes from Birmajer are my translation].

50 Ibid.: 89.

Tres mosqueteros contains very little action: essentially, Javier is tasked with writing a story about Elías Traúm for the newspaper for which he works. Traúm arrives from Israel, the two meet and drink and Javier asks him questions. Interspersed with the scenes with Traúm, Javier spends time by himself, at the newspaper office or in his apartment, longing for his wife who left him after he was unfaithful or lusting after other women. These snippets of Javier's everyday life establish him as an archetypical *schlemiel* narrator, as Giffney notes. While scenes of Javier drinking too much and feeling sorry for himself, or fantasizing graphically about his housekeeper, might grow tiresome for readers, this archetype is important insofar as it resembles other fictional Jewish characters in early twenty-first century Argentina, such as Ariel Makaroff, in the 2004 Daniel Burman film *El abrazo partido*, co-written by Birmajer and Burman. That is, the novel conveys a sense that Argentine Jews of Javier's generation are somewhat aimless because of unreconciled aspects of Jewish Argentine life that they inherited from the previous generation.

Like *Hija del silencio*, *Tres mosqueteros* also uses silences as a way of underscoring past trauma that affects both Jewish communities worldwide in the wake of the Holocaust and Argentines following the dictatorship. As Sarah Giffney has posited in her excellent reading of *Tres mosqueteros*, "socialism provided these boys with a path away from the theological doubt left by God's silence in the face of Nazi horrors."[51] In this regard, the novel is similar to *Hija del silencio*, whose protagonist is compelled to political militancy largely because of her mother's experience as a Holocaust survivor. In order to discuss the connection between Socialism and the aftermath of the Holocaust, Giffney notes that the "musketeers" established their own Socialist magazine, titled *Dios No Dice Nada*, (God says nothing), as teenagers. I would linger here on Giffney's characterization that Socialism allowed the boys to move away from their doubt left by God's silence, for the relationship between Judaism and Socialism casts the two in a complex and contradictory binary: if indeed the boys turn to Socialism as a response to their theological doubt, then Socialism is both in keeping with and a disruption of their Jewish identities. Their subsequent identification with the Montoneros is in many regards a continuation of this initial identification with Socialism and is also presented as a more general zeitgeist of youth culture in 1970s' Argentina. Yet, the move from Socialism to Marxism is too extreme from Traúm's perspective.

As part of the pending conversation between generations that Goldfine notes in the novel, a sense of fear surrounding Traúm overwhelms Javier at times. This dread is related to Traúm's association with members of a group

51 Giffney, "Argentina's Wandering Jews": 1565–69.

associated with death. Similar to the dream that Neuman's adolescent protagonist describes in *Una vez argentina*, Javier experiences a nightmarish vision of Traúm late one night when the latter enters Javier's apartment unexpectedly. Prior to that moment, he had lost track of Traúm and he worried that he was being persecuted: "Where was Traúm? Had death returned unbridled? Were they torturing him, did they have him locked in a *chupadero*, or in a clandestine jail, part 'town jail', part aristocratic dungeon?"[52] Javier thinks at first that he is seeing Traúm's ghost:

> Before my eyes the ghost of Traúm appeared. I saw him hanging from the ceiling, like a newspaper editor hanged in a Communist Party office during the military dictatorship. I saw him bloody, riddled with bullets, thrown in the Ezeiza woods in a nylon bag like the people killed by the Peronist Right during Perón's third administration.[53]

While Javier's vision of Traúm as a victim of state brutality is not presented in patently Jewish terms, when Javier asks him how he learned to enter people's homes like a thief, the latter responds, "not like a thief, like the prophet Elijah"[54] (149). Here, Traúm emphasizes a connection to the prophet whose name he bears and the belief among Jewish communities that Elijah sneaks into homes and drinks wine during Passover. The exchange between the two men thus links Jewish identity with a fear of repression at the hands of the military government, a dread that Javier registers now because of Traúm's past connection to the Montoneros through his friends.

Like the debate sparked by Oscar del Barco that I mentioned in the introduction to this chapter, Birmajer's novel criticizes the Montoneros' taking of human life. Thus, as noted, Birmajer's novel is more critical of the Montoneros than Fingueret's. While it does not explicitly present its critique of the Montoneros' violence within a Jewish or religious framework, the novel does suggest that having belonged to the Montoneros meant necessarily that one was culpable—at least by association—of the taking of human life. Javier asks Traúm whether Guidi and Benja killed people, to which Traúm responds, "I don't know, but I consider them to be murderers anyway, both of them."[55] Traúm announces this judgment as the two stand on the doorstep of Benja's parents' home, and Traúm explains that they have to sneak in rather than

52 Birmajer, *Tres Mosqueteros*: 147–48.

53 Ibid.

54 Ibid.: 149.

55 Ibid. 153.

ring the doorbell. Javier responds, “like the prophet Elijah,” again creating an exchange between the two characters in which Traúm’s namesake and its significance for Jewish belief is linked to his Montonero friends’ activities. Giffney interprets Javier’s characterization of his friends as murderers:

> The ultimate mistake that Guidi and Benjamin make is not to pursue socialist ideals. Rather, it is to forget their Jewish identity and their moral obligation to other human beings. In militant pursuit of social change, the boys ceased to prioritize human life over social goals. They became, in Traúm’s words, “asesinos.”[56]

Interestingly, Traúm does not dwell on the matter, but rather laconically states that he considers both of his deceased friends to be murderers. The brevity of his pronouncement suggests that armed violence was so obviously abhorrent that he thinks of his friends as murderers despite not knowing whether or not they took lives.

At other points, however, Traúm offers more prolonged reflections on the moral question of armed violence. He offers a judgment of the way that his friends died, stating:

> The way that a person dies changes everything … they didn’t die in peace, … not only because of the people who killed them: they didn’t know how to die. They ruined their entire lives with that death. The killers are one thing; but I’m talking about them: I couldn’t take care of them.[57]

Traúm also emphasizes Guidi’s reaction to his father’s death around the same time as his radicalization and increasing Marxist fervor. He harangued Traúm: “Do you know how many people died the same day as he did, of hunger, of repressors’ bullets, in dictators’ prisons?”[58] Guidi is focused on a vision of revolutionary violence that dovetails with Fanon’s points and with Fingueret’s narrator-protagonist’s belief that, while she does not personally engage in armed violence, it is necessary for the liberation of Argentina and other countries. Yet, through the conversations—decades later—between Traúm and Javier, Guidi’s dismissal of his father’s death because many other people died on the same day due to the injustices in the world is criticized by Traúm and the narrator. Nonetheless, Birmajer includes this recapitulation of

---

56 Giffney, “Argentina’s Wandering Jews”: 1575.
57 Birmajer, *Tres Mosqueteros*: 186.
58 Ibid.: 188.

Traúm's conversation with Guidi to show that, whether it is understandable from the perspective of a younger generation or from that of a contemporary (Traúm), who did not agree with the Montoneros, many Jewish revolutionaries of Guidi and Traúm's age did think like the former. *Tres mosqueteros* and *Hija del silencio* are similar in their tacit coupling of non-violence and Jewish beliefs.

Tres mosqueteros also presents political militancy as bound up in Jewishness through linking Guidi's life and death with the model of the wandering Jew. The novel ends with Javier's reconciliation with his estranged wife Esther, in Mar del Plata, positing his previous womanizing and lusting after other women as a short-lived experience of straying from his family. Shortly before this resolution, Javier spends time in Mar del Plata with Traúm. While there, he wanders the sandy beaches and panics over feeling lost as he thinks of what Traúm has told him about Guidi, evoking Jewish experiences of wandering the desert.[59] He reflects:

> But I was not Guidi: I knew my way back. Maybe the sky would fall on my head but I was not going to proclaim cheerfully that I was ready to give my life for the life of a stranger. Or to take a stranger's life.[60]

For Javier, this experience guides him back to family (his Jewish wife, in contrast to his non-Jewish lovers), unlike the wandering that he learns characterized Guidi's life and which led him, in Javier's view, astray from his Jewish roots and toward the Montoneros. Whereas Guidi believed that sacrificing his own life for the liberation of the country—in line with Che Guevara's and Frantz Fanon's beliefs regarding the necessity of armed violence in the colonized and formerly colonized nations of the world—Javier and Traúm reject this view. Importantly, they repudiate this idea in ways that are—at times patently and at times implicitly—coupled with Jewishness. Thus, unlike Rozitchner, who responded to Oscar del Barco's "Thou shalt not kill" by delineating a Jewish understanding of the commandments from what he perceived as del Barco's Christian interpretation as a way of defending revolutionary violence, Traúm and Javier both suggest that Guidi and Benjamín were led astray from their Jewish beliefs by the Montoneros and their embrace of armed violence.

---

59 Similarly, in *El abrazo partido*, the mother of the protagonist, Ariel, tells a story about him getting lost on the beach as a child and screaming when he could not find her, an anecdote the mother characterizes as an "ancestral scream," also evoking the biblical moment of Jews wandering the desert in Exodus.

60 Birmajer, *Tres Mosqueteros*: 185.

5 Conclusions

Both Fingueret and Birmajer connect Jewish political affinities of the 1970s to a need to make sense of their lives and the post-Holocaust world. Moreover, through association with the Montoneros, the characters come to relate more strongly to Argentine national identity (the "national destiny" of which Fanon speaks). Where the two novels differ is in that for Fingueret political militancy is in keeping with Jewishness, whereas Birmajer's characters disagree (at least Traúm and Javier do; readers are not privy to Benja and Guidi's own perspectives on their militancy). Another salient difference between the novels is that Fingueret emphasizes her militant character's relationship to the older generation (her mother), whereas Birmajer underscores the Montoneros' legacy for the younger generation (Javier). Yet for both novelists, the place of Jewishness vis-à-vis revolutionary politics remains a question worth exploring. In particular, the implications of the legacy of armed violence—both for the Argentine nation at large and for Jews in particular—continue to influence cultural understandings of how Jews could or could not belong to the Montoneros.

Bibliography

Bauman, Z. (2002) *Modernity and the Holocaust*, Ithaca, NY: Cornell University Press.

Birmajer, M. (2001) *Tres Mosqueteros*, Buenos Aires: Debate.

Birmajer, M. (2004) *Ser judío en el siglo XXI*, Buenos Aires: AMIA/Editorial Milá.

Brodsky, A., B. Gurwitz, and R. Kranson (2015) "Editors' Introduction: Jewish Youth in the Global 1960s," *Journal of Jewish Identities 8*, no. 2: 1–11.

Burman, D. (dir.) (2004) *El abrazo partido*, Buenos Aires: BD Cine.

Cucchetti, H. (2010) *Combatientes de Perón, herederos de Cristo: Peronismo, religión secular y organizaciones de cuadros*, Buenos Aires: Prometeo.

Fanon, F. (1963) *The Wretched of the Earth*, New York: Grove Press.

Fingueret, M., and D.B. Lockhart (2012) *Daughter of Silence*, Lubbock, TX: Texas Tech University Press.

Franco, M. (2014) "La 'teoría de los dos demonios': un símbolo de la posdictadura en la Argentina," *A Contracorriente: Una Revista de Estudios Latinoamericanos 11*, no. 2: 22–52.

Giffney, S. (2009) "Argentina's Wandering Jews: Judaism, Loyalty, Text, and Homeland in Marcelo Birmajer's *Tres Mosqueteros*," in David William Foster ed., *Latin American Jewish Cultural Production*, Nashville: Vanderbilt UP, 97–116.

Goldfine, D. (2013) "Birmajer, Burman, Winograd: Tres mosqueteros que no esperan al Mesías: (Trans)formaciones de la producción cultural judeo-argentina en el siglo XXI," in *Actas Del Congreso de La Asociación Internacional de Hispanistas: Hispanismos Del Mundo: Diálogos y Debates en (y Desde) El Sur VI*, Literatura Latinoamericana, 261–70.

Goldfine, D. (2015) "La postmemoria del silencio: Transmision truncada y elipsis tangibles en la obra de Manuela Fingueret y Sergio Chejfec," *Estudios Interdisciplinarios de America Latina y el Caribe* 26, no. 1: 91–111.

Goñi, U. (2003) *The Real Odessa: How Perón Brought the Nazi War Criminals to Argentina*, London: Granta Books.

Guevara, Che (1966) "Message to the Tricontinental," https://www.marxists.org/archive/guevara/1967/04/16.htm (accessed October 12, 2020).

Heker, L., and A. Labinger (2012) *The End of the Story*, Emeryville, ON: Biblioasis.

Kahan, E. (2014) *Recuerdos que mienten un poco: vida y memoria de la experiencia judía durante la última dictadura militar*, Buenos Aires: Prometeo.

Kaminsky, A. (2014) "Memory, Postmemory, Prosthetic Memory: Reflections on the Holocaust and the Dirty War in Argentine Narrative," *Hispanic Issues Online* 14 (2014): 104–17.

Lanusse, L. (2005) *Montoneros: El mito de sus 12 fundadores*, Buenos Aires: Vergara.

Mennell, D.J. (2003) "(Im)Penetrable Silence: The Language of the Unspeakable in Manuela Fingueret's *Hija Del Silencio*," *Revista Canadiense de Estudios Hispánicos* 27, no. 3: 485–507.

Neuman, A. (2004) *Una Vez Argentina*, 2nd ed., Narrativas Hispánicas, Barcelona: Editorial Anagrama.

Peiro, C. (n.d.) "Murió Alejandro Álvarez, Mítico Fundador de Guardia de Hierro—Infobae," https://www.infobae.com/politica/2016/06/06/murio-alejandro-alvarez-mitico-fundador-de-guardia-de-hierro/ (accessed October 12, 2020).

Pridgeon, S.M. (2020) *Revolutionary Visions: Jewish Life and Politics in Latin American Film*, Toronto: University of Toronto Press.

Ran, A. (2009) "'Israel': An Abstract Concept or Concrete Reality in Recent Judeo-Argentinean Narrative?" in David William Foster ed., *Latin American Jewish Cultural Production*, Nashville: Vanderbilt UP, 24–40.

Ran, A. (2011) *Made of Shores: Judeo-Argentinean Fiction Revisited*, Bethlehem, PA: Lehigh University Press.

Rein, R. (2010) *Argentine Jews or Jewish Argentines? Essays on Ethnicity, Identity, and Diaspora*, Leiden, Netherlands: Brill.

Rein, R. (2015) *Los muchachos peronistas judíos: Los argentinos judíos y el apoyo al Justicialismo*, Buenos Aires: Sudamericana.

Rozitchner, L. (1967) *Ser judío*, Buenos Aires: Ediciones de la flor.

Rozitchner, L. (2014) "Primero Hay Que Saber Vivir: Del Vivirás Materno al No Matarás Patriarcal" and "Intercambio Rozitchner-Del Barco," https://laempresadevivir.wordpress.com/2010/04/25/intercambio-rozitchner-del-barco/ (accessed October 13, 2020).
Strejilevich, N., and C. De la Torre (2002) *A Single, Numberless Death*, Charlottesville: University Press of Virginia.
Vezzetti, H. (2009) *Sobre la violencia revolucionaria: memorias y olvidos*, Buenos Aires: Siglo Veintiuno.

Index

www.ingramcontent.com/pod-product-compliance
Ingram Content Group UK Ltd.
Pitfield, Milton Keynes, MK11 3LW, UK
UKHW022224051026
15624UKWH00002B/336

9 789004 46253

SpringerBriefs in Petroleum Geoscience & Engineering

The SpringerBriefs series in Petroleum Geoscience & Engineering promotes and expedites the dissemination of substantive new research results, state-of-the-art subject reviews and tutorial overviews in the field of petroleum exploration, petroleum engineering and production technology. The subject focus is on upstream exploration and production, subsurface geoscience and engineering. These concise summaries (50–125 pages) will include cutting-edge research, analytical methods, advanced modelling techniques and practical applications. Coverage will extend to all theoretical and applied aspects of the field, including traditional drilling, shale-gas fracking, deepwater sedimentology, seismic exploration, pore-flow modelling and petroleum economics. Topics include but are not limited to:

- Petroleum Geology & Geophysics
- Exploration: Conventional and Unconventional
- Seismic Interpretation
- Formation Evaluation (well logging)
- Drilling and Completion
- Hydraulic Fracturing
- Geomechanics
- Reservoir Simulation and Modelling
- Flow in Porous Media: from nano- to field-scale
- Reservoir Engineering
- Production Engineering
- Well Engineering; Design, Decommissioning and Abandonment
- Petroleum Systems; Instrumentation and Control
- Flow Assurance, Mineral Scale & Hydrates
- Reservoir and Well Intervention
- Reservoir Stimulation
- Oilfield Chemistry
- Risk and Uncertainty
- Petroleum Economics and Energy Policy

Contributions to the series can be made by submitting a proposal to the responsible Springer contact, Charlotte Cross at charlotte.cross@springer.com or the Academic Series Editor, Prof Dorrik Stow at dorrik.stow@pet.hw.ac.uk.

More information about this series at http://www.springer.com/series/15391

Helei Liu · Raphael Idem
Paitoon Tontiwachwuthikul

Post-combustion CO_2 Capture Technology

By Using the Amine Based Solvents

Springer

Helei Liu
Process Systems Engineering
University of Regina
Regina, SK, Canada

Paitoon Tontiwachwuthikul
Process Systems Engineering
University of Regina
Regina, SK, Canada

Raphael Idem
Process Systems Engineering
University of Regina
Regina, SK, Canada

ISSN 2509-3126 ISSN 2509-3134 (electronic)
SpringerBriefs in Petroleum Geoscience & Engineering
ISBN 978-3-030-00921-2 ISBN 978-3-030-00922-9 (eBook)
https://doi.org/10.1007/978-3-030-00922-9

Library of Congress Control Number: 2018956589

This Springer imprint is published by the registered company Springer Nature Switzerland AG
The registered company address is: Gewerbestrasse 11, 6330 Cham, Switzerland

Contents

Chapter 1
Introduction and Background Information

Abstract Recently the worsening global warming issue caused by emissions and accumulation of greenhouse gases in the atmosphere has become a subject of public concern. Carbon dioxide (CO_2) is widely considered as the predominant contributor of greenhouse gases. The post combustion CO_2 capture technology is one of the commonly used technologies for dealing with CO_2 emissions. The absorption of CO_2 into aqueous amine solution is regarded to be one of the most promising technologies for post combustion CO_2 capture due to its maturity, cost effectiveness, and capacity to handle large amounts of exhaust streams. In this chapter, the general information about the CO_2 capture is present. The comparison of the several technologies for CO_2 capture is also discussed in order to give the detail about the advantages and disadvantages for each technology. In addition, the detailed introduction about the post combustion CO_2 capture by using the amine based solvent is provided in this chapter.

1.1 General

Recently the worsening global warming issue caused by emissions and accumulation of greenhouse gases in the atmosphere has become a subject of public concern. National and international governments and industries have collaborated for the purpose of formulating strategies to control GHGs. Also, a number of institutions and programs aimed at addressing this issue have emerged, including the Intergovernmental Panel on Climate Change (IPCC), the Global Climate Change Initiative and the United Nations Framework Commission on Climate Change.

Carbon dioxide (CO_2) is widely considered as the predominant contributor of greenhouse gases, with an annual global emissions having escalated by approximately 80% between 1970 and 2004 (D'Alessandro et al. 2010). Given that the main source of the huge demand for energy for mankind is still fossil fuel combustion (i.e. petroleum, coal and natural gas), there is an urgent need to develop strategies to mitigate CO_2 discharge in the atmosphere. According to IPCC, atmospheric concentrations of CO_2 equivalent must be limited to 450 ppm by 2100 to avoid global

H. Liu et al., *Post-combustion CO_2 Capture Technology*, SpringerBriefs in Petroleum Geoscience & Engineering, https://doi.org/10.1007/978-3-030-00922-9_1

warming greater than 2 °C (Miller et al. 2016). There are a number of means to achieve this target including energy production with higher efficiency, energy conservation, renewable energy and CO_2 capture and storage (CCS). After studying and evaluating a number of these approaches and its combinations, the IPCC concluded that without CCS as an option the costs of limiting temperature increase are significantly higher by 38% (Miller et al. 2016).

CO_2 capture can be applied to large-scale emissions processes, including coal and gas-fired power generation, natural gas processing and fertilizer production. Currently, it can be categorized into three main techniques in industry: pre-combustion capture, oxy-fuel combustion and post-combustion capture. Among these technologies, post-combustion process has the highest potential to be retrofitted to traditional pulverized coal power plants (Kenarsari et al. 2013; Liang et al. 2015). The major full commercial demonstrations are the TMC Mongstad in Norway (300,000 tonnes per year CO_2 captured) and BD3 SaskPower in Canada (1 million tonnes per year CO_2). Chemical solvent absorption, solid sorbent adsorption, cryogenic distillation and membrane separation are important techniques in post-combustion capture. In the chemical solvent scrubbing, an aqueous solution of some absorbent, e.g. alkanolamine or potassium carbonate, reacts reversibly with carbon dioxide. Absorbent for adsorption generally can be categorized into two groups: (i) physical absorbent, and (ii) chemical absorbent. Physical absorbent relies on clathrate that cages CO_2. Chemical absorbent utilizes the attraction with CO_2 molecule (Yang et al. 2008). Cryogenic distillation uses the principle of separation on the basis of cooling and condensation, thus the substantial energy requirement makes it less desirable for industrial application (Figueroa et al. 2008). Membrane separation is based on the different physical or/and chemical interactions between different gases and membrane material. The long-term stability of the membrane performance and expenditure for large size membranes are major hindrances for industrial implementations (Brunetti et al. 2010). The characteristics of each technology are summarized in Table 1.1.

Chemical absorption has been investigated extensively for post combustion CO_2 capture because it has shown the most potential for valid CO_2 control, and it has been commercialized for many decades. Alkanolamines are widely used as the absorbents for CO_2 capture, and can be classified into three types, namely, primary amines, secondary amines and tertiary amines. The condensed structural formulas are RNH_2, R_2NH and R_3N, respectively, in which the alkanolamines comprise of at least one OH and one amine groups. For one thing, primary and secondary amines, such as monoethanolamine (MEA) (Rochelle 2009) and diethanolamine (DEA) (Mandal et al. 2003), have high rates of reaction and high mass transfer toward CO_2 to form carbamate. However, MEA and DEA have several drawbacks, including lower CO_2 loading capacity at 0.5 mol CO_2/mol amine, and high energy consumption for loaded amine regeneration because the carbamate formed has relatively high heat of CO_2 absorption (Mondal et al. 2012). For tertiary amines such as methyl-diethanolamine (MDEA) (Benamor and Aroua 2007), which has a relatively lower reactivity towards CO_2 as compared with the primary and secondary amines, but the loading capacity of MDEA is high about 1 mol CO_2/mol amine, and exhibits a lower energy penalty for

Table 1.1 Characteristic of absorption, adsorption, distillation and membrane in Post-combustion (Kenarsari et al. 2013; Sema 2012)

Technology	Characteristic
Chemical absorption	• Most mature and been proven commercially • Easy to be retrofitted to the existing power plant • High selectivity, high absorption efficiency and low capacity cost • High regeneration energy requirement • Solvent management, i.e. corrosiveness, degradation
Chemical adsorption	• High thermal and chemical stability • No corrosiveness and easy to operate • High capacity and operation cost • Low adsorption and regeneration efficiency
Physical adsorption	• Low selectivity and low adsorption efficiency • High energy consumption for regeneration
Cryogenic distillation	• Can obtain liquid CO_2 with high purity • Significantly energy penalty
Membrane separation	• No need for regeneration process • High separation energy efficiency • Low permeation selectivity • Limited on the operating temperature

regeneration resulting from the relatively lower heat of CO_2 absorption associated with bicarbamate formation (Ramachandran et al. 2006; Vaidya and Kenig 2007).

In addition to the primary, secondary and ternary amines, the specialty amines like steric hindrance amine [e.g. 2-amino-2-methyl-1-propanol (AMP)] (Tontiwachwuthikul et al. 1991) and cyclic diamine [e.g. piperazine (PZ)] (Rochelle et al. 2011) have also been commercially employed. In the reaction of AMP and CO_2 the formed carbamate is not stable, and the stability constant of carbamate for AMP and MEA are 0.1 and 12.5, respectively at 303 K (Vaidya and Kenig 2007), and CO_2 mainly reacts to form carbonate and bicarbonate ions (Khan et al. 2015). AMP has outstanding absorption and desorption characteristics, such as lower energy consumption for desorption, higher degradation resistance and higher loading capacity of 1 mol CO_2/mol amine (Yeh et al. 2001). PZ is generally used as an activator to add into the other amine systems for CO_2 capture due to its rapid reaction rate with CO_2 (Cullinane and Rochelle 2005; Xu et al. 1998), and it can improve CO_2 mass transfer rates and improve kinetics, and blended amines such as MEA with PZ and MDEA with PZ are the typical representatives (Hilliard 2008). In addition, some researchers have reported the use of PZ as an absorbent for absorption/desorption system to capture CO_2, and the concentrated PZ showed better performance than MEA, including, higher absorption rate, lower equivalent work and lower degradation rate. Nevertheless, the solubility of PZ in water is very low, for example, the solubility is 14 wt% at 293 K, CO_2 capture by a high PZ content solution is thus needed to conduct at high temperatures (Yu et al. 2012). However, a comprehensive study of this solvent to include all the activity parameters (e.g. reaction kinetics, CO_2 equilibrium solubility, thermodynamic characteristics, heat duty) should be conducted before its commercial application (Park et al. 2002). In addition, the physicochemical properties of this solvents such as density, viscosity, and specific heat capacity should be measured because they are necessary requirements for determination of the reaction kinetics model and design of gas-liquid contactor columns. In order to apply the developed post-combustion CO_2 capture technology, the fundamental research and pilot plant study should be addressed in order to further understand the process of this technology.

1.2 Objectives of This Book

This book presents a comprehensive review of the most recent information available on all aspects of the PCC processes. It provides designers and operators of amine solvent-based CO_2 capture plants with an in-depth understanding of the most up-to-date fundamental chemistry and physics of the CO_2 absorption technologies using amine-based reactive solvents. Topics covered physical properties, chemical analysis, reaction kinetics, CO_2 solubility, and innovative configurations of absorption and stripping columns. The corrosion and degradation for the post combustion CO_2 capture process also be discussed. Also, the summary of the pilot plant and demonstration plant of this technology is presented as well. The insights will assist engineers, scientists, and decision makers working in academia, industry and government, to gain a better appreciation of the post combustion carbon capture technology.

References

Benamor, A., & Aroua, M. K. (2007). An experimental investigation on the rate of CO_2 absorption into aqueous methyldiethanolamine solutions. *Korean J Chem Eng, 24,* 16–23.

Brunetti, A., Scura, F., Barbieri, G., & Drioli, E. (2010). Membrane technologies for CO_2 separation. *J Membr Sci, 359,* 115–125.

Cullinane, J. T., & Rochelle, G. T. (2005). Thermodynamics of aqueous potassium carbonate, piperazine, and carbon dioxide. *Fluid Phase Equilib, 227,* 197–213.

D'Alessandro, D. M., Smit, B., & Long, J. R. (2010). Carbon dioxide capture: prospects for new materials. *Angew Chem Int Ed Engl, 49,* 6058–6082.

Figueroa, J. D., Fout, T., Plasynski, S., McIlvried, H., & Srivastava, R. D. (2008). Advances in CO_2 capture technology—The US Department of Energy's Carbon Sequestration Program. *Int J Greenhouse Gas Control, 2,* 9–20.

Hilliard MD (2008) A predictive thermodynamic model for an aqueous blend of potassium carbonate, piperazine, and monoethanolamine for carbon dioxide capture from flue gas

Kenarsari, S. D., Yang, D., Jiang, G., Zhang, S., Wang, J., Russell, A. G., et al. (2013). Review of recent advances in carbon dioxide separation and capture. *Rsc Adv, 3,* 22739–22773.

Khan, A. A., Halder, G. N., & Saha, A. K. (2015). Carbon dioxide capture characteristics from flue gas using aqueous 2-amino-2-methyl-1-propanol (AMP) and monoethanolamine (MEA) solutions in packed bed absorption and regeneration columns. *Int J Greenhouse Gas Control, 32,* 15–23.

Liang, Z. H., Rongwong, W., Liu, H., Fu, K., Gao, H., Cao, F., et al. (2015). Recent progress and new developments in post-combustion carbon-capture technology with amine based solvents. *Int J Greenhouse Gas Control, 40,* 26–54.

Mandal, B., Biswas, A., & Bandyopadhyay, S. (2003). Absorption of carbon dioxide into aqueous blends of 2-amino-2-methyl-1-propanol and diethanolamine. *Chem Eng Sci, 58,* 4137–4144.

Miller, D. C., Litynski, J. T., Brickett, L. A., & Morreale, B. D. (2016). Toward transformational carbon capture systems. *AIChE J, 62,* 2–10.

Mondal, M. K., Balsora, H. K., & Varshney, P. (2012). Progress and trends in CO_2 capture/separation technologies: a review. *Energy, 46,* 431–441.

Park, J.-Y., Yoon, S. J., Lee, H., Yoon, J.-H., Shim, J.-G., Lee, J. K., et al. (2002). Density, Viscosity, and Solubility of CO_2 in aqueous solutions of 2-Amino-2-hydroxymethyl-1, 3-propanediol. *J Chem Eng Data, 47,* 970–973.

Ramachandran, N., Aboudheir, A., Idem, R., & Tontiwachwuthikul, P. (2006). Kinetics of the absorption of CO_2 into mixed aqueous loaded solutions of monoethanolamine and methyldiethanolamine. *Ind Eng Chem Res, 45,* 2608–2616.

Rochelle, G., Chen, E., Freeman, S., Van Wagener, D., Xu, Q., & Voice, A. (2011). Aqueous piperazine as the new standard for CO_2 capture technology. *Chem Eng J, 171,* 725–733.

Rochelle, G. T. (2009). Amine scrubbing for CO_2 capture. *Science, 325,* 1652–1654.

Sema T (2012) Kinetics of carbon dioxide absorption into aqueous solutions of 4-(diethylamino)-2-butanol and blended monoethanolamine and 4-(diethylamino)-2-butanol. Faculty of Graduate Studies and Research, University of Regina

Tontiwachwuthikul, P., Meisen, A., & Lim, C. J. (1991). Solubility of carbon dioxide in 2-amino-2-methyl-1-propanol solutions. *J Chem Eng Data, 36,* 130–133.

Vaidya, P. D., & Kenig, E. Y. (2007). CO_2-alkanolamine reaction kinetics: a review of recent studies. *Chem Eng Technol, 30,* 1467–1474.

Xu, G.-W., Zhang, C.-F., Qin, S.-J., Gao, W.-H., & Liu, H.-B. (1998). Gas-liquid equilibrium in a CO_2-MDEA-H_2O system and the effect of piperazine on it. *Ind Eng Chem Res, 37,* 1473–1477.

Yang, H., Xu, Z., Fan, M., Gupta, R., Slimane, R. B., Bland, A. E., et al. (2008). Progress in carbon dioxide separation and capture: a review. *J Environ Sci, 20,* 14–27.

Yeh, J. T., Pennline, H. W., & Resnik, K. P. (2001). Study of CO_2 absorption and desorption in a packed column. *Energy Fuels, 15,* 274–278.

Yu, C.-H., Huang, C.-H., & Tan, C.-S. (2012). A review of CO_2 capture by absorption and adsorption. *Aerosol Air Qual. Res, 12,* 745–769.

Chapter 2
Solvent Property of Amine Based Solvents

Abstract In this chapter, the method about selection of amine solvent for CO_2 capture plant was proposed in order to improve the efficiency of the process. The properties of amines solvents related to CO_2 capture were discussed in order to design the plant based on the selected amines. The physical properties (i.e. density, viscosity, specific heat capacity, Henry's law constant, and diffusivity) are very key parameters for design of the plant or determination the performance of solvent. The measurement of those physical properties are also present in this chapter. Also, the present correlations for the experimental results of Henry's law constant and the diffusivity are discussed. In addition, some chemical properties related to the reaction of amines and CO_2 such as the solubility, reaction kinetics should be addressed before the further study. Those parameter determine the capacity of solvent or the reaction rate of the solvent, which are necessary for the process design or process stimulation. In order to understand the process of reaction, the ions speciation plots are studied by using the different methods, which are added in this chapter. The trend of the new solvent and the improvement of the current solvent are discussed in this chapter.

2.1 The Selection of Amine for CO_2 Capture

For the technology of CO_2 capture based on amines, the researchers are selecting the efficient amines solvents, which should have good performance for CO_2 capture (e.g. fast reaction kinetics, high absorption capacity, faster mass transfer, and low energy for regeneration) (Liu et al. 2014; Rochelle 2009). In our previous work (Liu et al. 2017a), the selection method or the guideline for the amines solvents for the post combustion CO_2 capture technology, which could provide direction for future design or screening of better or ideal amines for CO_2 capture. All detailed information could be founded in our previous work (Liu et al. 2017a).

H. Liu et al., *Post-combustion CO_2 Capture Technology*, SpringerBriefs in Petroleum Geoscience & Engineering, https://doi.org/10.1007/978-3-030-00922-9_2

2.2 Physical Properties

Physicochemical properties of solvents such as density, viscosity, and specific heat capacity are regarded as one of significant parameters for practical application. Densities and viscosities are necessarily required for determination of reaction kinetics model and design of gas-liquid contactor columns, because the data of densities and viscosities are essential for calculations (Park et al. 2002). Those data could be used for dimensioning the column diameter, velocities, and pressure drop in a column as described by Eckert (1970) and the calculation of mass transfer correlations and mass transfer area as described by Wang et al. (2005). Specific heat capacity plays an essential role in the industries of gas-treating, especially for CO_2 capture process, because the values of specific heat capacity are required to calculate the heat balance of absorber, stripper and heat exchanger.

2.2.1 Density, Viscosity and Specific Heat Capacity

Physicochemical properties of solvents such as density, viscosity, and specific heat capacity are regarded as one of significant parameters for practical application. Densities and viscosities are necessarily required for determination of reaction kinetics model and design of gas-liquid contactor columns, because the data of densities and viscosities are essential for calculations (Park et al. 2002). Those data could be used for dimensioning the column diameter, velocities, and pressure drop in a column as described by Eckert (1970) and the calculation of mass transfer correlations and mass transfer area as described by Wang et al. (2005). Density and viscosity have a significant effect on the mass transfer coefficient, hydrodynamic behavior, and diffusivity behavior. Both density and viscosity are necessary for the design of the absorber and stripper column (Gao et al. 2017a). Specific heat capacity plays an essential role in the industries of gas-treating, especially for CO_2 capture process, because the values of specific heat capacity are required to calculate the heat balance of absorber, stripper and heat exchanger.

For measurement of density, the densities were presented by using density meter with model of DMA 4500M (Anton Paar) with an accuracy of 0.00001 g/cm3. The temperature of the system was controlled with the accuracy of 0.01 K. Before conducting the measurement of density, the density meter was cleaned by acetone and methanol. In addition, the equipment was well dried by air. After that, the sample of solution was injected to measure the density. All the process (cleaning, drying, injection, measurement) were conducted automatically. In order to obtain the reliable result, all the samples were run two times and the average was achieved for the final result. For the measurement of viscosity, the viscosity data were obtained by using density meter with model of Lovis (Anton Paar) with an accuracy of 0.5%. The temperature of the system was controlled with an accuracy of 0.01 K. As in the case of density meter, the system was cleaned using the acetone and methanol and was dried by air. And then, the sampled was introduced to DMA 4500M and Lovis 2000 at the same time. The viscosity was measured at the desired temperature. The result of viscosity of the sample is given as an average of two measurements. The measurement of thermal conductivity, thermal diffusivity and specific heat capacity were by obtained by employing thermal conductivity meter (THW-LAMBDA). This performance of thermal conductivity meter was achieved using a transient. Firstly, the liquid sample was introduced into the stainless bottle. And then, a thin wire was fitted to the bottle and immersed into the sample. A current was conducted through the wire in order to heat the sample. The resistance as a function of time is obtained. Meanwhile, the temperature of sample with respect to temperature was also monitored. The thermal properties of a sample can be automatically generated based on the typical graph of versus time using THW-LAMBDA, which gives an accuracy of 5%.

2.2.2 *Henry's Law Constant and Diffusivity*

In addition, Physical solubility and physical diffusivity of CO_2 in aqueous amine solution under different temperature and solution concentration play an important role in developing the CO_2 absorption kinetic correlation models. Unfortunately, it is difficult to measure the physical solubility and physical diffusivity directly due to the accompanying chemical reactions between CO_2 and the aqueous amine solution.

Compared with CO_2, N_2O has a similar molecular weight, molecular space structure, molecular volume, and chemical configuration. Therefore, it can be assumed that the physical solubility and physical diffusivity of CO_2 and N_2O in water or in aqueous amine solution are similar at the same temperature. This concept was first proposed by Clarke (1964). This concept was applied for estimating the CO_2 physical solubility and physical diffusivity in aqueous amine solutions, according to Versteeg and Van Swaaij (1988), the physical solubility and physical diffusivity of CO_2 in aqueous amine solutions can be determined from N_2O diffusion coefficient in the same solution and the diffusivity in water at the same temperature. This method is called the N_2O analogy.

$$He_{CO_2-\text{amine}} = He_{N_2O-\text{amine}}\left(\frac{He_{CO_2-H_2O}}{He_{N_2O-H_2O}}\right) \tag{2.1}$$

$$D_{CO_2-\text{amine}} = D_{N_2O-\text{amine}}\left(\frac{D_{CO_2-H_2O}}{D_{N_2O-H_2O}}\right) \tag{2.2}$$

where $He_{CO_2\text{-amine}}$ and $He_{CO_2\text{-water}}$ are the physical solubility of CO_2 in amine solution and water, respectively. $He_{N_2O\text{-amine}}$ and $He_{N_2O\text{-water}}$ are the solubility of N_2O in amine solution and water, respectively. $D_{CO_2\text{-amine}}$ and $D_{CO_2\text{-water}}$ are the physical diffusivity of CO_2 in amine solution and water, respectively. $D_{N_2O\text{-amine}}$ and $D_{N_2O\text{-water}}$ are the diffusivity of N_2O in amine solution and water, respectively.

This analogy is an effective approach and has been widely applied into determining the physical diffusivity of CO_2 in aqueous amine solutions such as AMP, MIPA, DIPA, EDA, DETA, DMMEA, MEA-MDEA, MEA-AMP, MDEA-AMP, AMP-TMS, DIPA-AMP-PZ, MDEA-ionic liquid (Haghtalab et al. 2014; Haimour and Sandall 1984; Hartono et al. 2008; Li and Lai 1995; Munder et al. 2000; Park and Sandall 2001; Sada et al. 1978; Saha et al. 1993; Versteeg and Van Swaaij 1988; Wang et al. 1992; Xu et al. 1992).

2.2.2.1 Henry's Law Constant

Henry's law constant represents the gas concentration of gas-liquid interface, which is an important parameter for studying mass transfer of gas and reaction kinetics of gas absorption. Henry's law constants of CO_2 in aqueous amine solution under different temperature and solution concentration play an important role in developing the CO_2 absorption kinetic correlation models and mass transfer of CO_2 in the absorber. According to Versteeg and Van Swaaij (1988). Thus, the Henry's law constants of CO_2 in aqueous amine solutions can be determined from Henry's law constants of N_2O in the same solution and Henry's law constants of N_2O and CO_2 in water at the same temperature.

The Henry's law constant of N_2O in water is shown in Eq. 2.7. Henry's law constants of CO_2 in water could be obtained from the literature (Wang et al. 1992) as follows:

$$\mathrm{He}_{\mathrm{CO_2-H_2O}} = 2.82 \times 10^6 \exp\left(-\frac{2204}{T}\right) \tag{2.3}$$

To allow the experimental data to be readily utilized, many correlation models (Park and Sandall 2001) have been proposed such as the polynomial model, the semiempirical model, and the Redlich-Kister model (Hartono et al. 2008; Li and Lai 1995; Versteeg and Van Swaaij 1988; Wang et al. 1992), ENRT model (Haghtalab et al. 2014). According to Versteeg and Van Swaaij (1988), the polynomial model can be used successfully for predicting the N_2O solubility in DMMEA, DEA, TEA, MIPA, and DIPA solutions at the range of the temperatures and amine concentrations. Based on the work of Wang et al. (1992). and Li and Lai (1995), the semi-empirical model is applicable for single amine aqueous solution (MEA, DEA, MDEA, AMP, DIPA) and blend amine aqueous solution (MEA-MDEA, MEA-AMP, MDEA-AMP, AMP-TMS). Hatono et al. (2009) have pointed out that the Redlich-Kister model gives a good estimation of the solubility of N_2O in aqueous amine solutions considering the excess quantity of Henry's constant and is widely used for polyamine solution. Hartono et al. (2008) and Haghtalab et al. (2014) have observed that the ENRTL model presents a good correlation of the present experimental data with for the quaternary systems.

The experimental set-up used to measure Henry's law constant is the similar to the one used in the work of Sema et al. It consisted of two parts; reactor and controller.

The reactor was equipped with a rotary-type 600 mL stainless steel container (model Parr 5500, Parr Instrument Co. Moline, IL). The reactor was made up of a variable-speed impeller, a heating mantle, a cooling coil, a gas feed port, a thermocouple, and a pressure transducer. Another part was a controller with model Parr 4843 (Parr Instrument Co., Moline, IL), which was used to control the temperature of reactor and monitor the temperature and pressure of the reactor.

In order to correlate the obtained Henry's law constant in amines solution, some models [semi-empirical model (Wang et al. 1992), Redlich-Kister model (Hartono et al. 2008), and polynomial model (Versteeg and Van Swaalj 1988)] were employed to represent the experimental results.

Redlich-Kister Equation

Redlich-Kister equation is another correlation to represent the Henry's law constant of aqueous amines solutions, which also can be widely used to represent the physical properties of aqueous amines solutions (e.g. density, viscosity, and refractive index) (Maneeintr et al. 2008). As Redlich and Kister (1948) mentioned, the correlation equation for aqueous amines solutions should take the correlation equation of pure amines and water into account, because the aqueous amines solutions is a binary solution of the pure amine with water. In addition, an important factor, temperature, also needs to be added in the correlation equation. Thus, Redlich-Kister equation can finally be expressed as in the following equation:

$$\begin{aligned}\ln \mathrm{He}_{\mathrm{N_2O}-Solution} &= x_{\mathrm{H_2O}} \ln \mathrm{He}_{amine-\mathrm{H_2O}} + x_{amine} \ln \mathrm{He}_{\mathrm{N_2O}-amine} + x_{\mathrm{H_2O}} x_{amine} [A_0 \\ &+ A_1(x_{amine} - 1) + A_2(x_{amine} - 1)^2 + A_3(x_{amine} - 1)^3 + A_4(x_{amine} - 1)^4]\end{aligned} \tag{2.4}$$

where x_i is mole fraction of water and amine, respectively, A_i could be considered as a temperature dependent parameter.

Based on the work of Hartono et al. (2008), A_i is defined as a polynomial function of temperature, which is shown in the following equation:

$$A_i = a_0 + a_1 T + a_2 T^2 + a_3 T^3 + a_4 T^4 \tag{2.5}$$

where a_i are constants of the temperature coefficient equation, T is the temperature (K).

Semi-empirical Model

A semi-empirical model proposed by Wang et al. (1992) is commonly used to correlate the Henry's law constants of aqueous amines solutions (e.g. MEA, DEA, MDEA, AMP, and DIPA). According to their work, it was founded that this model could give a good correlation and predication. Similar to Redlich-Kister equation, the correlation equation of pure amines and water were also taken into account in the Semi-empirical model. Also, temperature is also introduced into the Semi-empirical model equation with a different expression. Thus, the semi-empirical model is expressed as in the following equation:

$$\ln He_{N_2O-Solution} = \emptyset_{H_2O}\emptyset_{amine}\left(k_1 + k_2 t + k_3 t^2 + k_4\emptyset_{H_2O}\right) + \emptyset_{H_2O}\ln He_{N_2O-H_2O} + \emptyset_{amine}\ln He_{N_2O-amine} \quad (2.6)$$

where $\emptyset_i$ represents volume fraction of water and amine, respectively, and t represents temperature (°C).

$He_{N_2O-H_2O}$ is the Henry's law constants of water, which could be obtained by using the following equation:

$$He_{N_2O-H_2O} = 8.55 \times 10^6 \exp\left(\frac{-2284}{T}\right) \quad (2.7)$$

Polynomial Model

One of the common models, the polynomial model, was applied to correlate and represent the experimental results of Henry law constant in 1DMA2P solution. The polynomial model mentioned by Versteeg and Van Swaaij (1988) is usually used to predict the Henry's law constant in an amine solution. This model was employed to correlate and predict the Henry's law constant in different amine solutions (Diethanolamine (DEA), Triethanolamine (TEA), dimethylmonoethanolamine (DMMEA), Monoisopropanolamine (MIPA), and (DIPA)). This model can be shown in the following equation:

$$He_{N_2O-Amine} = A_0 + A_1C + A_2C^2 + A_3C^3 + A_4C^4 \quad (2.8)$$

where C is the concentration, and A_i is the temperature coefficient.

The temperature coefficient was also correlated using a polynomial function (Hartono et al. 2008), which can be described as follows:

$$A_i = a_0 + a_1T + a_2T^2 + a_3T^3 + a_4T^4 \quad (2.9)$$

where a_i are constants of the temperature coefficient equation, T represents the temperature (K).

2.2.2.2 Diffusivity

Concerning the solubility of N_2O in aqueous amine solution, the diffusion of N_2O has also attracted much attention of researchers. Diffusivity of CO_2 in aqueous amine solution under different temperatures and solution concentrations plays an important role in developing the CO_2 absorption kinetics correlation models. Unfortunately, it is difficult to measure physical diffusivity directly due to the accompanying chemical reactions between CO_2 and the aqueous amine solution.

According to Versteeg and Van Swaaij (1988), the physical diffusivity of CO_2 in aqueous amine solutions can be determined from N_2O diffusion coefficient in the same solution and the diffusivity in water at the same temperature.

The physical diffusivity of N_2O in aqueous amine solution has been performed by various setups such as a single-sphere absorber, a wetted wall column, a stirred cell reactor and a laminar jet absorber. Until now, the diffusion of N_2O has been measured in various amine systems. (1) Single amines such as MEA, MDEA, DEA, TEA, AMP, DEEA, DIPA, and DMMEA. (2) Blend amines such as MEA-MDEA, AMP-MDEA, AMP-MEA, MDEA-DEA, DEA-AMP, PZ-MDEA (Ko et al. 2001; Li and Lai 1995; Li and Lee 1996; Versteeg and Van Swaaij 1988; Xu et al. 1992).

In addition, many correlation models also have been proposed for predicting the diffusion of N_2O in aqueous amine solution. The modified Stokes-Einstein model correlates N_2O diffusivity in aqueous amine solutions via viscosity. The model describes the dependence of N_2O diffusivity in terms of viscosity. Versteeg and Van Swaaij (1988) measured the physical diffusivity of N_2O in aqueous MEA, DEA, TEA, MIPA solution and pointed out that the modified Stokes-Einstein model was applicable for predicting the N_2O diffusivity into these aqueous amine solutions.

Modified Stokes-Einstein Model

The modified Stokes-Einstein model proposed by Versteeg and Van Swaaij (1988) can be used to estimate the N_2O diffusivity into aqueous amine solutions. By applying this modified model, the N_2O diffusivity in MEA, DEA, TEA, MIPA and DIAP were well correlated and predicted. Based on modified Stokes-Einstein model, the N_2O diffusivity in amines solution shows a relationship with viscosity of amines solution as follows:

$$D_{N_2O} \times \eta^{0.8} = Constant \tag{2.10}$$

where η is viscosity of 1DMA2P solution.

Semi-empirical Model

The semi-empirical model is another model, which could be used to represent the N_2O diffusivity of aqueous amine solutions. This model was proposed by Ko et al. (2001), and has been applied to predict the N_2O diffusivity in aqueous MEA, DEA, DIPA MDEA, AMP, TEA and AMP. According to the work of Ko et al. (2001), two parameters (concentration and temperature) are taken into account in the semi-empirical model. This is shown as follows:

$$D_{N_2O} = (b_0 + b_1 C + b_2 C^2)\exp\left(\frac{b_3 + b_4 C}{T}\right) \tag{2.11}$$

where C is the concentration of 1DMA2P solution, and T is temperature.

2.3 Chemical Property

2.3.1 Thermodynamic Model

The thermodynamic model for representing acid gas solubility in aqueous amine solutions is crucial. By far, a number of models (i.e. Kent-Eisenberg model, Modified Kent-Eisenberg model, UNIQUAC model, electrolyte-NRTL model) have been developed to describe the vapour-liquid equilibria.

Kent-Eisenberg model (KE model) and Modified Kent-Eisenberg (M-KE model) models are semi-empirical models used for represent CO_2-amine-H_2O system. It considers the chemical reaction in liquid phase, CO_2 concentration in gas phase concentration as well as a description of CO_2 relationship in liquid phase and gas phase. In the work of Kent and Eisenberg (1976), it predicted the solubility of acid gas and their mixtures in monoethanolamine (MEA) and diethanolamine (DEA) solutions. The equilibrium constants of amine protonation and carbamate formation were fitted to experimental data as functions of temperature. The model could give acceptable prediction result for CO_2-amine-H_2O system. To expand the application range of KE model, some modifications were done on the equilibrium constants of reactions and therefore developed M-KE model. Li and Shen (1993) applied M-KE models for predicting equilibrium solubility of CO_2 in aqueous amine solutions of blending MEA and methyldiethanolamine (MDEA). The chemical equilibrium constants were expressed as function of temperature, amine concentration and CO_2 loading. It showed satisfactory results for temperature ranging from 40 to 100 °C and for CO_2 partial pressure up to 2000 kPa. Hu and Chakma (1990) presented equilibrium solubility of CO_2 and H_2S in aqueous 2-amino-2-(ethoxy)ethanol (DGA) solutions using M-KE model. The equilibrium constant of carbamate formation was found as function of temperature, free acid gas concentration and DGA concentration as well. The mathematical model provided prediction results agreed well with experimental data. Tontiwachwuthikul et al. (1991) determined solubility of CO_2 in 2 and 3 M solutions of 2-amino-2methyl-l-propanol (AMP) at 20, 40, 60, 80 °C and for CO_2 partial pressure from approximately 1–100 kPa. The equilibrium constant of amine protonation was found to be function of temperature, CO_2 physical concentration, amine concentration as suggested by work of Chakma and Meisen. The M-KE model was found represent the experimental data quite accurately. However, since empirical model is limited by the experimental condition, it may fail to give reasonable prediction result under a wider range of temperature, CO_2 partial pressure or amine concentration. Therefore, some rigorous thermodynamic framework considering activity coefficients were developed such as electrolyte-NRTL model and UNIQUAC model. In vapour-liquid equilibria model, electrolyte-NRTL equation takes both long-rang electrostatic interactions and short rang-binary interactions to represent activity coefficients. In work of Austgen (1991), electrolyte-NRTL model was applied to predict H_2S and CO_2 solubility in mixtures of MDEA with MEA or DEA (David et al. 1991). The experimental data were obtained at 40 and 80 °C over a wide range of CO_2 partial pressure while it concluded that the representation of model is good. Faramariz et al. (2009) used extended UNIQUAC model to the thermodynamic representation of CO_2 absorption in aqueous MEA, MDEA and their mixtures. The experimental data used in this model covered CO_2 pressure range of 3–13,000 kPa and temperature of 25–200 °C. By using only one set of parameters for correlation of different thermodynamic properties, the model could represent the experimental data well.

2.3.2 *Ion Speciation of Amine-CO_2-H_2O Systems*

The VLE data of the systems of amine-H_2O-CO_2 are essential for the research and development of the theoretical models, which are very significant for the design of absorption and desorption process of CO_2 and simulation of the CO_2 treating plant (Holmes et al. 1998; Jakobsen et al. 2005). In order to obtain precise vapor–liquid equilibrium (VLE) models of the amine-H_2O-CO_2 systems, ion speciation plots of cations and anions at different CO_2 loadings (VLE plots) are required (Liu et al. 2015).

Based on the reports of the related researchers, with the exception of simulation by the models and computer, the ion speciation plots of the amine-H_2O-CO_2 systems can be obtained by three methods, they are the pH method, the pH + NMR method and the ^{13}C NMR method.

For the pH method, the ion speciation can be generated by the calculation based on the reaction equilibrium constants, which exist in the amine-H_2O-CO_2 system under the experimental environment. As reported by Liu et al. (2015). Two-dimensional (2D) VLE plots of the 1-dimethylamino-2-propanol (1DMA2P)-H_2O-CO_2 system, 3D VLE surface representing the concentrations of ions in the 1DMA2P solution and the equations used to predict CO_2 loading in 1DMA2P solution were generated by calculating the ion concentrations in the 1DMA2P solution using the pH method.

For the pH + NMR method, the ion concentrations of the carbamate of amine, HCO_3^- and CO_3^{2-} can be generated based on the ratio of the integrated peak areas and the chemical shift of the peaks, while the ion concentrations of free amine and protonated amine are obtained based on the reaction equilibrium constants. As reported by Fan et al. (Fan et al. 2009), the speciation studies in aqueous monoethanolamine (MEA) solution at different CO_2 loadings have been conducted by the pH + NMR method.

For the ^{13}C NMR method, the ion concentrations of free amine and protonated amine are generated based on the related protonation calibration curves and the mass balance of amine, while the ion concentrations of the carbamate of amine, HCO_3^- and CO_3^{2-} can be generated based on the ratio of the integrated peak areas, the chemical shift of the peaks and the mass balance of CO_2 absorbed into the amine solution, which is the same with those in the pH + NMR method. As shown in the research of Shi et al. (Shi et al. 2014a) the quantitative liquid phase speciation analysis of a complex quaternary CO_2 loaded blended amine solution such as MEA-DEAB-CO_2-H_2O at high and low temperatures respectively representing CO_2 stripping and absorption conditions have been conducted using the ^{13}C NMR method.

2.3.3 *Kinetics of CO_2 Absorption into Aqueous Amine Solutions*

The reaction kinetics of CO_2 absorption into aqueous amine solutions is a critical parameter to simulate the optimized absorption process and design the appropriate absorption column. To obtain the kinetics data and understand the kinetics behavior, reaction kinetics has been studied for several decades to aid the understanding of the reaction mechanism. In recent years, the kinetics of CO_2 absorption into a wide variety of aqueous amine solutions have been studied, such as MEA, DEA, MDEA, AMP, DETA, MDEA-MEA, MEA-AMP, MDEA-PZ, and AMP-PZ (Versteeg and Van Swaaij 1988; Saha and Bandyopadhyay 1995; Hartono et al. 2009; Mandal et al. 2001; Zhang et al. 2001; Sun et al. 2005; Edali et al. 2009).

The rate of CO_2 absorption into aqueous AMP solutions was measured by Saha and Bandyopadhyay (1995) using a wetted wall column over temperature and AMP concentration ranges of 294–318 K and 0.5–2.0 mol/L, respectively. In their work, based on the zwitterion mechanism, a graphical method was used to determine the reaction rate constant and the order of reaction with respect to concentration of CO_2 and AMP. Ramachandran et al. (2006) studied the kinetics of CO_2 absorption into an aqueous solution of CO_2 loaded, blended MEA-MDEA with a laminar jet absorber. Generally, the absorption rate data obtained from the CO_2 absorption equipment can then be interpreted using graphical methods, simplified kinetics models based on reaction mechanisms (i.e., zwitterion, termolecular or base-catalyzed hydration mechanisms) and comprehensive numerically solved reaction kinetics models that take into account the coupling between chemical equilibrium, mass-transfer and chemical kinetics of all possible reactions.

Hagewiesche et al. (1995) developed the very first numerical comprehensive reaction rate/kinetics models, which takes into account the coupling between chemical equilibrium, mass transfer, and chemical kinetics of all possible chemical reactions. The rate of CO_2 absorption into an aqueous solution of unloaded, blended MEA-DEAB was measured using a laminar jet absorber. The CO_2 absorption rate, the enhancement factor, and the rate coefficient of reaction between CO_2 and MEA can be predicted by their developed reaction rate/kinetics model. However, the experiment was performed with unloaded solution at only 313 K.

The kinetics of CO_2 absorption in aqueous unloaded solutions of MDEA and blended DEA-MDEA was studied using a wetted sphere absorber over a temperature range of 293–342 K and concentration range of 10–30% mass MDEA by Rinker et al. (1995 and 2000). Their comprehensive reaction rate/kinetics models also take into account the coupling between chemical equilibrium, mass transfer, and chemical kinetics of all possible chemical reactions. The PDEs system was developed and transformed into ODEs using the method of lines. The system of ODEs was solved numerically using the code DDASSL in FORTRAN. Their predicted rates of CO_2 absorption were found to agree reasonably well with the experimental results.

Aboudheir et al. (2003) measured the rates of CO_2 absorption into an aqueous solution of loaded MEA using a laminar jet absorber over a temperature range of 293–333 K, a concentration range of 3–9 M, and a CO_2 loading range of 0.1–0.49, developed the very more comprehensive reaction rate/kinetics model. Based on the Barakat-Clark scheme, the partial differential-nonlinear algebraic equations were solved using the finite difference method (FDM) in their study. They reported that the developed comprehensive reaction rate/kinetics model favorably predicted CO_2 absorption rates. Also, the absorption rate constant gave statistically robust results and in keeping with the literature's values.

The kinetics of DETA and CO_2 in an aqueous system was studied by Ardi Hartono et al. (2009). A string of discs contactor was used to measure the kinetics CO_2 absorption in unloaded aqueous solution with the diethylenetriamine (DETA) concentrations ranging between 1.0 and 2.9 kmol/m^{-3} and at temperatures ranging between 298.1 and 332.3 K. The reaction rates strongly increase as the increasing the concentrations and temperatures. Both the termolecular and the zwitterion models were applied to interpret the experimental data and gave identical results for all practical purposes. In this study, it was found that fitting of experimental data to the termolecular mechanism gave statistically more robust results than fitting to the zwitterion mechanism.

Recently, Edali et al. (2007 and 2009) studied blended MEA-MDEA and blended MDEA piperazine (PZ) system. In their studies, two numerical techniques were used to get the partial differential nonlinear algebraic equations solved: the FDM based on the Barakat-Clark scheme and the finite element method (FEM) based on COMSOL software. It was found that accurate predictions of CO_2 absorption rate results were obtained by both FDM and FEM. What's more, there are not significant difference between the results obtained from FDM and FEM.

Overall, kinetics of CO_2 absorption studies have been conducted to aid the investigation of how fast CO_2 reacts with an amine in terms of k_2 (the higher the k_2, the faster the reaction kinetics). For the generally used amines, the k_2 can be ranked as $k_{2,PZ} \ggg k_{2,DETA} > k_{2,AEEA} \gg k_{2,MEA} \gg k_{2,AMP} \approx k_{2,DEA} > k_{2,MDEA}$. In addition, the kinetics behavior can also be simply determined using carbamate stability (K_c) and free energy for carbamate formation (ΔG_c) for primary and secondary amines, and base strength of amine (pKa) for tertiary amines.

2.4 Novel Approaches and Future Trends

2.4.1 The Improvement of the Present Solvent

As well known, the recent technology for post combustion CO_2 capture has a big challenge, which is the high energy requirement for the stripper column. Given this scenario, a novel approach was proposed by Idem et al. (Idem et al. 2011; Srisang et al. 2017a), they pointed out that the introduction of solid acid catalyst into amine solution regeneration process greatly decreased the energy requirement. Different single solid acid catalysts have been reported, including molecular sieve catalysts HZSM-5, MCM-41 and SAPO-34 (Liang et al. 2016; Zhang et al. 2018); solid superacid catalysts SO_4^{2-}/TiO_2 and SO_4^{2-}/ZrO_2 (Liu et al. 2017b); metal oxide catalysts γ-Al_2O_3, MoO_3, TiO_2 and V_2O_5 (Bhatti et al. 2017). Among these catalysts, HZSM-5 (Brϕnsted acid catalyst) and γ-Al_2O_3 (amphoteric oxide) presented a better catalytic performance. Based on the literature (Liang et al. 2016; Shi et al. 2014b), HZSM-5 shows a superior catalytic CO_2 desorption performance to γ-Al_2O_3 in MEA solutions at the high CO_2 loading region, due to its remarkable Brϕnsted acid sites (BAS). Because BAS play an essential role in the CO_2 desorption process of carbamate ($AmineCOO^-$) breakdown reaction. However, γ-Al_2O_3 presents a better catalytic CO_2 desorption performance than HZSM-5 when the CO_2 loading is low, which could be attributed to the presence of prominent basic sites. The basic sites have a positive effect on the deprotonation reaction of protonated amine ($AmineH^+$) (Srisang et al. 2017b). Both of BAS and basic site can enhance the regeneration process in different way, respectively. Therefore, there is a great opportunity to improve the catalytic performance to further reduce the energy requirement by combing the BAS and basic sites to obtain bifunctional catalysts for the rich MEA regeneration process. To the best of our knowledge, no study has been reported in literature on the application of the bifunctional catalysts for this purpose.

Recently, bifunctional catalysts of metal oxide modified acid zeolites, such as, Ni-HZSM-5 (Jiang et al. 2012; Zhang et al. 2012), MgO-HZSM-5 (Saravanan et al. 2017), In_2O_3-HZSM-5 (Gao et al. 2017), and Na-Fe_3O_4-HZSM-5 (Wei et al. 2017) have attracted much attention in the organic chemistry reactions. These bifunctional catalysts showed a higher catalytic activity and catalyst stability than single zeolite catalysts, which resulted from the good synergistic effect between the zeolite and metal oxide (Balasundram et al. 2018). Furthermore, Shi et al. (2016) reported that the preparation methods of bifunctional catalyst also have a great influence on their catalytic activity. Khoshbin and Haghighi (2013) put forward an advanced method, namely, combined precipitation ultrasound for preparation of the hybrid catalyst to acquire the catalyst with a better performance. Because, the presence of ultrasound irradiation would generate the fast nucleation, limited particle growth, strong interaction between particles and fragmentation of particle aggregates in the solution. All of them led to obtain highly reactive surface area and porosity, enhance mass transfer and provide a well dispersed particles (Frusteri et al. 2015).

2.4.2 *Alternative Solvents for Post Combustion CO_2 Capture*

2.4.2.1 Amino Acid Salts

Amino acid salts are a class of chemical absorbents for CO_2 capture normally referred to as aqueous alkaline salts of amino acids which contain two important functional groups, namely amine ($-NH_2$) and carboxylic acid (–COOH). However, Taurine that contains a sulfonic acid group instead of the carboxylic acid can also be considered as an amino acid (Van Holst et al. 2006) and is one of the popular amino acids that have been tested for CO_2 capture.

2.4.2.2 Phase-Change Solvents

Phase change solvents such as the Alstom chilled ammonia (CAP) process (Kozak et al. 2009), the DMX process (Raynal et al. 2011a, b) and thermomorphic biphasic solvent (TBS) systems (Agar et al. 2008), constitute a new set of reactive solvents and have shown their prominent advantages for CO_2 capture. These solvents form two immiscible phases upon CO_2 loading, either precipitate/liquid phases or two liquid phases. The concept of using phase change solvents is to recycle the CO2-lean phase (liquid phase) back to the absorber without regeneration while only the CO_2-rich phase (either precipitate or liquid phase) is transported back to the stripper. In this way, the CO_2-lean liquid phase maintains a large driving force for CO_2 absorption to absorb more CO_2 while the CO_2-rich phase with its lower circulation rate and richer CO_2 loading can achieve lower consumption of sensible heat and vaporization heat for regeneration. The end result is higher efficiency for both CO_2 absorption and desorption.

2.4.2.3 Ionic Liquids

Ionic liquids are non-volatile liquids (salts), composed of a cation and an anion and are usually liquid at room temperature. Ionic liquids are called designer solvents as numerous combination of cation and anion are possible. It was shown that CO_2 is remarkably soluble in these solvents, which spurred interest in these solvents. Much research was done in the last fifteen years to develop competitive solvents based on ionic liquids. Conventional ionic liquids absorb CO_2 through physical absorption, and were found not competitive with conventional amines due to lower capacity and high viscosity. Later, task-specific ionic liquids were designed to enhance absorption capacity by chemical absorption incorporating a reactive functional group in the cation or the anion (Bates et al. 2002). However, the conventional aqueous amine system remained preferred overall due to lower cost, higher capacity, and low viscosity (Yusoff et al. 2013).

References

Agar, D., Tan, Y., & Zhang, X. (2008). *CO_2 removal processes by means of absorption using thermomorphic biphasic aqueous amine solutions*. Patent WO/2008/015217.

Balasundram, V., Ibrahim, N., Kasmani, R. M., Isha, R., Hamid, M. K. A., Hasbullah, H., et al. (2018). Catalytic upgrading of sugarcane bagasse pyrolysis vapours over rare earth metal (Ce) loaded HZSM-5: Effect of catalyst to biomass ratio on the organic compounds in pyrolysis oil. *Applied Energy, 220,* 787–799.

Bates, E. D., Mayton, R. D., Ntai, I., & Davis, J. H. (2002). CO_2 capture by a task-specific ionic liquid. *Journal of the American Chemical Society, 124,* 926–927.

Bhatti, U. H., Shah, A. K., Kim, J. N., You, J. K., Choi, S. H., Lim, D. H., et al. (2017). Effects of transition metal oxide catalysts on MEA solvent regeneration for the post-combustion carbon capture process. *ACS Sustainable Chemistry & Engineering, 5,* 5862–5868.

Clarke, J. (1964). Kinetics of absorption of cardon dioxide in monoethanolamine solutions at short contact times. *Industrial and Engineering Chemistry Fundamentals, 3,* 239–245.

David, M. A., Gary, T. R., & Chau-Chyun, C. (1991). Model of vapor-liquid equilibria for aqueous acid gas-alkanolamine systems. 2. Representation of H_2S and CO_2 solubility in aqueous MDEA and CO_2 solubility in aqueous mixtures of MDEA with MEA or DEA. *Industrial and Engineering Chemistry Research, 30,* 543–555.

Eckert, J. (1970). Selecting proper distillation column packing. *Chemical Engineering Progress, 66,* 39.

Fan, G.-J., Wee, A. G., Idem, R., & Tontiwachwuthikul, P. (2009). NMR studies of amine species in MEA–CO_2–H_2O system: Modification of the model of vapor–liquid equilibrium (VLE). *Industrial and Engineering Chemistry Research, 48,* 2717–2720.

Frusteri, F., Cordaro, M., Cannilla, C., & Bonura, G. (2015). Multifunctionality of Cu–ZnO–ZrO_2/H-ZSM5 catalysts for the one-step CO2-to-DME hydrogenation reaction. *Applied Catalysis, B: Environmental, 162,* 57–65.

Gao, H., Gao, G., Liu, H., Luo, X., Liang, Z., & Idem, R. O. (2017a). Density, viscosity, and refractive index of aqueous CO_2-loaded and-unloaded ethylaminoethanol (EAE) solutions from 293.15 to 323.15 K for post combustion CO_2 Capture. *Journal of Chemical and Engineering Data, 62,* 4205–4214.

Gao, P., Li, S., Bu, X., Dang, S., Liu, Z., Wang, H., et al. (2017b). Direct conversion of CO_2 into liquid fuels with high selectivity over a bifunctional catalyst. *Nature Chemistry, 6,* 1019–1024.

Haghtalab, A., Eghbali, H., & Shojaeian, A. (2014). Experiment and modeling solubility of CO_2 in aqueous solutions of diisopropanolamine + 2-amino-2-methyl-1-propanol + piperazine at high pressures. *The Journal of Chemical Thermodynamics, 71,* 71–83.

Haimour, N., & Sandall, O.C. (1984). Absorption of carbon dioxide into aqueous methyldiethanolamine. *Chemical Engineering Science, 39*, 1791–1796.

Hartono, A., Juliussen, O., & Svendsen, H. F. (2008). Solubility of N_2O in aqueous solution of diethylenetriamine. *Journal of Chemical and Engineering Data, 53,* 2696–2700.

Holmes, P. E., Naaz, M., & Poling, B. E. (1998). Ion concentrations in the CO_2-NH_3-H_2O system from 13C NMR spectroscopy. *Industrial and Engineering Chemistry Research, 37,* 3281–3287.

Hu, W., & Chakma, A. (1990). Modelling of equilibrium solubility of CO_2 and H2S in aqueous diglycolamine (DGA) solutions. *Canadian Journal of Chemical Engineering, 68,* 523–525.

Idem, R., Shi, H., Gelowitz, D., & Tontiwachwuthikul, P. (2011). *Catalytic method and apparatus for separating a gaseous component from an incoming gas stream*. Google Patents.

Jakobsen, J. P., Krane, J., & Svendsen, H. F. (2005). Liquid-phase composition determination in CO_2-H_2O-alkanolamine systems: An NMR study. *Industrial and Engineering Chemistry Research, 44,* 9894–9903.

Jiang, T., Wang, T., Ma, L., Li, Y., Zhang, Q., & Zhang, X. (2012). Investigation on the xylitol aqueous-phase reforming performance for pentane production over Pt/HZSM-5 and Ni/HZSM-5 catalysts. *Applied Energy, 90,* 51–57.

Kent, R. L., & Elsenberg, B. (1976). Better data for amine treating. *Hydrocabon process, 55,* 87–90.

Khoshbin, R., & Haghighi, M. (2013). Direct syngas to DME as a clean fuel: the beneficial use of ultrasound for the preparation of CuO–ZnO–Al_2O_3/HZSM-5 nanocatalyst. *Chemical Engineering Research and Design, 91,* 1111–1122.

Ko, J.-J., Tsai, T.-C., Lin, C.-Y., Wang, H.-M., & Li, M.-H. (2001). Diffusivity of nitrous oxide in aqueous alkanolamine solutions. *Journal of Chemical and Engineering Data, 46,* 160–165.

Kozak, F., Petig, A., Morris, E., Rhudy, R., & Thimsen, D. (2009). Chilled ammonia process for CO_2 capture. *Energy Procedia, 1,* 1419–1426.

Li, M.-H., & Lai, M.-D. (1995). Solubility and Diffusivity of N_2O and CO_2 in (Monoethanolamine + N-Methyldiethanolamine + Water) and in (Monoethanolamine + 2-Amino-2-methyl-1-propanol + Water). *Journal of Chemical and Engineering Data, 40,* 486–492.

Li, M.-H., & Lee, W.-C. (1996). Solubility and Diffusivity of N_2O and CO_2 in (Diethanolamine+ N- Methyldiethanolamine+ Water) and in (Diethanolamine+ 2-Amino-2-methyl-1-propanol+ Water). *Journal of Chemical and Engineering, Data 41,* 551–556.

Li, M.-H., & Shen, K.-P. (1993). Calculation of equilibrium solubility of carbon dioxide in aqueous mixtures of monoethyanolamine with methyldiethanolamine. *Fluid Phase Equilibria, 85,* 129–140.

Liang, Z., Idem, R., Tontiwachwuthikul, P., Yu, F., Liu, H., & Rongwong, W. (2016). Experimental study on the solvent regeneration of a CO_2-loaded MEA solution using single and hybrid solid acid catalysts. *AIChE Journal, 62,* 753–765.

Liu, H., Gao, H., Idem, R., Tontiwachwuthikul, P., & Liang, Z. (2017a). Analysis of CO_2 solubility and absorption heat into 1-dimethylamino-2-propanol solution. *Chemical Engineering Science, 170,* 3–15.

Liu, H., Liang, Z., Sema, T., Rongwong, W., Li, C., Na, Y., et al. (2014). Kinetics of CO_2 absorption into a novel 1-diethylamino-2-propanol solvent using stopped-flow technique. *AIChE Journal, 60,* 3502–3510.

Liu, H., Luo, X., Liang, Z., & Tontiwachwuthikul, P. (2015). Determination of Vapor-Liquid Equilibrium (VLE) Plots of 1-Dimethylamino-2-propanol Solutions Using the pH Method. *Industrial and Engineering Chemistry Research, 54,* 4709–4716.

Liu, H., Zhang, X., Gao, H., Liang, Z., Idem, R., & Tontiwachwuthikul, P. (2017b). Investigation of CO_2 regeneration in single and blended amine solvents with and without catalyst. *Industrial and Engineering Chemistry Research, 56,* 7656–7664.

Maneeintr, K., Henni, A., Idem, R. O., Tontiwachwuthikul, P., & Wee, A. G. (2008). Physical and transport properties of aqueous amino alcohol solutions for CO_2 capture from flue gas streams. *Process Safety and Environmental Protection, 86,* 291–295.

Munder, B., Lidal, H., & Sandall, O.C. (2000). Effect of carbon dioxide loading on the solubility of nitrous oxide in aqueous solutions of 2-(tert-butylamino) ethanol. *Journal of Chemical and Engineering Data, 45,* 1195–1200.

Park, J.-Y., Yoon, S. J., Lee, H., Yoon, J.-H., Shim, J.-G., Lee, J. K., et al. (2002). Density, Viscosity, and Solubility of CO_2 in Aqueous Solutions of 2-Amino-2-hydroxymethyl-1, 3-propanediol. *Journal of Chemical and Engineering Data, 47,* 970–973.

Park, M. K., & Sandall, O. C. (2001). Solubility of carbon dioxide and nitrous oxide in 50 mass methyldiethanolamine. *Journal of Chemical and Engineering Data, 46,* 166–168.

Raynal, L., Alix, P., Bouillon, P.-A., Gomez, A., de Nailly, M.l.F., Jacquin, M., Kittel, J., di Lella, A., Mougin, P., & Trapy, J. (2011a). The DMX™ process: An original solution for lowering the cost of post-combustion carbon capture. *Energy Procedia, 4,* 779–786.

Raynal, L., Bouillon, P.-A., Gomez, A., & Broutin, P. (2011b). From MEA to demixing solvents and future steps, a roadmap for lowering the cost of post-combustion carbon capture. *Chemical Engineering Journal, 171,* 742–752.

Redlich, O., & Kister, A. (1948). Algebraic representation of thermodynamic properties and the classification of solutions. *Industrial and Engineering Chemistry, 40,* 345–348.

Rochelle, G. T. (2009). Amine scrubbing for CO_2 capture. *Science, 325,* 1652–1654.

Saravanan, K., Ham, H., Tsubaki, N., & Bae, J. W. (2017). Recent progress for direct synthesis of dimethyl ether from syngas on the heterogeneous bifunctional hybrid catalysts. *Applied Catalysis, B: Environmental, 217,* 494–522.

Shi, H., Naami, A., Idem, R., & Tontiwachwuthikul, P. (2014a). 1D NMR analysis of a quaternary MEA–DEAB–CO_2–H_2O amine system: liquid phase speciation and vapor–liquid equilibria at CO_2 absorption and solvent regeneration conditions. *Industrial and Engineering Chemistry Research, 53,* 8577–8591.

Shi, H., Naami, A., Idem, R., & Tontiwachwuthikul, P. (2014b). Catalytic and non catalytic solvent regeneration during absorption-based CO_2 capture with single and blended reactive amine solvents. *International Journal of Greenhouse Gas Control, 26,* 39–50.

Shi, Y., Cao, Y., Duan, Y., Chen, H., Chen, Y., Yang, M., et al. (2016). Upgrading of palmitic acid to iso-alkanes over bi-functional Mo/ZSM-22 catalysts. *Green Chemistry, 18,* 4633–4648.

Srisang, W., Pouryousefi, F., Osei, P. A., Decardi-Nelson, B., Akachuku, A., Tontiwachwuthikul, P., et al. (2017a). Evaluation of the heat duty of catalyst-aided amine-based post combustion CO_2 capture. *Chemical Engineering Science, 170,* 48–57.

Srisang, W., Pouryousefi, F., Osei, P. A., Decardi-Nelson, B., Akachuku, A., Tontiwachwuthikul, P., & Idem, R. (2017b). Evaluation of the heat duty of catalyst-aided amine-based post combustion CO_2 capture. *Chemical Engineering Science.*

Tontiwachwuthikul, P., Meisen, A., & Lim, C. J. (1991). Solubility of CO_2 in 2-Amino-2-methyl-1-propanol Solutions. *Journal of Chemical and Engineering Data, 36,* 130–133.

Van Holst, J., Politiek, P. P., Niederer, J. P., & Versteeg, G. F. (2006). CO_2 capture from flue gas using amino acid salt solutions. In *Proceedings of 8th International Conference on Greenhouse Gas Control Technologies.*

Versteeg, G. F., & Van Swaalj, W. (1988). Solubility and diffusivity of acid gases (carbon dioxide, nitrous oxide) in aqueous alkanolamine solutions. *Journal of Chemical and Engineering Data, 33,* 29–34.

Wang, G., Yuan, X., & Yu, K. (2005). Review of mass-transfer correlations for packed columns*. *Industrial and Engineering Chemistry Research, 44,* 8715–8729.

Wang, Y., Xu, S., Otto, F., & Mather, A. (1992). Solubility of N_2O in alkanolamines and in mixed solvents. *The Chemical Engineering Journal, 48,* 31–40.

Wei, J., Ge, Q., Yao, R., Wen, Z., Fang, C., Guo, L., et al. (2017). Directly converting CO_2 into a gasoline fuel. *Nature Communications, 8,* 15174.

Yusoff, R., Aroua, M., Shamiri, A., Ahmady, A., Jusoh, N., Asmuni, N., et al. (2013). Density and viscosity of aqueous mixtures of N-methyldiethanolamines (MDEA) and ionic liquids. *Journal of Chemical and Engineering Data, 58,* 240–247.

Zhang, Q., Wang, T., Li, B., Jiang, T., Ma, L., Zhang, X., et al. (2012). Aqueous phase reforming of sorbitol to bio-gasoline over Ni/HZSM-5 catalysts. *Applied Energy, 97,* 509–513.

Zhang, X., Zhang, R., Liu, H., Gao, H., & Liang, Z. (2018). Evaluating CO_2 desorption performance in CO_2-loaded aqueous tri-solvent blend amines with and without solid acid catalysts. *Applied Energy, 218,* 417–429.

Chapter 3
Design, Modeling and Simulation of Post Combustion CO_2 Capture Systems Using Reactive Solvents

Abstract Design is the key step for the commercial process of post combustion CO_2 capture by using the amines based solvents. Also, the simulation of the CO_2 capture process could provide the guideline or more information for the design. The improvement or enhancement of the process configuration could be achieved by using the simulation. This chapter consists of two main sections. One is the methods for determination of column height for CO_2 absorber. The procedure for column design are present in order to get the final destination of column height. Four different method are discussed in this part. Another is about the developed process configurations for post combustion CO_2 capture. In this part, the present development for the process improvement of post combustion CO_2 capture are discussed in terms of absorber, stripper, and economizer modification. All detailed development are shown in this chapter in order to give the clear picture of the process configurations development for post combustion CO_2 capture.

3.1 Determination of Column Height for CO_2 Absorber

Determination of the column height of absorber for CO_2 capture process is an important part of the design a post-combustion CO_2 capture plant. Figure 3.1 is the diagram procedure for the absorber design. As shown in Fig. 3.1, the achievement for determination the column height of the absorber needs a large amount of data to support. From Fig. 3.1, there are four different methods, which could reach the final destination of column height. All of them needs the design conditions i.e. gas and liquid flow, gas compound. The gas-liquid contacting system are necessary. In addition, the physical information of the liquid and gas should be available in order to finish the design. Based on the provided information, the superficial velocities could be obtained. By using the obtained data, four methods as shown in Fig. 3.1 could be employed to determine the column height. There are several methods to design absorber height as given in Fig. 3.1. Route A is the empirical design method. For this method, the average overall volumetric mass transfer coefficient ($K_G a_v$) should be obtained from the experiments in order to calculate the height of column. Route

H. Liu et al., *Post-combustion CO_2 Capture Technology*, SpringerBriefs in Petroleum Geoscience & Engineering, https://doi.org/10.1007/978-3-030-00922-9_3

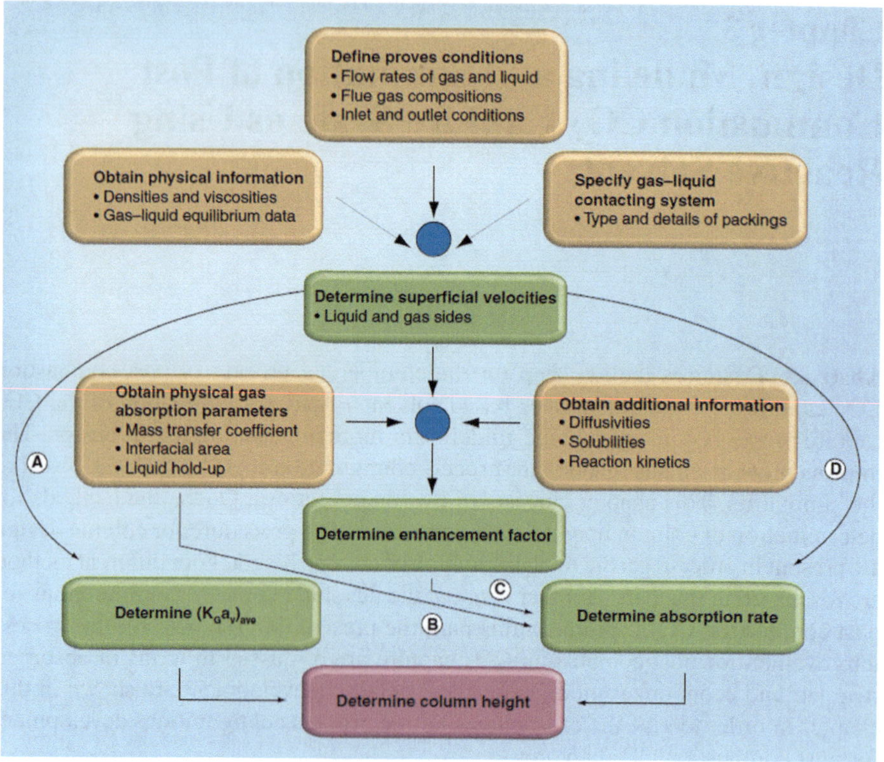

Fig. 3.1 The design procedure column height for CO_2 absorber. Modified from (Liang et al. 2011)

B is the theoretical design method. Route C is the Laboratory method. Route D is pilot plant technique method. All detailed information about those four method have been discussed and are available in our previous work (Liang et al. 2011).

3.2 Developed Process Configurations for Post Combustion CO_2 Capture

There are some issues which is closely related to the cost of the process. Solvent development, the efficiency of the facility, and the process flow sheet modification (Le Moullec and Kanniche 2011). For the modification of the process flow sheet, there are three main technology for modification: absorber modification, stripper modification, and economizer modification.

3.2.1 *Absorber Modification*

3.2.1.1 Intercooler

This technology is involved in the cooler in the absorber, which is named as the intercooler. The intercooler technology is related to draw off or apart of the solution flow from the absorber, cool it down and reject it at the same level in the absorber (Freguia and Rochelle 2003). The temperature in the absorber decrease due to the reaction of solvent with CO_2. The high temperature gives the limit of the CO_2 capacity of the solution, which leads to the more used solvent in the process. The intercooler can cool the solvent used in the column down, which result in the higher thermodynamic driving force in absorber. This technology will give the less used in the solvent.

3.2.1.2 Flue Gas Precooler

According to the diagram of flue gas precooler (Kvamsdal et al. 2009), it can be seen that the cooler was set up before the flue gas was injected into the column. The cooler was used to decrease the temperature of the flue gas. As the decrease of the temperature in the absorber column, the lower temperature increases the thermodynamic driving force and makes the contribution to the CO_2 capacity. This result in the consumption of solution in the process of CO_2 capture.

3.2.2 *Stripper Modification*

3.2.2.1 Rich Solvent Flash

In this process, the rich solution was flashed in the flash drum before it was injected into the stripper column for regeneration. The gas from the top of the flash drum was sent the stripper condenser. The solution from the bottom of the flash drum was then injected into the stripper (Tobiesen et al. 2007).

3.2.2.2 Stripper Intermediate Compression

In this part, the technology of Stripper intermediate compression was discussed. The principle of this technology is diving the stripper into several parts with different operation pressures (Oyenekan and Rochelle 2006; Rochelle and Oyenekan 2011). The Variations in pressure are obtained by the compressors. This technology can make the solvent boil at the lower temperature. And also, it can allow the recovery of all heat which was produced in the process of compression. In this case, it can save the energy consumption, which leads to the less cost of CO_2 capture process. However, this process required the different optimization of the pressure for the top and the bottom of stripper column. The different optimized pressure is closely to the number of the diving parts of stripper column.

3.2.2.3 Vacuum Stripper

The vacuum stripper technology was achieved by boiling the solvent at the lower pressure stream. In this process, the boiling solvent was obtain by using the low pressure steam. This leads to the decrease of the high pressure steam, which makes a contribution to the decrease of energy consumption.

3.2.3 Economizer Modification

3.2.3.1 Pre-condenser

This modification is made up of two parts: cooling the CO_2 and water at the top of the stripper column with rich cold solvent. The consumption of cool water decrease in this process. The decrease use of water will cut down the price of CO_2 capture. It can make a significant gain under the condition that the fraction of the cold rich solvent was operated. However, this technology is inefficient if this modification was used on the preheated rich solvent.

3.2.3.2 Split Flow

Split-flow is the generic term used to describe processes with two levels of solvent regeneration. One part was achieved in the stripper column; and another part appears in the Economizer (Chang and Shih 2005). The rich solution was draw off from the absorber. And then the lean solution was rejected into the stripper column. In this process, the part of regeneration was operated in the Economizer. As a result, the needed energy of regeneration process was decreased, which result in the price of CO_2 capture. However, there are a lot parameters which should be taken into consideration, i.e. the height of taking the rich solvent from absorber, the height of injection, the rich CO_2 loading, the flow of solution.

References

Chang, H., & Shih, C. M. (2005). Simulation and optimization for power plant flue gas CO_2 absorption-stripping systems. *Separation Science and Technology, 40,* 877–909.

Freguia, S., & Rochelle, G. T. (2003). Modeling of CO_2 capture by aqueous monoethanolamine. *AIChE Journal, 49,* 1676–1686.

Kvamsdal, H., Jakobsen, J., & Hoff, K. (2009). Dynamic modeling and simulation of a CO_2 absorber column for post-combustion CO_2 capture. *Chemical Engineering and Processing: Process Intensification, 48,* 135–144.

Le Moullec, Y., & Kanniche, M. (2011). Screening of flowsheet modifications for an efficient monoethanolamine (MEA) based post-combustion CO_2 capture. *International Journal of Greenhouse Gas Control, 5,* 727–740.

Liang, Z. H., Sanpasertparnich, T., Tontiwachwuthikul, P. P., Gelowitz, D., & Idem, R. (2011). Part 1: Design, modeling and simulation of post-combustion CO_2 capture systems using reactive solvents. *Carbon Management, 2,* 265–288.

Oyenekan, B. A., & Rochelle, G. T. (2006). Energy performance of stripper configurations for CO_2 capture by aqueous amines. *Industrial and Engineering Chemistry Research, 45,* 2457–2464.

Rochelle, G. T., & Oyenekan, B. A. (2011). *Regeneration of an aqueous solution from an acid gas absorption process by matrix stripping*. Google Patents.

Tobiesen, F. A., Svendsen, H. F., & Mejdell, T. (2007). Modeling of blast furnace CO_2 capture using amine absorbents. *Industrial and Engineering Chemistry Research, 46,* 7811–7819.

Chapter 4
Solvent Management

Abstract In the operational process for the post combustion CO_2 capture plant, there are several issues related to the cost of plant, which need be addressed. Among those issues, the corrosion and the degradation are the emergency problems when the plant is running. The corrosion and degradation could create the problems for the plant, which leads to the shutdown of the plant. The main parts are the corrosion and degradation. In this chapter, the corrosion is discussed in detail. The effect of corrosion and the solution of the corrosion are presented in this chapter in order to provide the more information for decrease the corrosion in the CO_2 capture plant. In addition, the degradation is discussed as well. The two different types of degradation as oxidative degradation and thermal degradation are covered in this chapter. The mechanism of degradation are presented in this chapter to understand the process of degradation. The solution to solve this issue is provided in order to exclude the degradation.

4.1 Corrosion

Corrosion in CO_2 absorption processes, which has an impact on the amine efficiency is one of the most severe operational problems. Typically, the incoming flue gas contains CO_2 with water, oxygen, some SOx and NOx, particulates (fly ash) and other contaminants. This composition is extremely corrosive to carbon steel wherever free water and/or amine are present. Due to the presence of acid gases and high temperature in the capture plant, corrosion occurs, and corrosion problems like production losses or reduction in equipment life are impossible to avoid (DuPart et al. 1993a, b). Corrosion occurs in the form of general, galvanic, crevice, pitting, intergranular, selective leaching, erosion and stress corrosion cracking (Soosaiprakasam and Veawab 2008). When acid gases dissolve in the amine, they can react with unrelated materials such as tubing lines, reactor vessels and other process compartments. The typical locations where corrosion can occur are at the bottom of the absorber, regenerator, heat exchanger, reboiler bundles, amine cooler and condenser (Kittel and Gonzalez 2014). Figure 4.1 shows the typical locations prone to corrosion. Secondly,

H. Liu et al., *Post-combustion CO_2 Capture Technology*, SpringerBriefs in Petroleum Geoscience & Engineering, https://doi.org/10.1007/978-3-030-00922-9_4

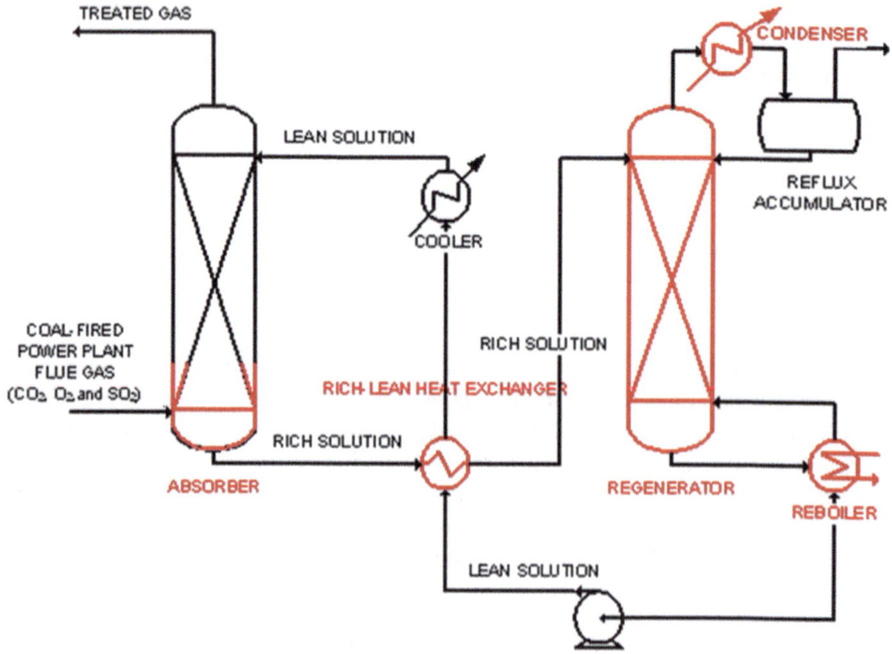

Fig. 4.1 Locations in amine-based CO_2 capture plant most susceptible to severe corrosion issues

the material corrosion creates unsafe and unreliability issues in the CO_2 capture system, which directly affects the safety of plant personal (Krzemień et al. 2016). In addition, corrosion problems cause high substantial expenditure. According to the work of Mogul (1999), each corporation in U.S. needs to set aside about \$170 billion each year as cost to combat corrosion. For the industrial workers, corrosion damage leads to a multitude of challenges concerning safety, replacement costs, lost production costs, maintenance scheduling, new coating and protection systems and non-traditional materials. This reflects an important effect of corrosion problems in plant operations.

In the past, treating gas plants arbitrarily used some species as oxidizing agents which could cause faulty corrosion control, so it was necessary to identify the oxidizing agents to minimize corrosion problems. In order to figure out the specific oxidizing agents, some investigators established a mechanistic model (Fig. 4.1) to simulate the corrosion process that takes place at the metal solution interface in the aqueous amine–CO_2 system (Veawab 2001). Based on that model simulation result, it was observed that hydrogen ion was not important in the reduction reaction. Instead, it was the bicarbonate ion and water that formed the main oxidizing agents. In order to minimize the corrosion of amines, plants usually use carbon steel as plant construction material for absorption and striping units. In specific key plant areas such as internals of reboiler, stripper, cross exchanger tubes and rich amine piping, stainless steel material is used to provide adequate corrosion protection. Even though carbon steel is a popular material used in the construction of gas treating plants, it is more vulnerable to corrosion (Saiwan et al. 2011). Therefore, it is necessary for researchers to study the factors which have effect on the corrosion rate of carbon steel. Some researchers used the electrochemical experimental methods to evaluated process parameters that impact the rate of corrosion of carbon steel in different aqueous solvents. Such factors include solution temperature, partial pressure of oxygen, solution velocity, CO_2 loading in solution, solvent concentration and metal surface condition.

The use of corrosion inhibitors to minimize corrosion has also received much attention in the past. Heavy metals such as arsenic and vanadium were used in the past (Kohl and Nielsen 1997). Even though heavy metal inhibitors provide excellent corrosion protection, especially for wetted areas, these inhibitors are not friendly to the environment and human life. In addition, the heavy metal inhibitors do not protect splash and vapor regions (DuPart et al. 1993b). Consequently, capture plants have recently started to use some non-toxic corrosion inhibitors such as carbohydraizide instead of heavy metal as corrosion inhibitors (Fytianos et al. 2016).

In addition, there are some mechanical design criteria that can be employed to minimize corrosion, some of which are based on common sense while others have been developed from past operating experience. Minimizing suction pressure is to prevent pump cavitation (Discenzo et al. 2003). Also, it is important to have adequate mechanical and carbon filtration to reduce corrosion in amine units. These filter equipment can remove all particles over 5 microns in size. This will effectively reduce particles that can promote erosion corrosion and abrasion of passive films (Campbell 1981). Meanwhile, all types of amine services should use stress relieve, especially at welding points (Reddy and Chin 1998). Using stress relieve, especially in MEA units, is important due to their poor cracking record. In order to control localized corrosion, some pilot plants use corrosion monitoring system. For instance, the ITC pilot plant uses corrosometer probes to test the corrosion rate in the pilot plant (Fig. 4.2). In the CO_2 capture process, nine probes were inserted into the system (Fig. 4.2), including the stripper overhead (IP1), rich amine to stripper (IP2), reflux and vapors (IP3), CO_2 product (IP4), stripper bottom (IP5), absorber bottom (IP6), lean amine to storage (IP7), absorber overhead (IP8), and absorber off-gas (IP9).At each location, the tail of the probe holder was connected to the remote data collectors to measure the current density, which subsequently was converted into a thickness reduction (Kittel et al. 2009).

4.1.1 *Effects of Corrosion*

Corrosion in amine-based post-combustion CO_2 capture plant includes general (or uniform) corrosion, localized CO_2 corrosion as well as erosion–corrosion, and various forms of stress-corrosion. Due to economic reasons most of the equipment and piping in amine-based post-combustion CO_2 capture plant is constructed of carbon

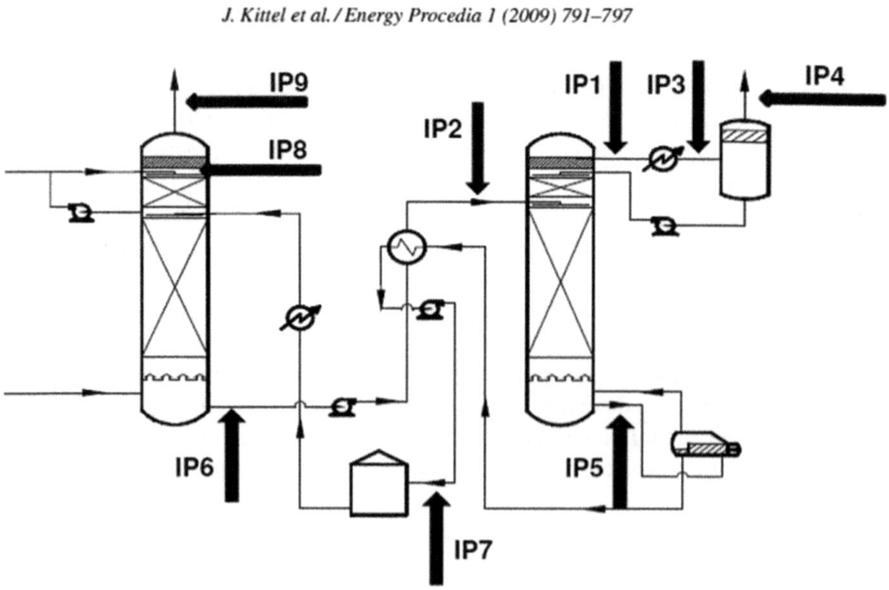

Fig. 4.2 Schematic ITC's CO_2 capture unit and corrosion monitoring insertion points (IP)

steel. These carbon steel parts of the installations may suffer from corrosion by several agents like wet acid gases, oxygen and products of the degradation of amines (Krzemień et al. 2016) as mentioned previously. The corrosion issue does not only occur in the case of carbon steel, but also in other types of materials used within the amine-based CO_2 capture plant (Krzemień et al. 2016). Consequently, polymers such as pump seals and heat exchanger linings are potentially susceptible to swelling and changes in physical properties due to absorption of CO_2 (Srisang et al. 2014).

4.1.1.1 Plant Failure and Unplanned Downtime

Corrosion causes the loss of mechanical integrity and loss of functionality of the affected part, which can lead to loss of containment of the circulating fluid (both gaseous and liquid phases). This can lead to unsafe situations which can compromise the health and safety of the staff. In severe situations it can lead to death. The other consequence of corrosion in a CO_2 capture plant is that it will lead to downtime with its attendant economic consequences. This is because the corroded parts of the plant need to be replaced to fix the plant in order to resume operations, and this is usually unplanned (Gunasekaran et al. 2013). Also, according to Cousins et al. (2013), in addition to unscheduled downtime, corrosion in amine gas treatment plants can result in production losses, reduced equipment life and even injury or death.

4.1.2 *Suggestions to Solve the Issues*

4.1.2.1 Use of Corrosion Inhibitors

Different types of corrosion inhibitors; how they act and blending the inhibitors to produce synergistic effects; choice of the right type of effective inhibitors (organic versus inorganic) (Srinivasan et al. 2013; Tontiwachwuthikul et al. 2011).The benefit is that it will minimize corrosion and allow cheap materials to be used for the construction of an amine based CO_2 capture plant. Because it will drastically reduce or eliminate corrosion, the use of a suitable corrosion inhibitor will allow the CO_2 capture plant owners to extract a long useful life out of the plant. According to Srinivasan et al. (2013), the use of a corrosion inhibitor is the most economical and flexible as compared to other methods since it can be applied to the existing process without requiring any major process modification (DuPart et al. 1993b; Kohl and Nielsen 1997; Veawab 2001). To date, a number of chemicals have been patented as corrosion inhibitors for amine-based acid gas treating plants. However, the effective ones including vanadium are toxic heavy metals. Their usage is restricted owing to more stringent environmental regulations particularly in the last few decades which consequently made their disposal and handling costly and difficult. This led to a shift in trend towards the use of low toxic corrosion inhibitors. The development of low toxic corrosion inhibitors such as sodium thiosulfate for inhibition of corrosion on carbon steel in amine solution in an amine-based CO_2 capture environment from the flue gases of a coal fired-electric power plant has been reported for applications at wide ranges of operating parameters (Srisang et al. 2014; Tontiwachwuthikul et al. 2011).

4.1.2.2 Use of Corrosion Resistant Materials for Absorber and Stripper Columns

This includes the lining of a carbon steel column with corrosion resistant plastics such as Teflon. The benefit is that it will allow the use of cheap carbon steel for the construction of the absorber and stripper vessels but at the same time providing a barrier that prevents the occurrence of corrosion.

4.1.2.3 The Use of Suitable Non Degradable and Corrosion Resistant Plastics and Elastomers as Pump Seals and Heat Exchanger Linings

This will maintain the integrity and good working conditions of the pumps and heat exchangers. The benefit is that it prevents downtime and unplanned stoppages due to pump and heat exchanger leaks leading to continuous operation of the plant.

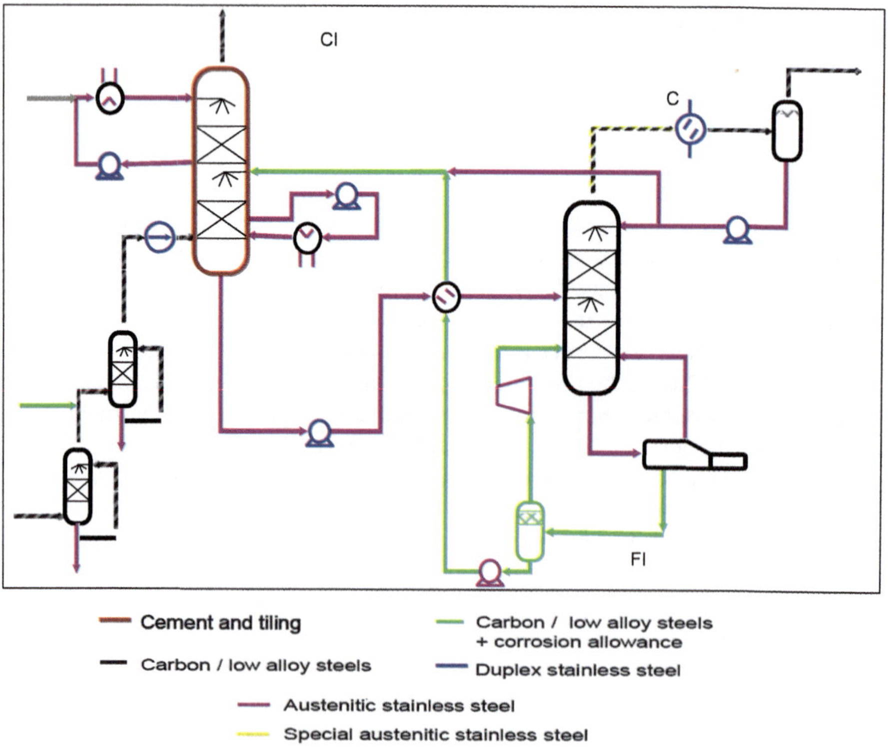

Fig. 4.3 Schematic materials selection diagram for post combustion capture on a coal-fired power plant (Billingham et al. 2011)

4.1.2.4 The Use of Specific Materials for Construction of Certain Parts of the CO_2 Capture Plant

The materials selected are as shown in Fig. 4.3 (Billingham et al. 2011). Also, the metallurgical guidelines for construction of an amine plant are provided in Table 4.1 by Dupart et al. (1993b). The benefit is that it will allow the use of cheap carbon steel for the construction of the pipes, vessels and units that are not prone to corrosion, and only uses specific corrosion resistant materials for areas that are prone to corrosion, thereby preventing the occurrence of corrosion in those areas.

4.1.2.5 Design of Piping to Reduce Velocity

As was indicated earlier (DuPart et al. 1993a), velocity of fluid flowing through pipes affects erosion corrosion. In designing an amine-based CO_2 capture plant, it is very important to follow strict guidelines for velocity in different piping and vessels.

Table 4.1 Metallurgical guidelines for construction of an amine plant (DuPart et al. 1993b)

Item	Material	Alternate material
Absorber		
Shell	Carbon steel	–
Internals	Carbon steel	Stainless steel
Lean/rich exchanger		
Tubes (rich)	Stainless steel	–
Shell (lean)	Carbon steel	–
Lean amine cooler		
Tubes	Carbon steel	Stainless steel
Shell	Carbon steel	–
Carbon bed		
Shell	Carbon steel	–
Internals	Stainless steel	–
Regenerators		
Shell	Carbon steel	–
Internals	Stainless steel	–
Reboiler		
Tubes	Stainless steel	–
Shell	Carbon steel	–
Tube sheet	Carbon steel	–
Steam side channels	Carbon steel	–
Reflux condenser		
Tubes	Carbon steel	Stainless steel
Shell	Carbon steel	–
Reflux drum	Carbon steel	–
Pumps		
Cases	Carbon steel	–
Impellers	Stainless steel	–
Piping		
Rich amine	Stainless	–
Lean amine	Carbon steel	Stainless steel

The benefit is that it will remove one of the conditions that is necessary for stress corrosion cracking to occur.

4.1.2.6 Post Welding Treatment

In the case of stress corrosion cracking, it was indicated that by Radgen et al. (2014) that residual stresses from welding are typically the highest localized stresses which induce stress corrosion cracking. It is therefore important to perform post welding stress relief wherever welding is done. The benefit is that it will remove one of the conditions that is necessary for stress corrosion cracking to occur.

4.2 Degradation

Amine degradation in the process is regarded as a big problem in the post-combustion CO2 capture. Because, this issue could result other issues such as solvent loss, formation of amine by products, foaming, fouling, corrosion and an increase in the solvent's viscosity (Zhou et al. 2012; Zoannou et al. 2013). All of them will increase the cost of the CO_2 capture. The cause of degradation are mainly because of thermally and oxidatively attributing to the presence of CO_2, and/or oxygen (O_2), NOx, and SOx in the flue gas stream. The degradation occurs in both of absorption process and desorption process, which leads to the vital effect of process operation. The formed products of degradation costs the amine molecular, which results in the less amines available for CO_2 capture. In this case, the CO_2 absorption capacity of amine are also cut down. Meanwhile, the products of degradation contain the heat stable salts, which contributed to the corrosion. In addition, some of the products are toxic. Those toxic products are toxic for both human being and environment, which have to be removed before the emission. The remove process of toxic products will also increase the cost of CO_2 capture. Considering all of listed issues, the degradation can constitute a major expense in the CO_2 capture process through solvent make up and reduction of absorption capacity (by up to 10%) (Freeman et al. 2010; Wang and Jens 2012b). Generally, amines degradation can be mainly categorized into two main types: thermal degradation at the high temperature and the present of CO_2; and oxidative degradation with presence of O_2 and other oxidants in the flue gas.

4.2.1 Oxidative Degradation

Oxidative degradation occurs in the CO_2 absorption process, which is caused by the oxygen (concentration varying from 3 to 5%) in the flue gas streams. The Oxidative degradation mainly happen in the absorber with at the temperature range of 40–70 °C and rich solvent at the heat-exchanger at the temperature range of 100–145 °C. (Rochelle, 2012) In order to understand the behavior of oxidative degradation, a large amount of amine-based solvent were investigated under different conditions by using the chlorine oxide as oxidant. By employing the advanced analytical techniques, the degradation products were identified to understand the mechanisms behind the degradation. Table 4.1 shows the summary of the work on the degradation for Oxidative degradation and thermal degradation. For the study of amine-based solvent degradation, the bath and semi-bath reactor are generally used to carry out the experiments in the laboratory. The concentration of O_2 was used as 2% (Sexton and Rochelle 2009). The experiments are operated at the operational absorber temperature as 55 °C. (Freeman et al. 2010) Generally, the gas flow was set as 100 mL/min. The high gas flow between 0.5 and 7.5 L/min were also investigated in the reference (Vevelstad et al. 2013). In comparison with the semi-batch reactor, the batch reactor exhibited the advantage in the precise control of the experimental condition. In addition, the new design for study the solvent degradation i.e. Solvent Degradation Apparatus have been proposed (Closmann and Rochelle 2011). The identification of the degradation products and the concentration of products could be achieved by using the Gas chromatography (GC), Ion chromatography (IC), Liquid chromatography (LC), High performance liquid chromatography (HPLC). In addition, other equipment (i.e. FTIR, Ultraviolet-vis spectrophotometry (UV-Vis), and NMR) have also been applied in the degradation study. There are two main mechanisms to explain the degradation behavior of amine-based solvent. The electron abstraction mechanism is proposed based on the electron extraction from the N of amine group. This extraction is motivated by a reactive free radical generated from the amine and aminium cation radical. All the process could be shown in Fig. 4.4. As shown in Fig. 4.5, hydrogen abstraction mechanism consists of an amine cyclic formation by hydrogen bonds, in which amine radical is transferred internally through the ring structure formed, resulting in ammonia, aldehydes and aldehydes radicals as main degradation products (Vega et al. 2014).

4.2.2 *Thermal Degradation*

Thermal degradation is usually found in the disrober with the desorption temperature of 120 °C. The present of CO_2 could catalyze the process of thermal degradation of amine-based solvents. (Kim and Sartori, 1984) The generation of thermal degradation products could lead to the decrease of the CO_2 cyclic capacity of amine, the increase of the viscosity, the formation of foaming. The formation of thermal degradation production also result in the corrosion (Martin et al. 2012). The analytical method for thermal degradation products are very similar as the oxidative degradation. The experimental set-up for thermal degradation products is similar as the oxidative degradation. The difference of two experiments is the operational condition, 50 °C for the oxidative degradation, and more than 100 °C for thermal degradation. The thermal mechanisms are proposed in order to explain the degradation behavior of amines with the present of CO_2. This pathway of ethanolamine could be founded in Fig. 4.6.

Fig. 4.4 The electron abstraction mechanism of MEA degradation (Chi and Rochelle 2002)

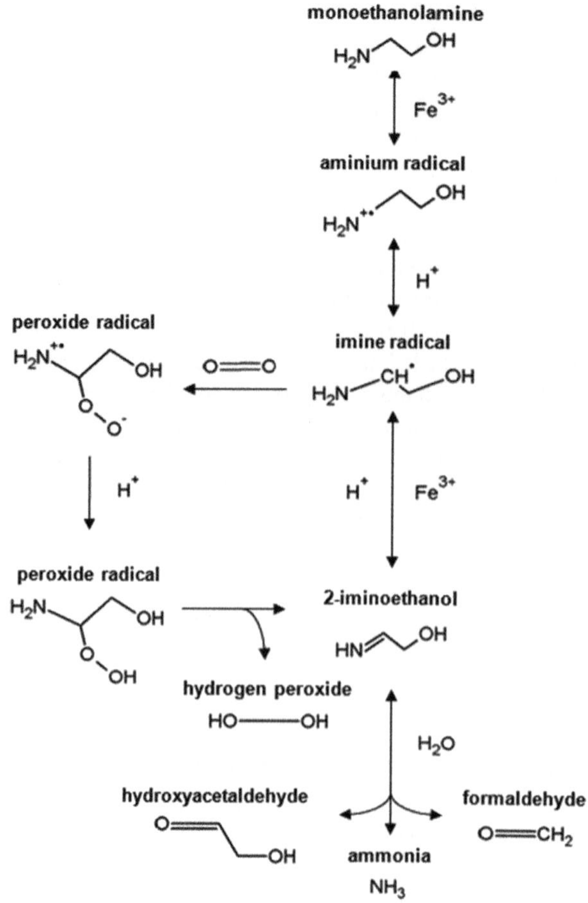

4.2.3 *Suggestions to Solve the Issues*

In order to decrease the impact of degradation on the system, there are a number of methods to be done. The removal of O_2, SO_2, fly ash before capture process. The present of O_2, SO_2, fly ash could result into the thermal and thermal degradation. The removal of those compounds plays a vital role in decrease the degradation. However, this total removal of those compounds are extremely difficult to achieve. Also, this process is very costly and time consuming, which could be applied in the industry.

4.2.3.1 Use of Inhibitors

As well known, the addition of inhibitor has been proposed to reduce the degradation. The introduction of inhibitors could be considered as the oxygen and radicals

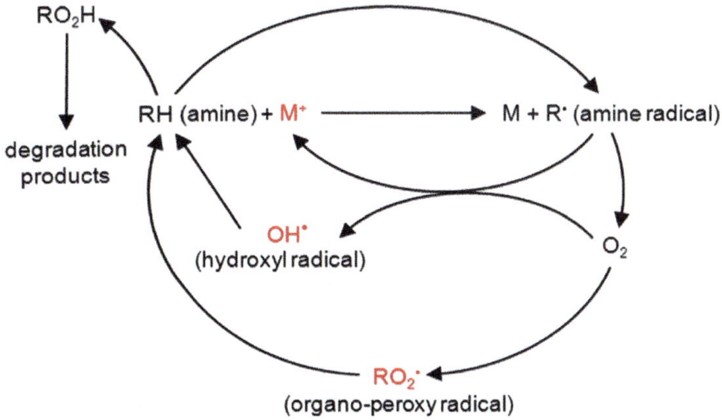

Fig. 4.5 The potential degradation process of hydrogen abstraction (Bedell 2009)

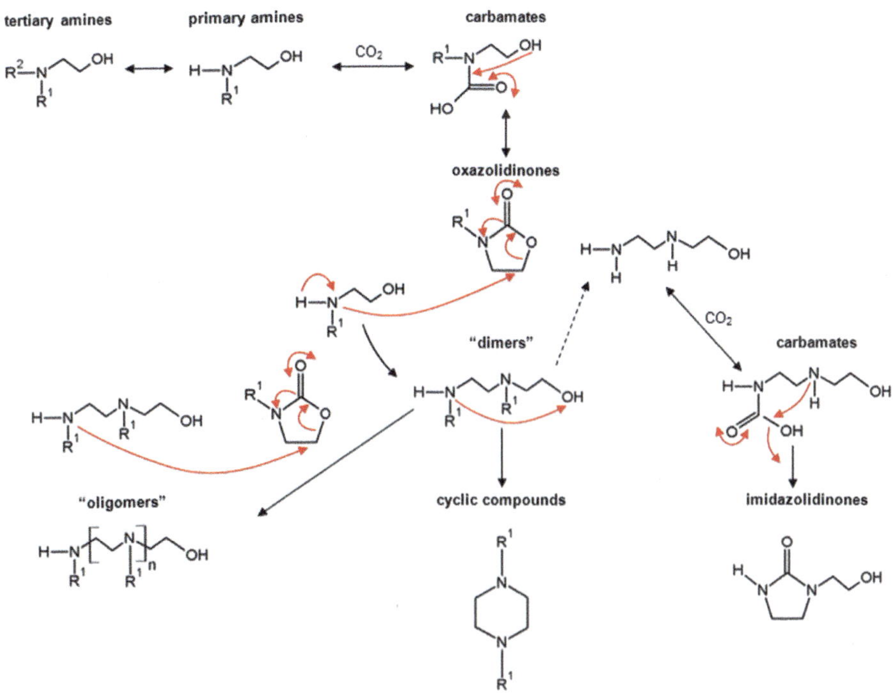

Fig. 4.6 The pathways of thermal degradation for ethanolamines (Lepaumier et al. 2009a)

Table 4.2 The comparison of radical inhibitor and oxygen inhibitor

	Radical inhibitor	O_2 inhibitor
Definition	Reaction with radicals to inhibit the initiation step or interrupt the propagation step of autoxidation	Inhibiting degradation by reacting with the dissolved O_2 before it can react with amine
Advantages	High rate of reaction	Preventing oxygen from degrading MEA
	Preventing oxygen from degrading MEA	
Dis-advantages	At elevated temperatures, the chain breaking antioxidants can lose their efficiency due to homolytic decomposition of the peroxides formed	Some of the oxygen scavengers give water soluble oxidation products, which have to be removed from the solution

scavengers. Thus, the inhibitors are mainly made up of two type as oxygen inhibitor and radical inhibitor. The commonly used radical inhibitor are sulfite and thiosulfate (Hakka and Ouimet 2006). The Ethylenediaminetetraacetic acid (EDTA) is considered as one of mostly used oxygen inhibitors (Goff 2005). For different inhibitors, the advantage and disadvantage are shown in this book as shown in Table 4.2.

Solvent	Main oxidative degradation products	References	Main thermal degradation products	References
MEA	Formic Acid Ammonia Acetic Acid 2-oxazolidone N-(2-hydroxyethyl)formamide N-(2-hydroxyethyl)imidazole N-(2-hydroxyethyl)piperazin-3-one N-(2-hydroxyethyl)-2-(2-hydroxyethylamino)acetamide N,N′-Bis (2-hydroxyethyl)oxamide	Goff (2005), Lepaumier et al. (2011a), Sexton (2008), Sexton and Rochelle (2010), Supap et al. (2006, 2009), Voice and Rochelle (2011)	1-amino-2-propanol 2-oxazolidone 2-Methylpiperazine 2-methyl-3-oxazolidine 4-methylmorpholine N-(2-hidroxyethyl)-ethylenediamine 1-(2-hydroxyethyl)-2-imidazolidone 1-piperazineethanol N,N′-bis(2-hidroxyethyl)urea N-(2-aminoethyl)-N′-(2-hydroxyethyl)imidazolidinone N,N′-bis(2-hydroxyethyl)imidazolidin-2-one N,N,N-tris(2-hydroxyethyl)ethylenediamine	Davis and Rochelle (2009), Eide-Haugmo et al. (2011a), Lepaumier et al. (2011a) Zhou et al. (2012)

(continued)

(continued)

Solvent	Main oxidative degradation products	References	Main thermal degradation products	References
DEA	Formic Acid Ammonia Acetic Acid Monoethanolamine Glycolic Acid N-methyldiethanolamine N-(carboxymethyl)diethanolamine N,N-bis(2-hydroxyethyl)piperazine N,N,N-tris(2-hydroxyethyl)ethylenediamine	Bedell (2009), Kennard (1983), Kennard and Meisen (1985), Lepaumier et al. (2009b), Rooney et al. (1998), Sexton (2008)	Monoethanolamine N-(2-hydroxyethyl)aziridine N-(2-hydroxyethyl)oxazolidin-2-one N,N-bis(2-hydroxyethyl)ethylenediamine Triethanolamine N,N-bis(2-hydroxyethyl)piperazine N,N,N-tris(2-hydroxyethyl)ethylenediamine Bis-(2-(2-hydroxyethylamino)ethyl)ether N-(2-hydroxyethyl)-N-(2-(2 hydroxyethy-lamino)ethyl)piperazine N-2-[bis(2-hydroxyethyl)-amino]ethylpiperazine N-2-[bis(2-hydroxyethyl)-amino]ethyloxazolidin-2-one N,N,N,N-tetrakis(2-hydroxyethyl)urea N,N,N,N-tetrakis(2-hydroxyethyl)ethylenediamine N-2-[bis(2-hydroxyethyl)-amino]ethyl-N-(2-hydroxyethyl)piperazine N,N,N,N-tetrakis(2-hydroxyethyl)diethylenetriamine N,N,N,N-tetrakis(2-hydroxyethyl)diethylenetriamine	Eide-Haugmo et al. (2011b), Lepaumier et al. (2009b)
MDEA	Methylamine Ethylene Oxide Dimethylamine Formic Acid Acetic Acid N-methylethanolamine Glycolic Acid N,N-(dimethyl)ethanolamine Oxalic Acida Diethanolamine N-methylmorpholin-2-one N-methylmorpholin-2,6-dione 2-[Methyl(2-hydroxyethyl)amino] Acetic Acid N,N,N-trimethyl-N-(2-hydroxyethyl)ethylenediamine triethanolamine N-(carboxymethyl)diethanolamine	Bedell (2009, 2011), Closmann and Rochelle (2011), Lawal et al. (2005), Lepaumier et al. (2009b), Rooney et al. (1998)	Methanol Ethylene oxide Formamide Formic Acid Trimethylamine Ethyleneglycol N,N-dimethylethylamine N-methylethanolamine 2–dimethylamino etanol Diethanolamine 1,4-dimethylpiperazine 3-(hydroxyethyl)-2-oxazolidone 1-(2-hydroxyethyl)-4-methylpiperazine Triethanolamine N-(carboxymethyl)diethanolamine N,N-bis-(2-hydroxyethyl)-piperazine N,N,N-tris(2-hydroxyethyl)ethylenediamine	Chakma and Meisen (1988, 1997), Closmann and Rochelle (2011), Davis and Rochelle (2009), Eide-Haugmo et al. (2011b), Lepaumier et al. (2009b)

(continued)

(continued)

Solvent	Main oxidative degradation products	References	Main thermal degradation products	References
APM	Ammonia Formic Acid Nitrous Acid Acetic Acid Nitric Acid Glycolic Acid Oxalic Acida 2,4-Lutidine 4,4-dimethyl oxazolidinone	Freeman et al. (2010), Lepaumier et al. (2009b), Sexton (2008), Strazisar et al. (2003), Voice and Rochelle (2011), Wang (2013), Wang and Jens (2012a)	N,2,2-trimethylethanolamine 4,4-Dimethyloxazolidin-2-one N,4,4-Trimethyloxazolini-2-one 4,4-Dimethyl-1-hydroxytertiobutylimidazolidin-2-one 2-[(2-amino-2-methylpropyl)amino]-2-methyl-1-propanol	Davis and Rochelle (2009), Eide-Haugmo et al. (2011a, b), Freeman et al. (2010), Lepaumier et al. (2009b), Wang 2013)
PZ	Formic Acid Nitrous Acid Acetic Acid Nitric Acid Ethylenediamine Glycolic Acid Oxalic Acida 2-oxopiperazine Formylpiperazine 1-mononitrosopiperazine	Alawode (2005), Freeman et al. (2010), Freeman and Rochelle (2012b), Li et al. (2013), Namjoshi et al. (2013), Rochelle (2013), Voice and Rochelle (2011)	Ethylenediamine Formic Acid N-methylpiperazine N-formylpiperazine N,N–dimethylpiperazine N-ethylpiperazine 1-(2-aminoethyl)piperazine N-(hydroxyethyl)piperazine 1,4-di(2-aminoethyl)piperazine	Freeman (2011), Freeman et al. (2010), Freeman and Rochelle (2012a), Wang (2013)
MMEA	Formic Acid Acetic Acid Monoethanolamine Ethylenglycol Glycolic Acid 1-methylimidazole 2.dimethylamino ethanol 1,x-dimethylimidazole N-methyloxazolidinone N-(2-hydroxyethyl)-N-methylformamide 1-(2-hydroxyethyl)imidazole N,N -dimethyl-N -(2-hydroxyethyl)ethylenediamine	Lepaumier et al. (2011b), Lepaumier et al. (2009b)	N-methyloxazolidinone N,N–dimethylpiperazine N,N′-dimethyl imidazolidinone N,N′-dimethyl-N′-(2-hidroxyethyl)ethylenediamine N-(-hydroxyethyl)-N′-methylpiperazine N,N′,N″-trimethyl-N″-(2-hydroxyethyl)diethylene triamine N,N′,N″,N″-tetramethyl-N″-(2-hydroxyethyl) triethylenetetramine	Davis and Rochelle (2009), Eide-Haugmo et al. 2011), Lepaumier et al. (2009b, 2011b)

References

Alawode, A. (2005). *Oxidative degradation of piperazine in the absorption of carbon dioxide*. University of Texas, Department of Chemical Engineering, Austin, TX, USA.

Bedell, S. A. (2009). Oxidative degradation mechanisms for amines in flue gas capture. *Energy Procedia, 1*, 771–778.

Bedell, S. A. (2011). Amine autoxidation in flue gas CO_2 capture—Mechanistic lessons learned from other gas treating processes. *International Journal of Greenhouse Gas Control, 5*, 1–6.

Billingham, M., Lee, C.-H., Smith, L., Haines, M., James, S., Goh, B., et al. (2011). Corrosion and materials selection issues in carbon capture plants. *Energy Procedia, 4*, 2020–2027.

Campbell, H. (1981). Aluminium bronze corrosion resistance guide. *Publication, 80*, 1–27.

Chakma, A., & Meisen, A. (1988). Identification of methyl diethanolamine degradation products by gas chromatography and gas chromatography-mass spectrometry. *Journal of Chromatography A, 457*, 287–297.

Chakma, A., & Meisen, A. (1997). Methyl-diethanolamine degradation—Mechanism and kinetics. *The Canadian Journal of Chemical Engineering, 75*, 861–871.

Chi, S., & Rochelle, G. T. (2002). Oxidative degradation of monoethanolamine. *Industrial and Engineering Chemistry Research, 41*, 4178–4186.

Closmann, F., & Rochelle, G. T. (2011). Degradation of aqueous methyldiethanolamine by temperature and oxygen cycling. *Energy Procedia, 4*, 23–28.

Cousins, A., Ilyushechkin, A., Pearson, P., Cottrell, A., Huang, S., & Feron, P. H. (2013). Corrosion coupon evaluation under pilot-scale CO2 capture conditions at an Australian coal-fired power station. *Greenhouse Gases: Science and Technology, 3*, 169–184.

Davis, J., & Rochelle, G. (2009). Thermal degradation of monoethanolamine at stripper conditions. *Energy Procedia, 1*, 327–333.

Discenzo, F.M., Chung, D., & Zevchek, J.K., (2003). System and method for controlling pump cavitation and blockage. *Google Patents*.

DuPart, M., Bacon, T., & Edwards, D. (1993a). *Understanding corrosion in alkanolamine gas treating plants: Part 1. Hydrocarbon Processing (United States)* (p. 72).

DuPart, M., Bacon, T., & Edwards, D. (1993b). *Understanding corrosion in alkanolamine gas treating plants: Part 2. Hydrocarbon Processing (United States)* (p. 72).

Eide-Haugmo, I., Lepaumier, H., da Silva, E. F., Einbu, A., Vernstad, K., & Svendsen, H. F. (2011a). A study of thermal degradation of different amines and their resulting degradation products. In *1st Post Combustion Capture Conference*, pp. 17–19.

Eide-Haugmo, I., Lepaumier, H., Einbu, A., Vernstad, K., da Silva, E. F., & Svendsen, H. F. (2011b). Chemical stability and biodegradability of new solvents for CO_2 capture. *Energy Procedia, 4*, 1631–1636.

Freeman, S. A. (2011). *Thermal degradation and oxidation of aqueous piperazine for carbon dioxide capture*.

Freeman, S. A., Davis, J., & Rochelle, G. T. (2010). Degradation of aqueous piperazine in carbon dioxide capture. *International Journal of Greenhouse Gas Control, 4*, 756–761.

Freeman, S. A., & Rochelle, G. T. (2012a). Thermal degradation of aqueous piperazine for CO_2 capture. 1. Effect of process conditions and comparison of thermal stability of CO_2 capture amines. *Industrial and Engineering Chemistry Research, 51*, 7719–7725.

Freeman, S. A., & Rochelle, G. T. (2012b). Thermal degradation of aqueous piperazine for CO_2 capture: 2. Product types and generation rates. *Industrial and Engineering Chemistry Research, 51*, 7726–7735.

Fytianos, G., Vevelstad, S. J., & Knuutila, H. K. (2016). Degradation and corrosion inhibitors for MEA-based CO_2 capture plants. *International Journal of Greenhouse Gas Control, 50*, 240–247.

Goff, G. S. (2005). *Oxidative degradation of aqueous monoethanolamine in CO_2 capture processes: iron and copper catalysis, inhibition, and O_2 mass transfer*. TX: University of Texas at Austin Austin.

Gunasekaran, P., Veawab, A., & Aroonwilas, A. (2013). Corrosivity of Single and Blended Amines in CO_2 Capture Process. *Energy Procedia, 37,* 2094–2099.

Hakka, L. E., & Ouimet, M. A. (2006). *Method for recovery of CO_2 from gas streams*. Google Patents.

Kennard, M. L. (1983). *Degradation of diethanolamine solutions*. University of British Columbia.

Kennard, M. L., & Meisen, A. (1985). Mechanisms and kinetics of diethanolamine degradation. *Industrial and Engineering Chemistry Fundamentals, 24,* 129–140.

Kim, C., & Sartori, G. (1984). Kinetics and mechanism of diethanolamine degradation in aqueous solutions containing carbon dioxide. *International Journal of Chemical Kinetics, 16,* 1257–1266.

Kittel, J., & Gonzalez, S. (2014). Corrosion in CO_2 post-combustion capture with alkanolamines—A review. *Oil & Gas Science and Technology-Revue d'IFP Energies nouvelles, 69,* 915–929.

Kittel, J., Idem, R., Gelowitz, D., Tontiwachwuthikul, P., Parrain, G., & Bonneau, A. (2009). Corrosion in MEA units for CO_2 capture: Pilot plant studies. *Energy Procedia, 1,* 791–797.

Kohl, A., & Nielsen, R. (1997). *Miscellaneous gas purification techniques. Gas Purification* (5th ed., pp, 1296–1373). Houston, TX: Gulf Professional Publishing.

Krzemień, A., Więckol-Ryk, A., Smoliński, A., Koteras, A., & Więcław-Solny, L. (2016). Assessing the risk of corrosion in amine-based CO_2 capture process. *Journal of Loss Prevention in the Process Industries, 43,* 189–197.

Lawal, O., Bello, A., & Idem, R. (2005). The role of methyl diethanolamine (MDEA) in preventing the oxidative degradation of CO_2 loaded and concentrated aqueous monoethanolamine (MEA)–MDEA blends during CO_2 absorption from flue gases. *Industrial and Engineering Chemistry Research, 44,* 1874–1896.

Lepaumier, H., da Silva, E. F., Einbu, A., Grimstvedt, A., Knudsen, J. N., Zahlsen, K., et al. (2011a). Comparison of MEA degradation in pilot-scale with lab-scale experiments. *Energy Procedia, 4,* 1652–1659.

Lepaumier, H., Grimstvedt, A., Vernstad, K., Zahlsen, K., & Svendsen, H. F. (2011b). Degradation of MMEA at absorber and stripper conditions. *Chemical Engineering Science, 66,* 3491–3498.

Lepaumier, H., Picq, D., & Carrette, P.-L. (2009a). New amines for CO_2 capture. I. Mechanisms of amine degradation in the presence of CO_2. *Industrial and Engineering Chemistry Research, 48,* 9061–9067.

Lepaumier, H., Picq, D., & Carrette, P.-L. (2009b). New amines for CO_2 capture. II. Oxidative degradation mechanisms. *Industrial and Engineering Chemistry Research, 48,* 9068–9075.

Li, L., Voice, A. K., Li, H., Namjoshi, O., Nguyen, T., Du, Y., et al. (2013). Amine blends using concentrated piperazine. *Energy Procedia, 37,* 353–369.

Martin, S., Lepaumier, H., Picq, D., Kittel, J., de Bruin, T., Faraj, A., et al. (2012). New amines for CO_2 capture. IV. Degradation, corrosion, and quantitative structure property relationship model. *Industrial and Engineering Chemistry Research, 51,* 6283–6289.

Mogul, M. (1999). Reduce corrosion in amine gas absorption columns. *Hydrocarbon processing, 78,* 47–56.

Namjoshi, O., Li, L., Du, Y., & Rochelle, G. (2013). Thermal degradation of piperazine blends with diamines. *Energy Procedia, 37,* 1904–1911.

Radgen, P., Rode, H., Reddy, S., & Yonkoski, J. (2014). Lessons learned from the operation of a 70 tonne per day post combustion pilot plant at the coal fired power plant in Wilhelmshaven, Germany. *Energy Procedia, 63,* 1585–1594.

Reddy, J., & Chin, C. (1998). Thermomechanical analysis of functionally graded cylinders and plates. *Journal of Thermal Stresses, 21,* 593–626.

Rochelle, G. T. (2012). Thermal degradation of amines for CO_2 capture. *Current Opinion in Chemical Engineering, 1,* 183–190.

Rochelle, G. T. (2013). *Aqueous piperazine/N-(2-aminoethyl) piperazine for CO_2 capture.*

Rooney, P., Dupart, M., & Bacon, T. (1998). Oxygen's role in alkanolamine degradation. *Hydrocarbon processing, 77,* 109–113.

Saiwan, C., Supap, T., Idem, R. O., & Tontiwachwuthikul, P. (2011). Part 3: Corrosion and prevention in post-combustion CO_2 capture systems. *Carbon Management, 2,* 659–675.

Sexton, A. J. (2008). *Amine oxidation in CO_2 capture processes*. University of Texas.
Sexton, A. J., & Rochelle, G. T. (2009). Catalysts and inhibitors for MEA oxidation. *Energy Procedia, 1,* 1179–1185.
Sexton, A. J., & Rochelle, G. T. (2010). Reaction products from the oxidative degradation of monoethanolamine. *Industrial and Engineering Chemistry Research, 50,* 667–673.
Soosaiprakasam, I. R., & Veawab, A. (2008). Corrosion and polarization behavior of carbon steel in MEA-based CO_2 capture process. *International Journal of Greenhouse Gas Control, 2,* 553–562.
Srinivasan, S., Veawab, A., & Aroonwilas, A. (2013). Low toxic corrosion inhibitors for amine-based CO_2 capture process. *Energy Procedia, 37,* 890–895.
Srisang, W., Supap, T., Idem, R., & Tontiwachwuthikul, P. (2014). Physical and chemical resistance of elastomers in aqueous monoethanolamine (MEA) and CO_2-loaded MEA solutions during post-combustion carbon dioxide capture processes. *Industrial and Engineering Chemistry Research, 53,* 5932–5940.
Strazisar, B. R., Anderson, R. R., & White, C. M. (2003). Degradation pathways for monoethanolamine in a CO_2 capture facility. *Energy & Fuels, 17,* 1034–1039.
Supap, T., Idem, R., Tontiwachwuthikul, P., & Saiwan, C. (2006). Analysis of monoethanolamine and its oxidative degradation products during CO_2 absorption from flue gases: A comparative study of GC-MS, HPLC-RID, and CE-DAD analytical techniques and possible optimum combinations. *Industrial and Engineering Chemistry Research, 45,* 2437–2451.
Supap, T., Idem, R., Tontiwachwuthikul, P., & Saiwan, C. (2009). Kinetics of sulfur dioxide-and oxygen-induced degradation of aqueous monoethanolamine solution during CO_2 absorption from power plant flue gas streams. *International Journal of Greenhouse Gas Control, 3,* 133–142.
Tontiwachwuthikul, P., Idem, R., Kladkaew, N., & Saiwan, C. (2011). *Corrosion inhibitors*. Google Patents.
Veawab, A. (2001). *Corrosion and corrosion control in CO_2 absorption process using aqueous amine solutions*.
Vega, F., Sanna, A., Navarrete, B., Maroto-Valer, M. M., & Cortés, V. J. (2014). Degradation of amine-based solvents in CO_2 capture process by chemical absorption. *Greenhouse Gases: Science and Technology, 4,* 707–733.
Vevelstad, S. J., Grimstvedt, A. M., Knuutila, H., Svendsen, H. F. (2013). *Thermal degradation on already oxidatively degraded solutions*.
Voice, A. K., & Rochelle, G. T. (2011). Oxidation of amines at absorber conditions for CO_2 capture from flue gas. *Energy Procedia, 4,* 171–178.
Wang, T. (2013). *Degradation of aqueous 2-Amino-2-methyl-1-propanol for carbon dioxide capture*.
Wang, T., & Jens, K.-J. (2012a). Oxidative degradation of aqueous 2-amino-2-methyl-1-propanol solvent for postcombustion CO_2 capture. *Industrial and Engineering Chemistry Research, 51,* 6529–6536.
Wang, T., & Jens, K.-J. (2012b). A study of oxidative degradation of AMP for post-combustion CO_2 capture. *Energy Procedia, 23,* 102–110.
Zhou, S., Wang, S., & Chen, C. (2012). Thermal degradation of monoethanolamine in CO_2 capture with acidic impurities in flue gas. *Industrial and Engineering Chemistry Research, 51,* 2539–2547.
Zoannou, K.-S., Sapsford, D. J., & Griffiths, A. J. (2013). Thermal degradation of monoethanolamine and its effect on CO_2 capture capacity. *International Journal of Greenhouse Gas Control, 17,* 423–430.

Chapter 5
Pilot and Demonstration Plants

Abstract The application of the post combustion CO_2 capture technology in the industry, a large amount of research need to be addressed in order to decrease the cost. The purpose of the pilot plants or demonstration plants is to find the issues which may not observed in the fundamental research. Another purpose of this study is to confirm the stimulation of process configuration. A pilot and demonstration plant study for the CO_2 capture process is a vital step to go to the commercial step. The critical information generated from those plants could provide the guideline to build the commercial plant. Thus, a lot of research on pilot have been conducted in the past decades for different amines solvents in the different process configurations. In this chapter, the summary of the recent study on the pilot plant are discussed in term of capacity, region, solvent technology, and status. In addition, the demonstration plants in terms of size, feedstock, solvent, status, and CO_2 fate are provided as well in this chapter.

A pilot and demonstration plant study for the CO_2 capture process is a vital step to go to the commercial step. The critical information generated from those plants could provide the guideline to build the commercial plant. In addition, the work on this scale plant also could provide the unique insights, which makes a contribution to the growth of a commercially competitive low carbon energy option through CCS. Based on the work of Abu-Zahra et al., the pilot plant was defined as the capacity <50 tons/day. The selected pilot plants were listed in table by capacity, region, solvent technology, and status. All information are shown as in Table 5.1. Figure 5.1 is the process diagram of pilot plant at the University of Regina. The detailed information could be founded in our previous work (Idem et al. 2006). In addition, the selected demonstration plants are shown in Table 5.2 in regards with size, feedstock, solvent, status, and CO_2 fate.

H. Liu et al., *Post-combustion CO_2 Capture Technology*, SpringerBriefs in Petroleum Geoscience & Engineering, https://doi.org/10.1007/978-3-030-00922-9_5

Table 5.1 The selected pilot plant for post combustion CO_2 capture [Modified from Idem et al. (2015)]

Project	Capacity	Technology/solvent used	Operational status	Reference
C2P3 University of Texas' Separations Research Program (SPR)	3 ton/day	MEA and 2 variants of PZ-Promoted K2CO3	2002–Ongoing	Wang et al. (2011)
University of Regina (UofR)		MEA and MEA/MDEA	1999–2000	Wang et al. (2011)
UofR/Consortium Saskpower station Boundary Dam (Saskatchewan)	4 ton/day	Fluor's Econamine FGTM Technology, MEA, RS-X solvents	2000–Ongoing	Wang et al. (2011)
CESAR/CASTOR (Dong Energy in Denmark)	24 ton/day	30 wt% MEA, CASTOR-1 and CASTOR-2, CESAR1 and CESAR2	2006–2012	Wang et al. (2011)
CO2CRC/International Power (Hazelwood coal-fired power plant, Australia)	50 ton/day	BASF PuraTreat™ and other new solvents	2009–Ongoing	Wang et al. (2011)
MHI-KEPCP Nanko Pilot Plant	2 ton/day	KS-1TM and other solvents	1991–Ongoing	Imai and Mitchell (2008)
Tampa electric/siemens big Bend station project (Florida)	1 ton/day	Amino acid salt	2013–Ongoing	Abu-Zahra et al. (2013)
CSIRO/China Huaneng Group Gaobeidian power station (Beijing)	3000 ton/year	Amine based solvents	2008–Ongoing	Wardhaugh (2010)
CSIRO/Delta electricity Munmorah power station (New South Wales)	3000 ton/year	Aqueous ammonia	2009–Ongoing	Wardhaugh (2010)
CSIRO Loy Yang power station (Victoria)	1000 ton/year	Amine based solvents including MEA	2008–Ongoing	Wardhaugh (2010)
SINTEF/NTNU pilot plant (Norway)	1.2 ton/day	New CO_2 capture solvent	– Ongoing	Abu-Zahra et al. (2013)

(continued)

Table 5.1 (continued)

Project	Capacity	Technology/solvent used	Operational status	Reference
Sigma Power Ariake Co. Ltd/Toshiba Mikawa Power plant (Japan)	10 ton/day	Amine based Toshiba solvent	2009–Ongoing	Suzuki (2012)
Hitachi/EERC pilot plant (University of North Dakota)	0.05 Nm^3/s	MEA and H3-1 Solvent	2010	Kikkawa et al. (2010)
TNO/E.ON. Maasvlakte pilot plant	0.69 kg/s	Amine acid salt	2008–Ongoing	Feron et al. (2010)
RWE/BASF/Linde RWE Coal Innovation Centre (Niederaussem)	12 ton/day	Amine solvents	2009–Ongoing	Feron et al. (2010)
Fluor/E.ON Power station (Wilhelmshaven, Germany)	72 ton/day	MEA solvent using Econamine FG	2010–Ongoing	Feron et al. (2010)
EdF (Le Havre, France)	25 ton/day	UCARSOLTM FGC 3000 (DOW solvent) using Alstom's advanced amine process technology	2012–Ongoing	Abu-Zahra et al. (2013)
Aker Clean Carbon Scottish power (Longannet)	4.8 ton/day	Amine solvents	2009–Ongoing	Feron et al. (2010)
China Huaneng Group (Shanghai)	100,000 ton/year	Amine solvents	2009–Ongoing	Feron et al. (2010)
Doosan/Emissions Reduction Test Facility (FRTF)	1 ton/day	MEA solvent, RS-2 solvent	2010	Elgarni and Aboudheir (2011)

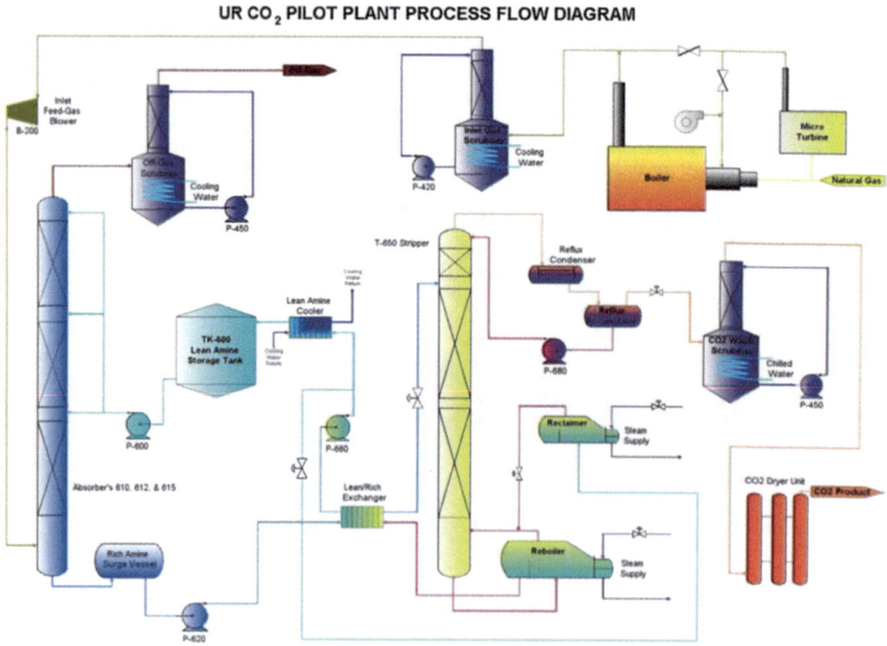

Fig. 5.1 The diagram of pilot plant for post-combustion plant at the University of Regina

Table 5.2 The selected demonstration plants for post combustion CO_2 capture [Modified from Idem et al. (2015)]

Project	Feedstock	Size	Technology/solvent	CO_2 fate	Status	Reference
AEP Mountaineer, AEP, WV, USA	Coal	30–600 ton/day	Chilled ammonia	Saline	Operated 2009–2011	Idem et al. (2015)
Karlshamn, E.ON, Sweden	Coal	5–100 ton/day	Chilled ammonia	Vented	Operated 2009–2010	Idem et al. (2015)
Aberthaw, RWE, Wales, UK	Coal	3–60 ton/day	Cansolv	N/A	Operational 2013–2014	Idem et al. (2015)
Shidongkou, Huaneng, China	Coal	0.1 Mt/year	Amine mix	Commercial use	Operational 2011	Idem et al. (2015)
IFP Energies nouvelles/EnEL Brindisi power plant (Italy)	Coal	54 ton/day	Different solvents including MEA (20 wt% to 40 wt%)	EOR	2011–ongoing	BOUILLON et al. (2011)
Wilhelmshaven, E.ON, Germany	Coal	3.5 or 70 ton/day	Fluor's Econamine FG PlusSM	Vented	Operational October 2012	Idem et al. (2015)
Mongstad, Statoil, Norway	Gas	0.1 Mt/year	Amines, chilled ammonia	Saline	Operational May 2012	Idem et al. (2015)
Boryeong, KEPCO, South Korea	Coal	10 MW	KEPCO wet amine	N/A	Operational May 2013	Idem et al. (2015)

References

Abu-Zahra, M. R., Abbas, Z., Singh, P., & Feron, P. (2013). Carbon dioxide post-combustion capture: solvent technologies overview, status and future directions. *Materials and processes for energy: Communicating current research and technological developments, 1,* 923–934.

Bouillon, P. A., Lemaire, E., Mangiaracina, A., Tabasso, C. (2011). *First results of the 2.25 t/h post-combustion CO_2 capture pilot plant of ENEL at the Brindisi coal power plant with MEA from 20 to 40% wt and HiCapt process.*

Elgarni, M., Aboudheir, A. (2011). Design of commercial post combustion CO_2 capture plant using pilot plant scale-up approach. In *1st Post Combustion Capture Conference.*

Feron, P., Hooper, B., Attalla, M., Dave, N., Kentish, S., Stevens, G., Wardhaugh, L., & Webley, P. (2010). *Research opportunities in post combustion CO_2 capture.*

Idem, R., Supap, T., Shi, H., Gelowitz, D., Ball, M., Campbell, C., et al. (2015). Practical experience in post-combustion CO_2 capture using reactive solvents in large pilot and demonstration plants. *International Journal of Greenhouse Gas Control, 40,* 6–25.

Idem, R., Wilson, M., Tontiwachwuthikul, P., Chakma, A., Veawab, A., Aroonwilas, A., et al. (2006). Pilot plant studies of the CO_2 capture performance of aqueous MEA and mixed MEA/MDEA solvents at the University of Regina CO_2 capture technology development plant and the boundary dam CO_2 capture demonstration plant. *Industrial and Engineering Chemistry Research, 45,* 2414–2420.

Imai, N., & Mitchell, R. (2008). *MHI's experience in post combustion CO_2 capture: An effective means to counter anthropogenic global warming.* Oslo, Norway: Mitsubishi Heavy Industries Ltd.

Kikkawa, H., Fukuda, Y., Takamoto, S., Katsube, T., Nakamoto, T., Oda, N., Kawasaki, T., Sugiura, T., Wu, S., & Eswaran, S. (2010). Hitachi's carbon dioxide scrubbing technology with new absorbent for coal-fired power plants. HITACHI.

Suzuki, K., (2012). Toshiba's activity in clean coal and carbon capture technology for thermal power plants. In *APEC Clean Fossil Energy Technical and Policy Seminar* (Feb 22, 2012).

Wang, M., Lawal, A., Stephenson, P., Sidders, J., & Ramshaw, C. (2011). Post-combustion CO_2 capture with chemical absorption: A state-of-the-art review. *Chemical Engineering Research and Design, 89,* 1609–1624.

Wardhaugh, L. (2010). *CSIRO post-combustion capture research—Latest developments and the next steps in NSW*. CSIR

GPSR Compliance

The European Union's (EU) General Product Safety Regulation (GPSR) is a set of rules that requires consumer products to be safe and our obligations to ensure this.

If you have any concerns about our products, you can contact us on ProductSafety@springernature.com

In case Publisher is established outside the EU, the EU authorized representative is:

Springer Nature Customer Service Center GmbH
Europaplatz 3
69115 Heidelberg, Germany

Batch number: 10682325

Printed by Printforce, the Netherlands